Marine Turbulence

FURTHER TITLES IN THIS SERIES

Elsevier Oceanography Series, 28

Marine Turbulence

PROCEEDINGS OF THE 11th INTERNATIONAL LIÈGE COLLOQUIUM
ON OCEAN HYDRODYNAMICS

Edited by

JACQUES C.J. NIHOUL

*Professor of Ocean Hydrodynamics,
University of Liège,
Liège, Belgium*

ELSEVIER SCIENTIFIC PUBLISHING COMPANY
Amsterdam — Oxford — New York   1980

ELSEVIER SCIENTIFIC PUBLISHING COMPANY
335 Jan van Galenstraat
P.O. Box 211, 1000 AE Amsterdam, The Netherlands

Distributors for the United States and Canada:

ELSEVIER/NORTH-HOLLAND INC.
52, Vanderbilt Avenue
New York, N.Y. 10017

ISBN 0-444-41881-4 (Vol. 28)
ISBN 0-444-41623-4 (series)

Printed in The Netherlands

FOREWORD

The concepts of mixing of water masses, dispersion of anomalies, transport of momentum and vorticity and the dissipation of kinetic energy by turbulent motions in the ocean, play a central role in almost every aspect of Oceanography. The action of turbulence appears not only in the circulation models of physical oceanographers but equally in the models of chemical distributions, of biological production and of sedimentation.

The interaction between atmosphere and ocean - with its important consequences for climate variation - can only be modelled by taking into account the transport of heat, salt and momentum by turbulent motions in the wind-mixed surface layer and the underlying thermocline.

The major advances that have been made in the past decade have revolutionized our thinking about turbulence in the ocean. The bulk effects of turbulence that were known fifty years ago can now for the first time be discussed in terms of models of the responsible flow patterns based on an understanding of the underlying dynamics and physics.

The IAPSO General Assembly in Grenoble (1976) accepted a proposal submitted by the British National Committee to sponsor a second Symposium on Turbulence in the Ocean, at which the emphasis would be on seeking a synthesis across the whole spectral band from millimetres to megametres. This proposal was subsequently discussed by SCOR which agreed to co-sponsor the meeting and by the IOC which offered support. A number of member states recommended that the Symposium should be recognised as a IDOE meeting. It was therefore designated the IAPSO-SCOR-IDOE Second Symposium on "Turbulence in the Ocean", with support from the IOC.

It was decided to make the first week an open meeting, during which both invited and contributed papers could be presented and discussed in open sessions to be attended by all comers. The plan was to follow this by a closed session during the second week. In order to accomodate the idea of an open first week, it was agreed that this would be combined with the annual Colloquium on Ocean Hydrodynamics held in Liège University.[*]

[*]J.D. Woods : Report to the IOC.

The international Liège Colloquia on Ocean Hydrodynamics are organized annually. Their topics differ from one year to another and try to address, as much as possible, recent problems and incentive new subjects in physical oceanography.

Assembling a group of active and eminent scientists from different countries and often different disciplines, they provide a forum for discussion and foster a mutually beneficial exchange of information opening on to a survey of major recent discoveries, essential mechanisms, impelling question-marks and valuable suggestions for future research.

The papers presented at the Eleventh International Liège Colloquium on Ocean Hydrodynamics report theoretical and experimental research and they address such different scales of motions as synoptic eddies, fronts, mesoscale blinis, and three-dimensional microscale fluctuations. Their unity resides in a common approach to the variability of the seas, based on the profound understanding of non-linear processes which the theory of turbulence provides.

Jacques C.J. NIHOUL.

The Scientific Organizing Committee of the Eleventh International Liège Colloquium on Ocean Hydrodynamics and all the participants wish to express their gratitude to the Belgian Minister of Education, the National Science Foundation of Belgium, The University of Liège, the Intergovernmental Oceanographic Commission and the Division of Marine Sciences (UNESCO) and the Office of Naval Research for their most valuable support.

LIST OF PARTICIPANTS

ADAM, Y., Dr., Ministère de la Santé Publique et Environnement, Belgium.

BAH, A., Prof., Dr., Institut Polytechnique de Conakry, Rép. de Guinee.

BELYAEV, V.S., Prof., Dr., Institute of Oceanology, Moscow, U.S.S.R.

BERGER, A., Prof., Dr., Université Catholique de Louvain, Belgium.

BOWDEN, K.F., Prof., Dr., University of Liverpool, U.K.

BURT, W.V., Prof., Dr., Office of Naval Research, London, U.K.

BUTTI, C.H., Dr., Rijkswaterstaat, 's Gravenhage, The Netherlands.

CALDWELL, D.R., Prof., Dr., Oregon State University, U.S.A.

CHABERT d'HIERES, G., Ir., Institut de Mécanique de Grenoble, France.

COANTIC, M., Prof., Dr., Institut de Mécanique Statistique de la
    Turbulence, Marseille, France.

COLIN DE VERDIERE, A., Dr., C.O.B., Brest, France.

DELCOURT, B., Dr., Von Karman Institute for Fluid Dynamics,
    Rhode St. Genèse, Belgium.

DERENNE, M., Mr, Université de Liège, Belgium.

DESAUBIES, Y.J.F., Dr., University of Washington, Seattle, U.S.A.

DILLON, T.M., Ir., Oregon State University, U.S.A.

DISTECHE, A., Prof., Dr., Université de Liège, Belgium.

FEDOROV, K., Prof., Dr., Institute of Oceanology, Moscow, U.S.S.R.

FERGUSON, S.R., Mr., University of Liverpool, U.K.

GIBSON, C.H., Prof., Dr., University of California, U.S.A.

GOODMAN, L., Dr., O.N.R., Washington, U.S.A.

GORDON, C.M., Dr., Naval Research Laboratory, Washington, U.S.A.

GREGG, M.C., Prof., Dr., University of Washington, Seattle, U.S.A.

HAUGUEL, A., Ir., Electricité de France, Chatou, France.

HOPFINGER, E., Dr., Institut de Mécanique de Grenoble, France.

JOSSERAND, M., Ir., Institut de Mécanique de Grenoble, France.

KITAIGORODSKII, S.A., Prof., Dr., University of Copenhagen, Denmark.

KRUSEMAN, P., Dr., K.N.M.I., De Bilt, The Netherlands.

KULLENBERG, G.E.B., Prof., Dr., University of Copenhagen, Denmark.

LEBON, G., Prof., Dr., Université de Liège, Belgium.

LE PROVOST, C., Prof., Dr., Institut de Mécanique de Grenoble, France.

LEWALLE, A., Ir., Université de Liège, Belgium.

LOFFET, A., Ir., Université de Liège, Belgium.

LUMLEY, J.L., Prof., Dr., Cornell University, Ithaca, U.S.A.

MAHRT, K.H., Dr., Universität Kiel, W. Germany.

MICHAUX, T., Ir., Université de Liège, Belgium.

MITCHELL, J.B., Dr., U.K. Meteorological Office, Bracknell, U.K.

MULLER, P., Dr., Harvard University, Cambridge, U.S.A.

NIHOUL, J.C.J., Prof., Dr., Université de Liège, Belgium.

NIZET, J.L., Ir., Université de Liège, Belgium.

OAKEY, N.S., Dr., Bedford Institute of Oceanography, Dartmouth, Canada.

OLBERS, D., Dr., Universität Kiel, W. Germany.

ORLANSKI, I., Dr., Princeton University, U.S.A.

OSBORN, T.R., Prof., Dr., University of British Columbia, Vancouver, Canada.

OSTAPOFF, F., Dr., Sea-Air Interaction Laboratory, Miami, U.S.A.

OZER, J., Ir., Université de Liège, Belgium.

OZMIDOV, R.V., Prof., Dr., Institute of Oceanology, Moscow, U.S.S.R.

PANCHEV, S., Prof., Dr., University of Sofia, Bulgaria.

PASMANTER, R., Dr., Rijkswaterstaat, Den Haag, The Netherlands.

PIACSEK, S.A., Dr., NORDA, Bay St. Louis, U.S.A.

PICHOT, G., Ir., Ministère de la Santé Publique et de l'Environnement, Belgium.

RENIER, I., Miss, Université de Liège, Belgium.

REVAULT d'ALLONNES, M., Dr., Museum d'Histoire Naturelle, Paris, France.

RHINES, P.B., Dr., Woods Hole Oceanographic Institution, Woods Hole, U.S.A.

ROISIN, B., Ir., Université de Liège and Florida State University, U.S.A.

RONDAY, F.C., Dr., Université de Liège, Belgium.

RUNFOLA, Y., Ir., Université de Liège, Belgium.

SMITZ, J., Ir., Université de Liège, Belgium.

STROSCIO, M.A., Dr., Johns Hopkins University, U.S.A.

SWIFT, M.R., Prof., Dr., University of New Hampshire, U.S.A.

THOMASSET, F., Ir., IRIA Laboria, Le Chesnay, France.

TURNER, J.S., Prof., Dr., Australian National University, Camberra, Australia.

VETH, C., Dr., N.I.O.Z., Texel, The Netherlands.

WOODS, J.D., Prof., Dr., Universität Kiel, W. Germany.

WOLF, J., Mrs, I.O.S., Bidston, U.K.

CONTENTS

XII

THE TURBULENT OCEAN

Jacques C.J. NIHOUL[1]

Mécanique des Fluides Géophysiques, Université de Liège, Sart Tilman B6, B-4000 Liège, (Belgium).

[1]Also at the Institut d'Astronomie et de Géophysique, Université de Louvain (Belgium).

ABSTRACT

The variability of the ocean over a wide range of scales, from the megameter to the millimeter, is examined in the light of turbulence theory.

The geophysical constraints which arise from the Earth's rotation and curvature and from the stratification are discussed with emphasis on the role they can play at different scales in inducing instabilities and a transfer of energy to other scales of motion.

INTRODUCTION

Turbulence in the ocean is still a very controversial subject. An innocent physicist, emboldened by a solid background in the theory of turbulence, who would approach the subject relying on his good understanding of, say, turbulent channel flow, might soon find himself confronted with a maze of conflicting experimental data and a nightmarish farrago of theories where he recognizes very little of what he regards as "turbulence".

Assuming that he is able to muster enough of his high school latin and greek to find his way among iso-halines, iso-pycnals and other proliferating peculiar surfaces which seem to fill the ocean with complexity, he may still require some time to adjust to new cabalistic concepts such as enstrophy, red energy cascade, eddies "which do not overturn", double diffusion or fossil turbulence.

Used to regard a turbulent flow as the superposition of a mean motion and turbulent fluctuations, he must face the fact that, if such things exist in the ocean, they are several thousands orders of magnitude apart in scales and apparently separated by a jungle of complicated movements which one refers to as the "variability of the ocean".

One of the most intriguing aspect of ocean variability is that, while the macroscale dynamics is claimed to be governed by a cascade of enstrophy to smaller scales, studies of mesoscale and microscale variability seem totally unconcerned with it as if it had disappeared somewhere on the way. (Planetary oceanographers talk about enstrophy "dissipation" but surely they mean something else like "annihilation"; even if one of them, answering a question at the Eleventh Liège Colloquium on Ocean Hydrodynamics, expressed the somewhat surprising view that "dissipated" enstrophy turns into heat).

For somebody trained in classical turbulence theory, it is not immediately obvious that the variability of the ocean is a form of turbulence.

If it is turbulence, it is clearly very different from the type of turbulence with which mechanical engineers, say, are familiar.

The reason why it should be so is found in the work of those ocea-nographers who, taking a quite opposite view, insist on describing ocean hydrodynamics in terms of non-turbulent theories such as linear wave propagation or molecular effects.

That they succeed in explaining some of the observations - after removing turbulence from the experimental signals in an operation which they gallantly call "decontamination" - is an indication of the mechanisms which are particular to geophysical flows and which are liable to modify, more or less drastically, geophysical turbulence.

The Coriolis force and the density stratification - which allow wave motions that do not exist in non-rotating non-stratified fluids -, the variations of density not only with temperature but also with salinity, - with different molecular diffusivities for momentum, heat and salt -, and their interference in other processes related to bottom or coastal topography and air-sea interactions, provide a great variety of mechanisms which eventually combine with mechanical effects to determine the stability or the instability of oceanic motions and the subsequent constraints on turbulence at different scales.

Including such geophysical constraints, it becomes possible to understand some essential characteristics of ocean variability and to build an image of what turbulence in the ocean may be.

TIME SCALES, LENGTH SCALES AND LINEAR WAVE THEORY

The equations of Geophysical Fluid Dynamics admit linear wave solutions of different kinds (e.g. Monin et al, 1977). Whether these
waves can be observed in the ocean depends on a series of factors.
Very small amplitude waves will not be affected by non-linear interactions but, on the other hand, they may be masked by stronger motions
and remain unnoticed. Interactions between larger amplitude waves may
create an intricated field of waves and wave packets of all scales,
wave breaking and turbulence, from which the ideal individual wave of
linear theory cannot be sorted out.

The theory of linear waves is however always a useful mathematical
exercise as it helps to identify the dominant length scales (wave
numbers) and time scales (frequencies) of motions.

From this point of view, it is convenient to divide ocean waves
into three categories (Nihoul, 1979) :

A. Macroscale waves

These waves have frequencies (ω) in the range

$$10^{-8} s^{-1} \lesssim \omega \lesssim 10^{-5} s^{-1} \tag{1}$$

and operate over a range of horizontal wave-numbers (k)

$$10^{-6} m^{-1} \lesssim k \lesssim 10^{-3} m^{-1} \tag{2}$$

(Waves of larger periods could be considered but the limiting frequency 10^{-8} corresponding to periods larger than 10 years, they are presumably rather irrelevant to the present discussion). Macroscale
waves are directly related to the spatial variations of the Coriolis
parameter f (f = 2 Ω sin ϕ where Ω is the angular velocity of the
earth's rotation and ϕ the latitude) i.e. to the parameter

$$\beta = \| \nabla f \| \sim 10^{-11} m^{-1} s^{-1} \tag{3}$$

Macroscale waves include (e.g. Rhines, 1977 ; Nihoul, 1979)

(i) very slow baroclinic waves ($\omega k \ll \beta$) for which

$$\omega \sim \frac{N^2 \beta H^2}{f^2} k \tag{4}$$

where H is the depth of the ocean and N the Brunt-Väisälä frequency.

(The condition $\omega k \ll \beta$ yields $L \sim k^{-1} \gg R$ where $R = NHf^{-1} \sim 10^5$ m is the so-called "Rossby internal scale". These waves are thus very slow large scale small amplitude waves which, if excited, are likely to break rapidly under the effect of bottom slope and general baroclinic instability).

(ii) barotropic Rossby waves ($\omega k \sim \beta$) for which

$$\omega \sim \beta k^{-1} \sim f \gamma H^{-1} k^{-1} \tag{5}$$

where γ is the non-dimensional mean bottom slope chosen here of the order $\gamma \sim 10^{-4}$ such that $f \gamma H^{-1} \sim \beta \sim 10^{-11}$.

(In a forced problem, with wind blowing accross the ocean surface, these waves constitute the most important mode at "weatherlike" time scales over horizontal scales greater than R).

(iii) fast baroclinic waves confined within a layer of thickness $f L N^{-1}$ above the sloping bottom, for which

$$\omega \sim \gamma N \tag{6}$$

(If, in the dispersion relation of topographic Rossby waves $\omega \sim f \gamma H^{-1} k^{-1}$, one replaces the depth H by the "penetration height" $f L N^{-1}$, one obtains the dispersion relation (6). These waves may thus be regarded as topographic Rossby waves where density stratification provides a lid for vortex stretching).

B. Mesoscale waves

These waves have frequencies in the range

$$10^{-5} s^{-1} \lesssim \omega \lesssim 10^{-2} s^{-1} \tag{7}$$

and include gyroscopic waves and inertial oscillations, tides, and internal gravity waves (e.g. Tolstoy, 1963 ; Monin et al, 1977).

The effect of the Earth's curvature becomes here negligible and the essential factors in the dispersion relations are the Coriolis parameter f and the Brunt-Väisälä frequencies N_{min} and N_{Max} .

C. Microscale waves

These waves have frequencies

$$10^{-2} s^{-1} \lesssim \omega \tag{8}$$

They are essentially surface waves and acoustic waves. The former only affect the upper layer of the ocean and may be regarded as an indispensable - unfortunately rather complicated - way of transferring momentum and energy directly from the wind to the sea ; the latter are marginally important in Ocean Hydrodynamics from which they are customarily excluded by the Boussinesq approximation.

MACROSCALE TURBULENCE

Macroscale motions in the ocean include large scale currents (gyres) and quasi-geostrophic or "synoptic" eddies which appear, from observational studies, to contain a large fraction of the ocean's kinetic energy.

The dynamics of the synoptic eddies is dominated by the earth's curvature - parameterized in terms of β - and their horizontal length scale is of the order of the Rossby internal scale R.

The spectral characteristics of the synoptic eddies, wave number κ_β, frequency ω_β and energy level $\kappa_\beta E(\kappa_\beta)$ where $E(\kappa)$ is the horizontal kinetic energy spectral density, can be estimated by turbulence similarity arguments. One finds

$$\kappa_\beta \sim R^{-1} \sim 10^{-5} m^{-1} \tag{9}$$

$$\omega_\beta \sim \beta \kappa_\beta^{-1} \sim 10^{-6} s^{-1} \tag{10}$$

$$\kappa_\beta E(\kappa_\beta) \sim \beta^2 \kappa_\beta^{-4} \sim 10^{-2} m^2 s^{-2} \tag{11}$$

These estimates appear to be in good agreement with the observations (e.g. Koshlyakov and Monin, 1978).

The use of the term "synoptic" emphasizes the physical analogy between these eddies and the synoptic eddies of the atmosphere (cyclones and anticyclones, quasi-geostrophic motions at the Rossby scale).

The synoptic variability of the atmosphere, however, has time scales of the order of a week and is shaped by pressure lows and highs with characteristic horizontal scales of the order of the thousand of kilometers and one must exclude the hypothesis of the generation of synoptic ocean eddies by direct resonant interactions. Atmospheric disturbances - lows and highs - generate large-scale currents in the ocean and it is the barotropic and essentially baroclinic instability of these currents which provide the energy for the synoptic eddies (Kosklyakov and Monin, 1978).

(The kinetic - and approximately equal potential-energy of the synoptic eddies is essentially higher than the kinetic energy of the

large scale currents and at the same time much smaller than the available potential energy of the latter. This constitutes strong experimental evidence of eddy generation through baroclinic instability
of the large scale oceanic currents).

Large scale ocean experiments (Polygon, Mode, ...) give evidence
of synoptic eddies of two kinds, "frontal eddies" produced by the
cut-off of meanders from such frontal currents as the Gulf Stream and
the Kuroshio, and much weaker "open-ocean eddies". The kinetic energy
of the frontal eddies can be two orders of magnitude larger than the
kinetic energy of the typical ocean eddies, the rotation velocity in
the upper part of frontal eddies can reach meters per second. (e.g.
Kosklyakov and Monin, 1978 ; Nihoul, 1979).

The vertical length scale of the synoptic eddies is of the order
of the depth (e.g. Rhines, 1977 ; Woods, 1977 ; Nihoul, 1979) and it
is very tempting to regard them as constituting a form of two-dimensional turbulence.

Potential vorticity would be conserved in such motion and, in the
words of Gill and Turner (1979), "patches of marked particles would
be teased out, into spindly shapes, leading to an enstrophy ('mean
square vorticity') cascade to smaller scales".

Such a cascade predicted by the mathematical theory of homogeneous
two-dimensional turbulence (e.g. Kraichnan, 1967 ; Batchelor, 1969)
implies that the flow of kinetic energy is from smaller to larger
scales (the "red cascade").

There must be however processes - presumably different at different levels and in different regions - which are not quasi-geostrophic
and which limit the extension of potential vorticity contours.

Turbulent energy transfer to small scale may jump over an eventual
synoptic valley via boundary turbulence, intermittent internal turbulence or non-local cascade into internal waves. Bottom roughness provides a permanent mechanism for the conversion from large to small
scales. An initial cluster of eddies, surrounded by quiet fluid may
cascade to longer scales but eventually the energetic patch will
contain too few eddies to act as turbulence. Another obstacle to the
2D red cascade is the restoring force provided by the β-effect or its
topographic equivalent. No matter how intense or how small the initial eddies, the red cascade carries the flow into the regime of linear waves. The red cascade is then not only blocked by Rossby wave
propagation but it is reversed near western boundaries as fast long
weak westward-propagating Rossby waves reflect at a western boundary
into slow short strong eastward-propagating waves (Rhines, 1977).

Instabilities of fronts could play an important part. Such fronts can be formed by the same mechanisms which produce atmospheric fronts but according to Woods (1977, 1978), it is possible that they reach a limiting equilibrium form well before the larger scale velocity field which produced them has changed significantly.

The transfer of energy to internal waves has been strongly advocated. According to Müller (1976), for instance, internal waves could extract energy from synoptic eddies at about the same rate as that at which they gain energy from baroclinic instability of the wind - generated Sverdrup flow . Desquieting evidence against such a scheme was presented by Ruddick and Joyce (1979) from direct measurements of the vertical eddy momentum flux, due to internal waves, with moored current-meters and temperature sensors. They found no significant correlation with the mean shear and estimated an upper bound for the vertical eddy viscosity more than one order of magnitude smaller than Müller's suggestion (Garrett, 1979).

Following Panchev (1976), one can estimate the rate of energy transfer ε_β from the synoptic eddies to smaller turbulent oceanic scales as being of the order of

$$\varepsilon_\beta \sim 10^{-9} \, m^2 s^{-3} \tag{12}$$

i.e. of the same order as the atmospheric energy input into the largest oceanic scales (e.g. Ozmidov, 1965) but apparently one or two orders of magnitude smaller than the rate of energy transfer to larger scales through the red cascade (Panchev, 1976).

In the macroscale range, one expects - as a result of the two-dimensional turbulence enstrophy cascade - the energy spectral function $E(\kappa)$ to fall off as κ^{-3} i.e.

$$\kappa^3 E(\kappa) \sim \kappa_\beta^3 E(\kappa_\beta) \sim 10^{-12} s^{-2} \tag{13}$$

Hence, at a scale of a few kilometers ($\kappa_w \sim 3 \; 10^{-4} m^{-1}$, say) characteristic of eddies and intrusive layers which may emanate from fronts (Woods, 1978), the energy level would be, in the mean (all space and time intermittencies taken into account)

$$\kappa_w E(\kappa_w) \sim \kappa_\beta^3 E(\kappa_\beta) \kappa_w^{-2} \sim 10^{-5} m^2 s^{-2} \tag{14}$$

Now, the rate of energy transfer from these scales, - which belong to the frontier districts between macro- and mesoscales -, into mesoscale turbulence can be estimated from turbulence similarity arguments. One finds

$$\varepsilon \sim \kappa_w \, E(\kappa_w) \omega_m \tag{15}$$

where ω_m is the frequency of energy transfer from macroscale turbulence to mesoscale turbulence. Using eqs. (12) and (14), one gets

$$\omega_m \sim 10^{-4} \, s^{-1} \tag{16}$$

This is precisely the characteristic frequency of the longuest mesoscale waves and one may speculate that the macroscale energy cascade is passed on to mesoscale turbulence through the instability of frontal eddies and intrusive layers transmuted by inertial, gyroscopic and tidal oscillations. The energy transferred into the mesoscales is found by eq. (12) to be only a small fraction of the energy which is apparently continuously recycled over the largest scales ($\kappa \sim 10^{-6}$, $10^{-5} \, m^{-1}$) via transformations between kinetic and potential forms and exchanges between synoptic eddies, Rossby wave motions and gyres ; a very exclusive society, it would seem, which allows just enough bottom friction and limited leakage through physical space and Fourier space to maintain the energy balance with the atmospheric forcing.

MESOSCALE TURBULENCE

In the range of frequencies $10^{-5} \lesssim \omega \lesssim 10^{-2}$ one expects turbulent motions to be deeply intermingled with linear and non-linear waves related to tides, inertial oscillations and the general stratification of the ocean.

On the basis of turbulence, the ocean can be divided into three layers : (i) an upper mixed layer with a thickness $\sim 10^2 \, m$ which is continuously filled with turbulence generated by atmospheric factors working through the breaking of surface waves, drift currents and convection, (ii) an internal layer (practically the entire thickness of the ocean) in which only intermittent turbulence appears in the form of isolated patches or "blinis", (iii) a turbulent bottom layer with a thickness $\sim 10 \, m$ which is presumably similar to the atmospheric boundary layer.

Under stable stratification, turbulence loses energy in working against the buoyancy forces and turbulent mixing may become so difficult that, under natural conditions, it cannot extend to the whole water column and remains confined in distinct (well-mixed) "layers" separated by "sheets" where abrupt changes occur in temperature, velocity etc...

This is confirmed by experimental data and vertical profiles of temperature and other variables, almost everywhere in the ocean, show vertical step-like inhomogeneities generally referred to as the "fine structure" of the ocean.

As described by Woods (1977), the intermittent turbulence observed in the internal layers may be associated with trains of internal waves which, by locally increasing the vertical shear and reducing the Richarson number, allow turbulent patches to develop.

As a result of turbulent mixing, the water density becomes fairly homogeneous in a patch and in stably stratified surroundings, the density at the top becomes larger, and the density at the bottom smaller, than that of the ambient fluid. Under the action of buoyancy forces, the turbulent patch will then tend to flatten while it spreads aside by continuity forming a blini-shape intrusive layer contributing to the formation of the fine vertical structure.

Taking into account that there is an important input of energy in the ocean at the low frequency - low wave number end of the mesoscale range, through tidal and inertial oscillations, it is possible to conceive a coherent model of mesoscale motions based on the cohabitation of a chaotic field of internal waves and turbulent blinis.

Initially long waves will form large scale turbulent patches resulting in layers of great thickness. In such layers, internal waves of smaller periods and wave-lengths will develop forming turbulent patches of smaller dimensions and layers of smaller thickness and the process will continue, producing smaller and smaller scale motions down to the smallest waves and Kelvin-Helmoltz billows merging into three-dimensional turbulence.

The cascade "tidal-inertial waves → turbulent patches → fine structure layers → internal waves → turbulent patches → and so forth" is consistent with the observed spectra of ocean variability in the mesoscale range.

Another mechanism which could produce a fine vertical structure is double-diffusive convection.

The density of sea water being essentially a function of temperature and salinity, a given density distribution may mask important - but more or less compensating - variations of temperature and salinity.

Since the rates of diffusion of heat and salt are different, this situation allows potential energy to be released from the heavy component at the top. This type of "instability" can break smooth density gradients into a series of layers and interfaces with vertical transports accross the interfaces much larger than could be effected by classical diffusion down the mean gradients (e.g. Turner, 1973 a,b).

Double diffusive layering can be most effective near fronts when large anomalies of temperature and salinity may occur even with little net density differences.

According to Turner (e.g. Gill and Turner, 1979), quasi-vertical fluxes associated with double-diffusion processes can produce local density anomalies and so drive intrusive layers accross a front. Such layers could have vertical scales up to hundreds of meters and horizontal scales of several kilometers.

It is also possible that intrusions are produced by the dynamical instability of fronts and, unfortunately, experimental data are not yet sufficient to discriminate between the two mechanisms and assess their relative importance.

In any case, intrusive motions may be - as much as internal wave straining - an important cause of the observed fine vertical structure (Fedorov, 1978).

The role played by mesoscale fronts in ocean turbulence, has recently been emphasized by Woods (e.g. Woods, 1977 ; 1978) who suggested that frontogenesis is the inevitable outcome of the macroscale enstrophy cascade and that mesoscale fronts take over from the synoptic eddies to transfer enstrophy to microscales.

Mesoscale fronts are formed by the synoptic scale deformation field ($\sim 10^5$m) acting on the gyre scale baroclinicity ($\sim 10^6$m). Hydrodynamic instability of a mesoscale front produces meanders with wave-lengths λ ranging from a few tens of kilometers to a few kilometers (corresponding to typical wave-numbers $\kappa = 2\pi/\lambda$ of the order of $\kappa_w \sim 3 \ 10^{-4}m^{-1}$) breaking into intrusive layers and eddies.

As mentioned before, such layers and eddies in the frontier districts between macroscale and mesoscale motions might provide the missing link between the synoptic eddies and the mesoscale blinis under direct influence of the stratification and the earth's rotation.

There is however an input of energy in the same range of scales and it must be taken into account.

According to Ozmidov (1965), this input corresponds to a rate of energy transfer of the order of

$$\varepsilon \sim 10^{-7} m^2 s^{-3}$$

How much of this energy affects the whole water column is still an open question.

One expects the effects of the tidal forcing to be felt at all depths but the amount of tidal energy which can be included in a turbulent cascade in the interior of the ocean appears rather uncertain (e.g. Garrett, 1979). An important fraction of the tidal energy is indeed dissipated on the continental shelf and in shallow coastal seas. Perhaps a gross estimate of $2 \ 10^{-9} \ m^2 s^{-3}$ (Monin et al, 1977) may be retained for later comparisons.

(Bell (1975) has suggested a similar value for the energy flux into internal tides).

The effect of inertial oscillations is certainly not restricted to the upper ocean layers. Webster (1969) pointed out that inertial motions may ack like a kind of energy flywheel and explain the important and fairly permanent peak of kinetic energy about the inertial frequency.

Evidence of fairly energetic oscillations with frequencies close to the inertial frequency f, extending to the deep sea, have been given by several authors (e.g. Webster, 1968 ; Brekhovskikh et al, 1971 ; Perkins, 1972 ; Monin et al, 1977).

Knowing that this value is liable to be revised, one may perhaps write down an estimate

$$\varepsilon_f \sim 10^{-8} \ m^2 s^{-3} \tag{17}$$

for the rate of energy transfer through the mesoscales in the ocean interior (with a higher value of $10^{-7} \ m^2 s^{-3}$ in the upper layers).

Many experimental data seem to be reasonably well explained with values of ε of that order (e.g. Webster, 1969 ; Nihoul, 1979).

Frequency spectra of horizontal kinetic energy derived from long series of oceanic data show a marked narrow spectral peak at a frequency of the order of the inertial and tidal frequencies ($f \sim 10^{-4} s^{-1}$), followed by a gentle slope frequently fairly close to the classical - 5/3 line (e.g. Fofonoff, 1969 ; Webster, 1969 ; Nihoul, 1979).

If one considers, the largest scales on the gentle slope, at a frequency close to $f \sim 10^{-4}$ one may associate to them, by similarity arguments an energy level $\kappa_f E(\kappa_f)$ and a wave number κ_f of the order of

$$\kappa_f E(\kappa_f) \sim \varepsilon_f \ f^{-1} \sim 10^{-4} \ m^2 s^{-2} \tag{18}$$

$$\kappa_f \sim \varepsilon_f^{-1/2} f^{3/2} \sim 10^{-2} \ m^{-1} \tag{19}$$

for the value of ε given by eq. (17).

The corresponding eddies are not however the "energy containing eddies", corresponding to the spectral peak, which have about the same frequency but a level of energy about two orders of magnitude larger (e.g. Webster, 1969 ; Nihoul, 1979).

By similarity arguments, the energy containing eddies are characterized by a wave number κ_m given by

$$\kappa_m^3 \, E(\kappa_m) \sim \omega_m^2 \tag{20}$$

i.e.

$$\kappa_m \sim \frac{\omega_m}{\left(10^2 \kappa_f E(\kappa_f)\right)^{1/2}} \sim 10^{-3} \, m^{-1} \tag{21}$$

(With the higher value $\varepsilon \sim 10^{-7} \, m^2 s^{-3}$ (applicable to the upper layers), one finds, by the same calculation, $\kappa_m \sim 3 \; 10^{-4} \, m^{-1} \sim \kappa_w$) .

In the range of frequencies between the Coriolis frequency and the maximum Brunt-Väisälä frequency ($10^{-4} \lesssim \omega \lesssim 10^{-2}$) turbulent eddies are intermingled with inertial-internal waves and although it should be possible to "see" fast internal waves passing through the turbulent eddies, internal waves are often so intense, interacting and diversified that it becomes difficult to separate the waves from the turbulence in a spectral analysis of current data. In fact, interactions and diversification of sources produce, in the internal wave field, an intricated collection of motions of various scales, a continuum of Fourier modes which appear as waves, wave packets or turbulent eddies depending on the distance they can propagate during their life-time (Nihoul, 1972) and which could very well be classed as turbulence if turbulence is defined as a "field of chaotic vorticity" (Saffman, 1968).

MICROSCALE TURBULENCE

In the range of frequencies $10^{-2} \lesssim \omega$, larger than the maximum Brunt-Väisälä frequency, the turbulence is no longer constrained by buoyancy and may be considered as three-dimensional.

Three-dimensional turbulence can be generated in the ocean by shear instability of local currents or by convection in layers with unstable density stratification.

The second mechanism would seem rather exceptional in the ocean where, unlike the atmosphere, the density stratification is always globally stable. However unstable layers may occasionally be produced by cooling of the ocean surface in the winter or by salt accumulation in the sub-surface waters during periods of intensive evaporation and, in deeper waters, by lateral instrusions (e.g. Fedorov, 1978).

One knows very little about convective turbulence in the ocean. It has been suggested that it might contribute something of the order of $10^{-8} m^2 s^{-3}$ to the rate of turbulent energy production ε , in some places.

Mechanical energy production is likely to be more widespread and to occur in all ocean currents whenever the local conditions of stability are not fulfilled. This process can be very important in boundary layers like the bottom boundary layer of the ocean where it may account for a rate of energy production ε larger than $10^{-8} m^2 s^{-3}$ (e.g. Nihoul, 1977), shallow continental seas where amplified tidal currents yield very high dissipation rates and in the upper mixed layer in association with drift currents generated by the wind.

In the upper mixed layer, however, the breaking of surface waves provides a generally more powerful mechanism for the generation of turbulence with rates of turbulent energy production of the order of 10^{-5} - 10^{-6} $m^2 s^{-3}$.

In the deep ocean, local mechanical production of turbulence in macroscale and mesoscale currents is certainly less efficient and the largest estimate is a value of $\varepsilon \sim 10^{-7}$ $m^2 s^{-3}$ for tidal and inertial non-stationary currents, with scales of the order of tens of kilometers (Lemmin et al, 1974).

Excluding boundary layers, the generation of three-dimensional turbulence in the ocean may thus be regarded as being largely the end product of the meso-scale cascade interpreted in terms of non-linear interactions of turbulent blinis or random, ultimately breaking, internal waves.

Microscale ocean turbulence will in general comprise an inertial range, where the energy spectral density $E(\kappa)$ is a function only of the wave number κ and of the energy ε transferred per unit time from one scale to the next, and a molecular range where molecular diffusivities play an essential role.

In the inertial range, similarity arguments predict the spectral law

$$E(\kappa) \sim \varepsilon^{2/3} \kappa^{-5/3} \tag{22}$$

and a similar $\kappa^{-5/3}$ dependence for the temperature fluctuations spectrum.

The inertial range exists for turbulent "frequencies"

$$\omega_\kappa \sim \kappa \, v_\kappa \sim \left(\kappa^3 E(\kappa) \right)^{1/2} \tag{23}$$

smaller than the maximum Brunt-Väisälä frequency N_{Max} and larger than the viscous dissipation frequency

$$\omega_{v\kappa} \sim \nu \kappa^2 \tag{24}$$

i.e.

$$N_{Max} \ll \varepsilon_3^{1/3} \kappa^{2/3} \ll \nu \kappa^2 \tag{25}$$

The local rate of energy transfer - and ultimate dissipation rate - ε_3 in patches of microscale turbulence should be considerably larger than the mesoscale value but it is expected to average to the same order of magnitude once vertical intermittency is taken into account.

Taking a conservative value of 1 % for the intermittency factor, one estimates

$$\varepsilon_3 \sim 10^{-6} m^2 s^{-3} \tag{26}$$

Then if $N_{Max} \sim 10^{-2} s^{-1}$ and $\nu \sim 10^{-6} m^2 s^{-1}$ eq.(25) is equivalent to

$$\kappa_N \sim 1 \, m^{-1} \ll \kappa \ll \kappa_v \sim 10^3 m^{-1} \tag{27}$$

so that the inertial range should be expected between typical scales of 1 metre and 1 millimetre.

Beyond κ_v, $E(\kappa)$ falls off rapidly as a result of viscous dissipation but the spectra of temperature and salinity fluctuations continue to higher wave numbers (with a κ^{-1} slope) before diffusion becomes significant and produces and exponential cut-off (Batchelor, 1959).

The predictions of the similarity analysis are confirmed by the observations in the ocean (e.g. Grant et al, 1968).

The persistence of temperature and salinity fluctuations of significant level beyond the viscous cut-off wave number κ_v is obviously a result of the smaller diffusivities of heat and salt.

The same persistence is likely to be observed in time, i.e., a patch of decaying turbulence will retain temperature and salinity heterogeneities of a given scale ℓ longer than velocity fluctuations of the same scale. The possibility of such "fossil turbulence" should not be overlooked when interpreting data based on temperature fluctuations.

THE CLIMATOLOGY OF OCEAN TURBULENCE

According to Munk (1966) a vertical eddy diffusivity of some $10^{-4} m^2 s^{-1}$ would be necessary to explain the vertical heat flux required in the ocean interior by global balances. Garrett (1979), reviewing the results of direct measurements, found that the experimental values were as a rule considerably smaller than Munk's theoretical estimate and suggested that perhaps isolated regions of very intense mixing existed which, somehow, the instruments had missed .

The discrepancy between the theoretical estimate and the experimental values cannot, according to Garrett (1979) be explained by the intermittency *in the vertical* (i.e. the unlikelihood of detecting a mixing event on a cast) but it might be related to the intermittency of mixing *in the horizontal* and *in time*.

The space and time intermittency of ocean turbulence is the inevitable consequence of the multiplicity of sources and physical processes acting over a wide range of scales. If one takes, for instance, the fine structure characteristic of the mesoscales, one finds that three different mechanisms at least may have generated the systems of layers and sheets. It is conceivable that all three operate in the ocean in different regions, or perhaps in the same region at different times, often under very different conditions. The question arises then whether it is possible to predict the mesoscale structure as a function of space and time or if one must regard the mesoscale turbulent blinis as appearing at haphazard with random sizes, locations and durations.

The same is true for energy dissipation or mixing intensity. Is it possible to survey the ocean, theoretically and experimentally and resolve the ocean dynamics, both in space and in time, with sufficient precision to associate numbers with given regions of the ocean and given periods of time ? Is this impossible and, then, is a statistical approach based on a limited number of indicative information the only feasible one ?

The answer may be yes to both questions.

To a certain extent, it should be possible to identify distinct ocean situations (semi permanent currents, intrusive plumes, down- and up-welling areas, regions of frontogenesis...), to ascertain the mechanisms that are actually operating and see that they are properly parameterized and to chart orders of magnitude showing the "climatology" of ocean turbulence.

To a large extent, however, this type of coarse climatology will
not provide enough information on local instantaneous events to allow
a deterministic approach. This means, to take a specific example,
that it will not be possible to model the dispersion of a pollutant
over scales going from 10^2m to 10^4m in the mesoscale range, taking
into account the actual mechanisms which are responsible for the stirring, the mixing and the diffusion of the contaminant. Apart from
measuring them simultaneously on the spot - in which case, it would
be much simpler to measure the pollutant's concentrations directly -
one can indeed have only statistical information of what they may be
and what they may do. If these information are properly included
into the parameterization, the model will be able to predict the extent of the contamination but not to reproduce the detailed dispersion pattern one might actually observe. Repeating the experiments
many times in the same conditions and superposing the observations,
however, a "mean" pattern would soon emerge which would look more and
more like the model's predictions.

MODEL TURBULENCE FOR A MODELLED OCEAN

Modellers deal with averages taken over ensembles of identical
oceans and, in doing so, they approach ocean variability with a re-
solute turbulence point of view.

Thinking in terms of turbulence, one likes to associate energy to
each scale of motion and one tends to regard the transfer of energy
from scale to scale as the essential mechanism of turbulence dynamics.
The rate of downscale energy transfer ε - which may have very little
to do with the rate of energy dissipation, at least its local value -
is looked upon as the cogent factor which, associated with the wave
number κ (the inverse of the length-scale), determines the strength
of the turbulence and the efficiency of mixing.

This is of course nothing but the classical Kolmogorov theory of
three-dimensional turbulence and many will undoubtedly object, to its
application to the ocean, that the conditions of its validity are far
from being satisfied in macroscale and mesoscale turbulence.

It is true that Kolmogorov's theory is essentially designed for
mechanical turbulence in which large turbulent eddies are hydrodynamically unstable and disintegrate into smaller eddies transferring to
them their kinetic energy (at a constant rate ε) until the eddies
are small enough to be stabilized by viscous dissipation.

Ocean turbulence is evidently more complex but modellers will argue that the complicated processes which have been described as occuring in the macroscale and mesoscale ranges, simply explain, something which is self-evident in three-dimensional mechanical turbulence, i.e. why and how large eddies are unstable and finally transfer energy downscale to the dissipation range. The mechanisms of instabilities, subject to the geophysical constraints on the ocean, are obviously rather sophisticated. They involve conversion between kinetic and potential energy, energy transport by wave motions in physical space, trapped oscillations at the epidermis of turbulent blinis and, possibly, a form of "Echternacht procession" in the macroscales where energy goes from the large scale currents to the synoptic eddies and from them backwards - but not entirely - to the gyre circulation, via Rossby waves or more direct mechanisms.

But the detailed machinery is perhaps not important.

Macroscale quasi-twodimensional turbulence is overwhelmed by the enstrophy cascade to smaller scales, but the arresting result is the subsequent generation of fronts which generate large mesoscale eddies, - either directly or, indirectly, through the formation of intrusive layers -, of just the right scale to be grappled by inertial and tidal waves and initiate the cascade "layers - long waves - turbulent patches - layers - internal waves - turbulent patches ... microscale turbulence".

An individual turbulent blini, in the mesoscale range, does not "overturn", but turbulent blinis occur with random sizes, locations orientations and durations and an ensemble average is likely to reflect more the global randomness than any individual stiffness.

Internal waves and mesoscale turbulence are deeply intermingled, but the random field of interacting non-linear rotational internal waves and wave packets differs very little from turbulence and presumably less and less so as individual realizations of such fields are superimposed by ensemble averaging.

On the average, the ocean may turn out to be more turbulent than any particular oceanographic situation might suggest.

Admittedly, the values $\varepsilon_f \sim 10^{-8} m^2 s^{-3}$ and $\varepsilon_3 \sim 10^{-6} m^2 s^{-3}$ for the rates of kinetic energy transfer in the mesoscale and microscale ranges, respectively,are intentionally higher than current estimates trying to conciliate turbulent theories and observations. One must remember however that they are not intended to describe a particular, observed, situation but rather as ensemble averages appropriate to the parameterization of turbulence. The subjacent idea is that oceanic turbulence is highly *intermittent in time* and that the ocean cannot be adequately sampled in this respect ; the most intensive events being associated with severe weather conditions proscribing most of the oceanographic research.

Admittedly also, this is a wager.

REFERENCES

Batchelor, G.K., 1959. Small-scale variation of convected quantities like temperature in turbulent fluid. J. Fluid Mech., 5: 113-139.
Batchelor, G.K., 1969. Computation of the Energy Spectrum in Homogeneous Two-Dimensional Turbulence, Phys. of Fluids, 12 (supplement II): 233-239.
Bell, T.H., 1975. Momentum and energy transport by mesoscale processes in the deep ocean. In: J.C.J. Nihoul (Editor), Proceedings of the Sixth International Liège Colloquium on Ocean Hydrodynamics, April 29 - May 3, 1974. Mémoires de la Société des Sciences de Liège, 7: 241-251.
Brekhovskikh, L.M., Ivanov-Frantskevich, G.N., Koshlyakov, M.N., Fedorov, K.N., Fomin, L.M. and Yampol'skiy, A.D., 1971. Certain results of a hydrophysical experiment on a test range in the tropical Atlantic, Izv. Akad. Nauk. SSSR, Fizika atm. i okeana, 7: 511-528.
Fedorov, K.N., 1978. The thermohaline finestructure of the Ocean, Pergamon, Oxford, 170 pp.
Fofonoff, N.P., 1969. Spectral characteristics of internal waves in the ocean, Deep Sea Res. Supplement to vol. 16: 58-71.
Garrett, Ch., 1979. Mixing in the Ocean Interior, Dynamics of Atmospheres and Oceans, 3: 239-265.
Gill, A.E. and Turner, J.S., 1979. Second Symposium on Turbulence in the Ocean, Ocean Modelling, 25: 1-3.
Grant, H.L., Hughes, B.A., Vogel, W.M. and Moilliet, A., 1968. Some observations of turbulence in and above the thermocline. J. Fluid Mech., 34: 443-448.
Grant, H.L., Moilliet, A. and Vogel, W.M., 1968. The spectrum of temperature fluctuations in turbulent flow. J. Fluid Mech., 34: 423-442.
Koshlyakov, M.N. and Monin, S.A., 1978. Synoptic eddies in the ocean, Ann. Rev. Earth. Planet. Sci. 6: 495-523.
Kraichman, R.H., 1967. Inertial sub-ranges in two-dimensional turbulence, Phys. of Fluids, 10: 1417-1423.
Lemnin, U., Scott, J.T. and Czapski, U.H., 1974. The development from two-dimensional to three-dimensional turbulence generated by breaking waves. J. Geoph. Res., 79: 3442-3448.
Monin, A.S., Kamenkovich, K.M. and Kort, V.G., 1977. Variability of the oceans, Wiley Publ., New York, 241 pp.
Müller, P., 1976. On the diffusion of momentum and mass by internal gravity waves, J. Fluid Mech., 77: 789-823.
Munk, W.H., 1966. Abyssal recipes. Deep Sea Res., 13: 707-730.
Nihoul, J.C.J., 1972. Proper and improper turbulence in the deep sea, Symposia Mathematica, 9: 433-446.
Nihoul, J.C.J., 1975. Modelling of Marine Systems. Elsevier Publ. Amsterdam, 272 pp.
Nihoul, J.C.J. (Editor), 1977. Bottom Turbulence, Elsevier Publ. Amsterdam, 306 pp.
Nihoul, J.C.J., 1979. Turbulence in the Ocean. In: W. Kollmann (Editor), Prediction Methods for Turbulent Flows, Von Karman Institute for Fluid Dynamics Publ., pp. 1-25.

Ozmidov, R.V., 1965. Certain features of the oceanic turbulent energy spectrum, Dokl. Akad. Nauk. SSSR, 161: 828-832.

Panchev, S., 1976. Inertial range spectra of the large scale geophysical turbulence, Bulg. J. of Geophys., 2: 3-11.

Perkins, H., 1972. Inertial oscillations in the Mediterranean. In: J.C.J. Nihoul (Editor), Proceedings of the Third International Liège Colloquium on Ocean Hydrodynamics, May 3-8, 1971, Mémoires de la Société des Sciences de Liège, 2: 43-50.

Rhines, P.B., 1977. The dynamics of unsteady currents. In: E.D. Goldberg, I.N. McCave, J.J. O'Brien and J.M. Steele (Editors), The Sea, Wiley Publ. New York, pp. 189-318.

Ruddick, B.R. and Joyce, T.M., 1979. Observations of interaction between the internal wavefield and low frequency flows in the North Atlantic, J. Phys. Oceanogr., 9: 498-517.

Saffman, P.G., 1968. Lectures on homogeneous Turbulence. In: N. Zabusky (Editor), Topics in Nonlinear Physics, Springer-Verlag.

Tolstoy, I., 1963. The theory of waves in stratified fluids, including effects of gravity and rotation, Rev. Mod. Phys., 35: 207-230.

Turner, J.S., 1973a. Geophysical examples of layering and micro-structure ; interpretation and relation to laboratory experiments. In: J.C.J. Nihoul (Editor), Proceedings of the Fourth International Liège Colloquium on Ocean Hydrodynamics, 20-24 March 1972, Mémoires de la Société des Sciences de Liège, 4: 11-39.

Turner, J.S., 1973b. Buoyancy Effects in Fluids, Cambridge University Press. 368 pp.

Webster, F., 1968. Observations of inertial-period motions in the deep sea, Rev. Geophys., 6: 473-490.

Webster, F., 1969. Turbulent spectra in the ocean. Deep Sea Res. Supplement to vol. 16: 357-368.

Woods, J.D., 1977. Parameterization of unresolved motion. In: E.B. Kraus (Editor), Modelling and Prediction of the upper layers of the ocean, Pergamon Press, Oxford, pp. 118-140.

Woods, J.D., 1978. Fronts in the ocean - a review of physical aspects, SCOR Symposium on Fronts in the Ocean, Brest, France Nov. 1978.

THE INFLUENCE OF MERIDIONAL BOUNDARIES UPON ROSSBY WAVE RECTIFICATION PROCESSES

A. COLIN DE VERDIERE

Centre Océanologique de Bretagne, Brest (France)

ABSTRACT

The conditions under which fluctuating transients can be rectified to give rise to mean currents are investigated for the case of an homogeneous, flat bottommed ocean on a mid latitude beta plane. The role of meridional boundaries is assessed by making use of a simple analytical model in which the various fields are expressed as perturbation expansions in powers of the small non linearity.

A detailed comparison of both bounded and unbounded oceans is made by computing the responses to a wide range of forcing parameters. The strength and distribution of the mean currents in the bounded ocean are very similar to that found in the periodic case, at short forcing scale. Eastward currents are found in the forced area with westward return flows in free regions. Differences arise at long forcing scale : meridional boundaries severely reduce the strength of the mean currents at high frequencies and modify the spatial distribution of the mean flows at low frequencies through the effect of western intensification. Finally, the model indicates that the level of rectified currents induced by the atmospherically generated transients in the synoptic cyclones frequency band is much smaller than the canonical value of the general circulation in mid ocean regions.

INTRODUCTION

The high level of eddy kinetic energy in the world ocean has prompted the study of oceanic models driven by time dependent forces. Pioneering works by Veronis and Stommel (1967), Phillips (1966) have shown that for the energetic frequency range of the wind spectrum, the response of flat bottommed ocean was largely depth independent, i.e. barotropic. Although there is also a significant amount of baroclinic eddy kinetic energy in nature, it is still useful to consider homogeneous ocean models to investigate the possibility of generating mean currents through nonlinear interactions of the transient motions.

Both analytical (Pedlosky (1965) and numerical (Veronis (1970) techniques have been used to look at the pattern of mean currents set up by oscillatory wind stresses in squarebox ocean.

In the former study in which viscous effects exceeded nonlinearity it was found that the structure and the sense of the mean circulation were strongly dependent upon the parameters of the forcing chosen to be harmonic in both space and time. Numerical runs carried out by Veronis at higher amplitude levels showed that cyclonic mean gyres (respectively anticyclonic) appeared in the northern (respectively southern) part of the basin.

Laboratory experiments by Whitehead (1975), Colin de Verdière (1979) have shown that in geometries which keep the geostrophic contours essentially unblocked, strong zonal mean flows were generated, westward in free regions, and eastward over the forced area. Rhines (1977) was able to show that these circulations were the consequence of an irreversible down gradient mixing of potential vorticity. In a bounded geometries (the so called sliced cylinder) similar experiments were carried out by Colin de Verdière (1979). With spatially homogeneous, oscillatory, forcing action, the strength of the mean circulations (if any) were an order of magnitude smaller than eddy velocities preventing evaluation of their structures. Under local forcing action, mean zonally elongated gyres were found, the main feature being a cyclonic circulation lying to the west of the driving at the forced latitudes.

The obvious advantages of these laboratory experiments, the possibilities of observing strongly nonlinear flows, are, however, mitigated by measurement difficulties and friction limited parameter ranges. To make a direct comparison of rectified currents in bounded and unbounded geometries, it was therefore thought that the use of a mathematical model might prove itself more rewarding.

A purely numerical approach was not chosen as computational procedures being strongly geometry dependent, might have obscured the issues. To test the sensitivity of the processes to frequency‾ wavenumber tuning of the forcing and to dissipation, the quasilinear approach of Pedlosky (1965) was found most convenient. Consequently, the present study provides no insight into the dependence of the rectified circulations upon the forcing amplitude as the latter are computed by perturbation expansions in powers of the nonlinearity.

THE MODEL

Only meridional boundaries are introduced in the model ocean to avoid resonance problems and also to represent more closely the geometries of mid latitude beta planes of the real ocean. The details of the model are now described. The equations governing horizontally non divergent, hydrostatic flow in a rotating homogeneous fluid are the following :

$$\underline{u}_t + \underline{u}.\nabla\underline{u} + f \underline{k} \times \underline{u} = -\frac{\nabla p}{\rho_o} - R\underline{u} + \frac{\tau}{\rho_o H}$$

$$\nabla.\underline{u} = 0$$

$$-\rho_o g - P_z = 0$$

Viscous effects are introduced through Ekman type friction, R being the inverse of the spin down time, while $\underline{\tau}$ represents the forcing stress. Because emphasis is placed upon quasigeostrophic motions on a beta plane, fittering of the inertia-gravity waves is necessary. Non dimensionalizing length by L, time by $(\beta L)^{-1}$, speed by U, pressure by $\rho_o f_o LU$ and stress by $U\rho_o H\beta L$ one obtains :

$$\underline{u}_t + M \underline{u} . \nabla \underline{u} + \frac{f}{\beta L} \underline{k} \times \underline{u} = - \frac{f_o}{\beta L} \nabla p - K \underline{u} + \underline{\tau}$$

where $M = U/\beta L^2$ and $K = R/\beta L$.

We will now examine the following limit of this 2 parameters model :

$M \ll 1$

and

$MK^{-1} \ll 1$ (or $U/RL \ll 1$)

A perturbation expansion of the field quantities in powers of M may now be considered:

$$\underline{u} = \underline{u}_o + M\underline{u}_1 + \ldots$$

Introducing a stream function, the vorticity equation of the 0th order problem becomes :

$(1) \nabla^2 \psi^0_t + \psi^0_x = -K\nabla^2\psi^0 + \nabla \times \underline{\tau}$

The first order steady problem gives rise to the following equation :

$(2) K\nabla^2\psi_1 + \psi_{1x} = -\nabla . < q_o' \underline{u}_o >$

In the above, the brackets denotes a suitable time average, and q_o represents the potential vorticity $\mathbf{q}_o + y$. The procedure is then to solve equation (1) for the waves, to compute the wave induced forcing term arising in (2) and finally to solve (2) for the mean ψ_1 field.

As we will consider both periodic and meridionally bounded beta planes it is useful to derive some integral relationships valid in these geometries. Evaluation of the surface integral of equation (2) over the semi-infinite domain bounded by the two north south walls and the arbitrary latitude contour y (pictured on figure 1) yields :

$(3) \oint_\Gamma \underline{u}_1 \ d\underline{s} = < \overline{q_o' v_o'} > /K$

Γ denotes the contour of the semi-infinite domain and the upper bar represents zonal averaging at latitude y. Use has been made of the continuity equation to show that the zonally averaged north south mass transport vanishes. In a periodic domain a similar relation holds, the left hand side of (3) being now simply the zonal average of the zonal component of the flow. The main interest of (3) is obvious as considerable information about both the strength and structure of the mean circulations is available from a simple computation of the zonal average of the meridional flux of potential vorticity induced by the waves. Although we have often illustrated the complete ψ_1 field, repeated use has been made of (3) to compare quantitatively the effects of the different geometries.

With meridional boundaries, the problem now remains to find a suitable forcing function both from physical and computational point of views. Clearly the important forcing aspect is to model its localness as we expect different behaviour in free and forced regions. This leads in turn to a richer scale content than the one used by Pedlosky (1965). Hence the curl of the wind stress is taken here as a periodic, zonally propagating wave whose energy is limited in a narrow band of latitude. This is the first step toward more realistic modelling of the rather stochastic nature of the atmospheric forcing. We now present the solution of (1) with the forcing function defined therefore as :

$$\nabla \underline{x\tau} = G(y)e^{i(\omega t + \alpha x)} \qquad \text{with } G(y) = 1 \text{ for } y < a$$
$$G(y) = 0 \text{ for } y > a$$

The infinite beta plane

Looking for a solution varying like $\phi(y) \, e^{i(\omega t + \alpha x)}$ one finds that $\phi(y)$ satisfies :

$$(4) \frac{d^2\phi}{dy^2} + \ell^2\phi = \frac{G(y)}{i\omega'}$$

where ω' now a complex frequency is equal to : $\omega' = \omega - ik$, while $\ell^2 = \frac{\alpha}{\omega'} - \alpha^2$

The solution of (4) is found after use of the Sommerfeld radiation condition according to which far field motions must decay. It is implemented by choosing the proper complex square root of the meridional wavenumber ℓ . The complete solution may be written for positive y as :

$$(5)\,\psi_o = \mathcal{R}e \left[\frac{e^{i(\omega t + \alpha x)}}{\omega'} \cdot \begin{cases} -\frac{1}{\ell^2} \, e^{i\ell y} \sin\ell a, \text{if } y > a \\ -\frac{i}{\ell^2}(1-e^{i\ell a} \cos\ell y), \text{if } 0 \leqslant y \leqslant a \end{cases} \right.$$

with $\text{Im}(\ell) > 0$ for $y > 0$

The bounded beta plane

With the same forcing function, the added constraint to impose is that :

$$\psi_o = 0 \text{ at } x = 0 \text{ and } x = 1$$

By setting :

$$\psi_o = \psi_o^1 \, e^{i(\omega t + \alpha x)}$$

th equation (1) becomes :

$$(6)\nabla^2\psi_o^1 + \gamma^2\psi_o^1 = \frac{G(y)}{i\omega'} \, e^{i(\alpha-\gamma)x}$$

where : $\gamma = 1/2\omega'$ and $\omega' = \omega - ik$

The x dependence of ψ_o^1 may now be represented in terms of a Fourier series :

$$\psi_o^1 = \sum_1^\infty \varphi_n(y) \sin n\pi x \, ,$$

and φ_n now checks the following equation :

$$(7)\frac{d^2\varphi_n}{dy^2} + \ell^2\varphi_n = G_n \frac{G(y)}{i\omega'}$$

where :

$$G_n = 2n\pi \frac{[e^{i(\alpha-\gamma)}(-1)^n - 1]}{(\alpha-\gamma)2 - n^2\pi^2}$$

and $\ell^2 = \gamma^2 - n^2\pi^2$

As equation (4), (7) is now solved in the infinite y plane using the radiation condition. The complete solution is :

$$(8)\psi_0 = R_e \left| \frac{e^{i(\omega t+\gamma x)}}{\omega'} \sum_1^\infty 2n\pi \frac{[e^{i(\alpha-\gamma)}(-1)^n - 1]}{(\alpha-\gamma)2 - n^2\pi^2} \right| \sin n\pi x \left|
\begin{array}{l}
- \frac{1}{\ell 2} e^{i\ell y} \sin \ell a \text{ for } y > a \\
- \frac{i}{\ell 2}[1 - e^{i\ell a}\cos\ell y] \text{ for } 0 < y < a
\end{array}
\right.$$

with $\text{Im}(\ell)$ for $y > 0$

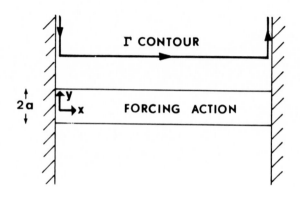

Fig. 1. The geometry of the bounded ocean and the contour Γ used to evaluate the mean circulation.

GENERAL RESULTS OF THE MODEL

Both solutions (5) and (8) have been evaluated numerically for various combinations of the three parameters α, ω and k. It is to note that the Fourier series in (8) is rather slowly converging and sometimes more than 100 modes were necessary to achieve an accuracy of 10^{-4} in the near field. From the stream function fields, variability, Reynolds stresses, potential vorticity fluxes and mean circulation could be evaluated. As oceanic interest is primarily focussed upon inviscid dynamics, values of K/ω much smaller than unity were chosen. The sensitivity to (α, ω) tuning was explored by varying the forcing spatial scale for given high and low frequency runs. In the next two sections, the forcing is taken as moving westward the right direction to excite Rossby waves. A separate section will be presented at the end in which eastward moving disturbances will be considered.

The high frequency case

For large values of the forcing frequencies, Rossby waves may exist only at very large scale and one may expect a priori that the boundary influences will be most severe in such cases. Table (1) summarizes the parameters and bulk statistics of several of these high frequency runs going from large to small scale forcing.

TABLE 1

A few high frequency runs going from large to small forcing scale. Common parameter values are : $\omega = .628$, $\alpha_c = 1.57$, $K = 0.01$

α	α/α_c	KE_o	A	$MAX(\overline{\frac{\partial u_o Y_o}{\partial y}})$	$MAX(\overline{q_o v_o})$	$\overline{U}_1^2$	$\overline{U}_1^2/KE_o$	Geometry[1]
6.210^{-2}	5.910^{-2}	2.410^{-3}	0.6	1.110^{-3}	8.910^{-4}	1.8610^{-3}	0.77	B
		4.710^{-3}	21.6	2.310^{-2}	2.310^{-2}	1.2	2.510^{2}	P
7.910^{-1}	0.5	2.410^{-3}	.58	1.210^{-3}	9.810^{-4}	2.2710^{-3}	0.95	B
		9.510^{-3}	.91	0.12	0.12	21.	2.210^{3}	P
1.57	1.	2.410^{-3}	.58	1.310^{-3}	1.110^{-3}	2.9410^{-2}	1.23	B
		.24	0.018	0.97	0.97	9.410^{-2}	4.10	P
6.28	4.	1.510^{-3}	.41	1.810^{-3}	1.810^{-3}	8.5410^{-3}	5.88	B
		1.7510^{-3}	.4	2.210^{-3}	2.210^{-3}	1.1210^{-2}	6.66	P
31.4	20.	$.10^{-4}$	0.11	2.110^{-3}	2.110^{-3}	4.810^{-3}	50.	B
		.10	0.10	2.10^{-3}	2.10^{-3}	1.810^{-3}	18.1	P

[1] B : Bounded
P : Periodic

In these runs, α_c the high wavenumber cut off ω^{-1} is order one so that the ocean width is too narrow for free waves to propagate. Values of the area averaged kinetic energy KE_o, and of the anisotropy factor A (ratio of zonal over meridional kinetic energy) are given. The two next columns represent the maximum values of the zonally averaged quantities $\overline{\frac{du_o v_o}{dy}}$ and $\overline{q_o v_o}$. Estimates of the mean kinetic energy $\overline{U}_1^2$ contained in the first order mean circulation have also been computed by meridional average of $<\overline{q_o v_o}>^2/K^2$. The ratio of the mean over the fluctuating kinetic energy which is a measure of the efficiency of the rectification processes is shown in the last column.

Looking at first into the runs for which α/α_c is smaller than unity, the unbounded case shows a strong dependence upon α : as α increases, the overall energy level increases as wave crests become oriented more meridionally. In the mean time the strength of the mean circulation and the rectification efficiency also build up. The bounded ocean does not appear to be sensitive to the choice of the forcing scale, as only forced waves are excited. The presence of the boundaries restricts strongly the rectification processes as the mean kinetic energy drops by a factor of at least 10^3 compared to the unbounded case. It is worth looking at the details of the ocean response for a particular run, namely that corresponding to $\alpha/\alpha_c = 0.5$. The zeroth order stream function fields are shown on figure 2.

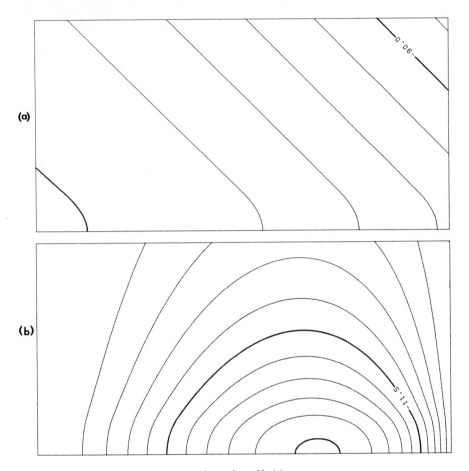

Fig. 2. The zeroth order stream function fields :
a) The periodic ocean
b) The bonded ocean
Non dimensional parameter values are : $\omega = 0.628$, $\alpha/\alpha_c = 0.5$, $K = 0.01$

North of the forced region of the unbounded ocean the shape of the wave crests implies the usual negative $\langle u_o v_o \rangle$ correlation. This well known signature of rectification is not apparent when boundaries are present. In that latter case, the forced field looks more isotropic and exhibits also an interesting flow intensification in the eastern basin. The associated vorticity fluxes are shown in figure 3.

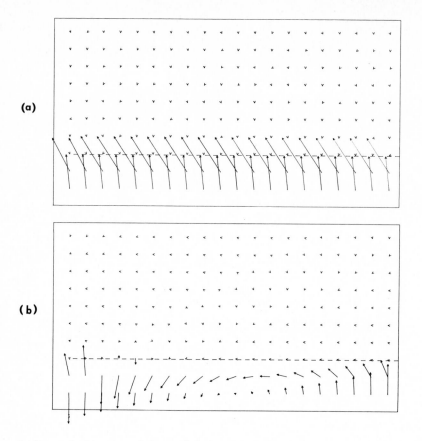

Fig. 3. Maps of the potential vorticity fluxes $\langle q \, u_o \rangle$
a) The periodic ocean
b) The bounded ocean
Parameters values are as in figure 2. The dashed horizontal line is the limit of
the forcing action.

The picture of the unbounded ocean is simple enough : large up gradient transport
of potential vorticity are found in the forced regions whereas weak down gradient
transports occur in the far field. A definite zonal structure appears over the forced
latitudes of the bounded ocean : significant down gradient (respectively up gradient)
transport of potential vorticity now occurs in the western basin (respectively eastern)
while the amplitudes are fully comparable with the preceding case. However,
after zonal averaging the net potential vorticity fluxes are strongly reduced over
the forced area, yet again comparable in free regions. The meridional structure of
these statistics can be appreciated on figure 4(a). The ultra long free scale of
the unbounded cas is now replaced by a much shorter forced scale.

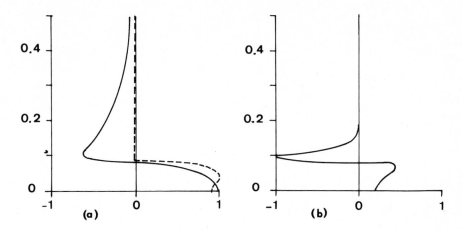

Fig. 4. The zonally averaged potential vorticity fluxes $\overline{\langle q_o v_o \rangle}$ at high frequency ($\omega = .628$).

a) Long scale forcing $\alpha/\alpha_c = 0.5$. Both profiles are normalised by their maximum values. The dashed curve corresponds to the periodic case and the plain one to the bounded case.
b) Short scale forcing $\alpha/\alpha_c = 20$. The profiles are identical in either geometries.

As α/α_c becomes greater than one, no free Rossby waves can be excited in either geometries and the statistics show little sensitivity to the variation in the spatial scale of the forcing. The influence of the boundaries appears negligible and limited to narrow boundary layers occurring on both eastern and western walls. The vorticity fluxes and the rectification efficiency are now much smaller in the periodic case as the motion fields mirrors closely the forcing pattern. The zonally averaged vorticity fluxes are shown on figure 4(b). The short e folding meridional scale is a forced scale of the order of the zonal forcing scale.

As a summary, high frequency waves do rectify mean circulations in both geometries. The distribution of the zonally averaged vorticity fluxes indicate that mean eastward jets will be found over forced regions with westward return flows at free latitudes. The complete mean circulation maps which can be obtained in the bounded ocean by solving (2) indicate a zonally elongated, zonally symmetrical cyclonic gyre (respectively anticyclonic) centered around the northern forcing discontinuity (respectively southern). The rectification efficiency is much stronger in the periodic case for long wave excitation while for short waves the efficiency drops severely and becomes comparable in both geometries.

The low frequency case

This is the most interesting parameter range. When illustrating the solutions given by (8) some care must be exercised to ensure proper field representation. If the forcing combines long spatial scale with low frequency, the bounded ocean reacts by generating short scale waves which must be properly resolved. We present in table 2 the results of 4 runs carried out at the same low frequency for increasing forcing scale.

TABLE 2

A few low frequency runs going from large to small forcing scale. Common parameters values are $\omega = 2.51 \ 10^{-2}$, $K = 4.10^{-4}$

α	α/α_c	KE_o	A	$MAX(\overline{u_o v_{oy}})$	$MAX(\overline{q_o v_o})$	$\overline{U}_1^2$	$\overline{U}^2/KE_o$	Geometry[1]
3.14	$7.8 \ 10^{-2}$	16.9	9.10^{-2}	49.	4.7	5.10^8	2.910^7	B
		2.17	10.15	18.	18.	2.410^8	1.310^7	P
19.8	0.5	1.97	.61	40.	41.	8.610^8	4.310^6	B
		1.69	.77	39.	39.	7.710^8	4.5619^8	P
39.75	1.	14.8	5.10^{-2}	120.	160.	2.6810^{10}	1.810^9	B
		55.	10^{-2}	260.	260.	1.0710^{11}	1.9410^9	P
62.8	1.57	0.128	4.410^{-2}	3.2	3.1	4.9510^6	3.8710^7	B
		.11	3.10^{-2}	3.1	3.1	8.110^6	7.410^7	P

[1]
B : Bounded
P : Periodic

Discussing the gross well known features such as kinetic energy, anisotropy and dominant scales is easy in the inertial limit.

From inspection of the solutions, the kinetic energy grows definitely with increasing meridional scale. In the periodic ocean, the dispersion relation is : $\ell^2 + \alpha^2 = \alpha\alpha_c$ where α_c is the wavenumber cut off. In the bounded ocean, the zonal scale is not the forcing scale. That scale is chosen internally when values of n are such that :

$$\left| (\alpha-\gamma)^2 - n^2\pi^2 \right| \ll 1 \text{ with } \gamma = \alpha_c/2$$

For such values the kinetic energy grows correspondingly. The associated meridional scale is found from the dispersion relation which is now :

$$\gamma^2 = n^2\pi^2 + \ell^2$$

These simple considerations show that the kinetic energy will be large when α is much smaller or of the order of α_c, parameter ranges for which the meridional scale will be very large. Above the cut off scale energy levels drop off quickly. The only important difference between the two geometries occurs for long forcing scale ($\alpha/\alpha_c \ll 1$) : the bounded regime exhibits the short scales of reflected waves at the western boundary and the kinetic energy is correspondingly enhanced. This leads

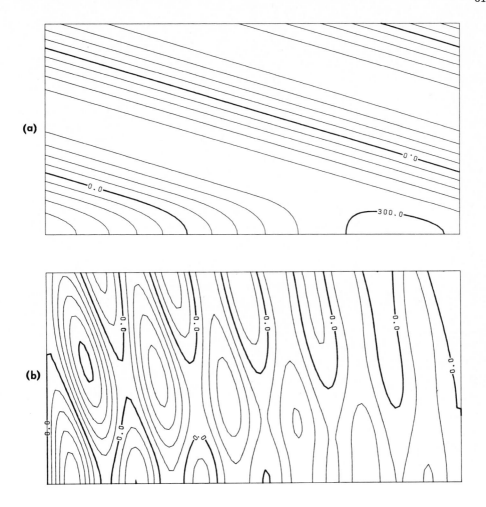

Fig. 5. The zeroth order stream function fields at long forcing scale.
(a) The periodic ocean
(b) the bounded ocean
Non dimensional parameter values are : $\omega = 2.51 \ 10^{-2}$, $\alpha/\alpha_c = 7.8 \ 10^{-2}$, $K = 4.10^{-4}$

to very different wavecrest orientations (zonally elongated in the periodic ocean,
meridionally elongated in the bounded one).

Looking at the parameters in table 2 which gives an indication about the strength
of the first order mean currents, one observes a great increase of the vorticity
fluxes at low frequencies. It is interesting to compare with the high frequency
runs of Table 1. In both cases, the ratio of K/ω has been kept the same, and thus
lowering of the frequency increases the rectification efficiency by at least 5 orders
of magnitude. Before describing further the results it is useful to visualize the
wave fields. To characterize the fields statistically we have also constructed maps

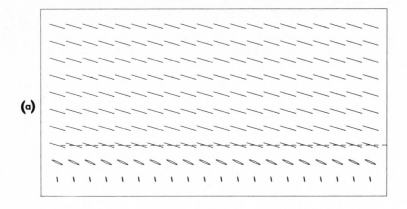

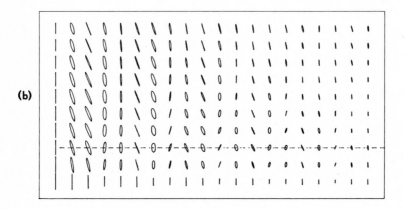

Fig. 6. Map of the variability ellipses
(a) the periodic ocean
(b) the bounded ocean
Parameter values are as in figure 5. The dashed horizontal line is the limit of
the forcing action.

of the variability ellipses whose characteristics are found by diagonalisation of
the tensor :

$$\begin{bmatrix} \langle u_o^2 \rangle & \langle u_o v_o \rangle \\ \langle u_o v_o \rangle & \langle v_o^2 \rangle \end{bmatrix}$$

Figure 5 and 6 depicts the results for long wave excitation. Western intensification
introduced by the slow, dissipating short waves is now evident and the corresponding
fine structure appears in the ellipse orientations and sizes in the bounded ocean.
At smaller forcing scale (figure 7) western intensification disappears whereas with
forcing excitation at the cut off wavenumber (figure 8) a peculiar eastern intensification of the wave field occurs.

The zonally averaged potential vorticity fluxes for these 4 runs are shown in
figure 9.

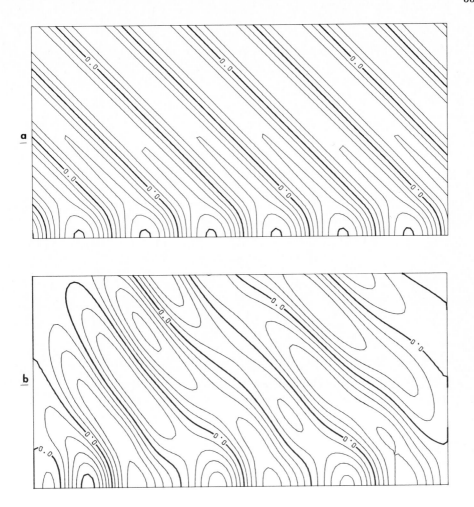

Fig.7. The zeroth order stream function fields :
(a) the periodic ocean
(b) the bounded ocean
Non dimensional parameter values are : $\omega = 2.51 . 10^{-2}$, $\alpha/\alpha_c = 0.5$, $K = 4 \ 10^{-4}$

At forcing scales smaller than the cut off scale the influence of the boundaries is small. Most of the rectification occurs in the near field. At the cut off scale stronger far field westward currents are now excited. It is for this tuning that rectification processes are at a maximum. The boundaries however reduce signifi-cantly the mean flows while rectification efficiencies are comparable. This is occur-ring through the long waves which are generated in the bounded ocean to satisfy the no energy flux requirement at the walls. These must also be responsible for the eastern instensification of the wave field as seen on figure 8. This latter effect leads to a net up gradient (positive) vorticity flux over the basin. The

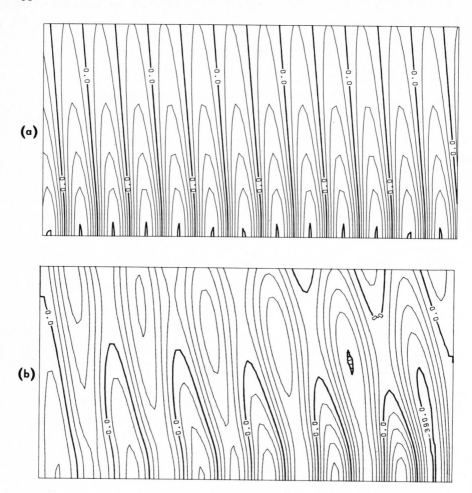

Fig. 8. The zeroth order stream function fields at the cut off scale.
(a) the periodic ocean
(b) the bounded ocean
Non dimensional parameter values are : $\omega = 2.51 \ 10^{-2}$, $\alpha/\alpha_c = 1.$, $K = 4.10^{-4}$

relation between the vorticity flux and the Reynolds stresses is :

$$(9) \ \overline{q_o v_o} = \frac{1}{2} \cdot \frac{\partial}{\partial x} (\overline{v_o^2} - \overline{u_o^2}) - \frac{\partial \overline{u_o v_o}}{\partial y}$$

It is the first term on the right hand side of (9) which starts playing a role in a bounded basin as zonal assymetry appears.

With forcing scale increasing further the amplitude of the mean and wavy currents decreases. When α/α_c reaches 1/2 it goes through a minimum. When this occurs, the wall influence appears again negligible. This is not surprising when one recalls from the Rossby wave dispersion relation that as the zonal energy radiation is zero in this parameter regime, the meridional walls should not induce new waves

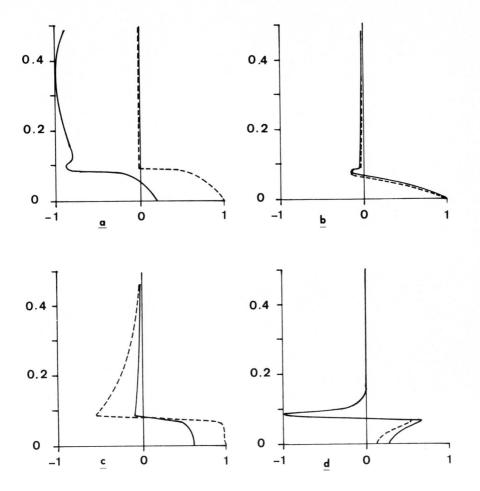

Fig. 9. The zonally averaged potential vorticity fluxes $\overline{\langle q_0 v_0 \rangle}$ at low frequencies
($\omega = 2.51\ 10^{-2}$, $K = 4.10^{-4}$)
(a) $\alpha/\alpha_c = 7.8\ 10^{-2}$
(b) $\alpha/\alpha_c = 0.5$
(c) $\alpha/\alpha_c = 1.$
(d) $\alpha/\alpha_c = 1.57$
The profiles are normalized by the largest values occurring in either geometries.
The dashed curves correspond to periodic oceans and the plain curves to bounded ones.

through reflexion processes. For illustrative purposes we show in figure 10 a
map of the vorticity fluxes in both geometries. Although the zonally averaged values
are similar, much more structure is present in the bounded ocean.

As α/α_c gets much smaller than unity, the amplitude of the motions increases
again specially in the bounded case for reason noted earlier. As noticed on figure 8
the fluxes are now widely different in both geometries. Much more meridional structure
appears now in the bounded case. However more drastic is the western signification

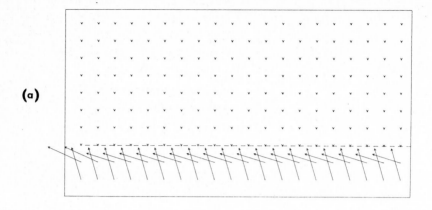

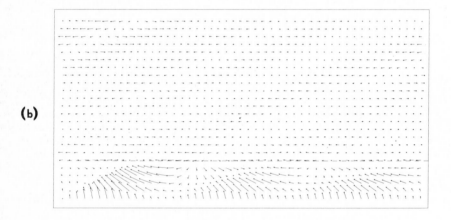

Fig. 10. Maps of the potential vorticity fluxes $\langle q\,u \rangle$
(a) The periodic ocean
(b) The bounded ocean
Parameter values are as in figure 7.

effect which shows up as inducing a net down gradient (negative) vorticity over
the basin as indicated by (9). Clearly the interior anticyclonic circulation will
be much more pronounced in this instance as compared to the periodic case. This
appears to be caused by the short Rossby waves, slowly propagating and with meridional
wavecrests, which are generated at the western wall.

Although the zonally averaged vorticity fluxes gives in a simple way quantitative
information about the mean flows, it is useful to describe the complete circulations
which arise. The periodic case is simple as the zonally averaged velocity is directly
related to the vorticity flux. Eastward currents are found in the forced area while
westward currents are found in the interior. In the bounded case, however, we need
to solve (2). At short forcing scale compared to the width of the basin, the circu-

lation always consists of a zonally elongated western intensified cyclonic (resp. anticyclonic) gyre centered at the northern forcing discontinuity (resp. southern). Clearly most of the divergence of the vorticity fluxes occurs there. At forcing scale long compared to the cut off scale, with western intensification dominant the circulation is very different. Along the northern forcing limit a zonally elongated, western intensified cyclonic gyre is found; to the south (over the forced area) isotropic gyres of alternating signs and of decreasing intensity to the east are observed. Corresponding cells but meridionally elongated exist north of the forcing limit, the scale of which appears closely related to the scale of the underlying Rossby waves. It has been noticed in several similar runs that the sense of the most westward of these gyres is always anticyclonic. This is consistent with the large and negative interior values of the zonally averaged potential vorticity fluxes.

To summarize at low frequency forcing the presence of the boundaries does not appear to modify drastically the bulk energy of the mean circulation found in the periodic case. The structure however are different : zonally elongated, western intensified circulations (cyclonic to the north of the forcing) are found for short scale excitation while for scales largely exceeding the cut off scale, shorter scale structure appears in the mean circulations schemes which tend to be western intensified and meridionally elongated. However "on the average", the senses of the circulations found in periodic beta plane oceans, namely eastward currents over the forced area and westward currents in the far field, free regions are not modified by the presence of the boundaries.

OCEANIC APPLICATIONS

Although the present study is only of qualitative relevance to the real ocean which is much more non linear than viscous, it is interesting to compute the ocean response to characteristic atmospheric forcing. A few runs are therefore presented when the driving pattern moved east, a more common occurrence for the atmospheric synoptic cyclone whose time scales occur in the 3-15 days band and with spatial scales of the order of the basin scale. With a typical ocean width of 5000 km, and a beta plane centered at 45°, the response to forcing with periods of 3.6 and 14.5 days respectively and half wavelength of 2500 km have been computed. We have chosen for a friction coefficient a value equals to Phillips's (1966) historical value. This implies an Ekman decay time of 260 days.

The non dimensional parameters and statistical indicators of the runs are shown in Table 3.

38

TABLE 3

A few runs with atmospheric forcing characteristics. The geometry is indicated by B (bounded) or P (periodic) whereas the forcing sense is indicated by W (westward) or E (eastward). The friction coefficient is $K = 4.10^{-3}$

ω	α	α/α_c	KE_o	A	$\overline{U}^2_1$	$\overline{U}^2_1/KE_o$	Forcing/Geometry
0.251	6.28	1.57	$2.23 10^{-2}$	0.24	9.12	$4.1 10^2$	B - W
0.251	6.28	1.57	$2.75 10^{-2}$	0.22	15.1	$5.5 10^2$	P - W
0.251	6.28	1.57	$.37 10^{-2}$	.66	.61	$1.66 10^2$	B - E
0.251	6.28	1.57	4.10^{-3}	.7	.71	$1.78 10^2$	P - E
$6.28 10^{-2}$	6.28	.39	.58	.84	$6.1 10^5$	$1.05 10^6$	B - W
$6.28 10^{-2}$	6.28	.39	.55	1.29	$6.25 10^5$	$1.13 10^6$	P - W
$6.28 10^{-2}$	6.28	.39	3.10^{-1}	.84	75	$2.5 10^3$	B - E
$6.28 10^{-2}$	6.28	.39	$.22 10^{-1}$	1.14	112	$5.2 10^3$	P - E

As expected the prograde (eastward) sense of forcing propagation reduces the wave energy level by a factor of 10 and in a manner rather independent of meridional boundaries. At high frequencies the mean values are decreased similarly and this leads to rectification efficiencies virtually unchanged. However at lower frequencies for which possible free Rossby waves may be excited it is seen that the mean flows are much weakened if the forcing moves eastward. In that latter case the rectification efficiency drops by 3 orders of magnitude. The zonally averaged vorticity fluxes are shown on figure 11 for the prograde case.

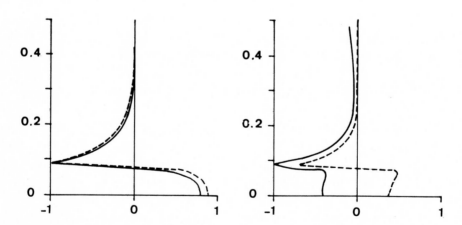

Fig. 11. The zonally averaged potential vorticity fluxes $\langle \overline{q_o v_o} \rangle$ for westerly forcing propagation.
(a) high frequency : $\omega = .251$ (3.6 days)
(b) low frequency : $\omega = 6.28 \ 10^{-2}$ (14.5 days)
The dashed curves correspond to periodic oceans and the plain curves to bounded ones. The profiles are normalized by the largest values occurring in either geometries

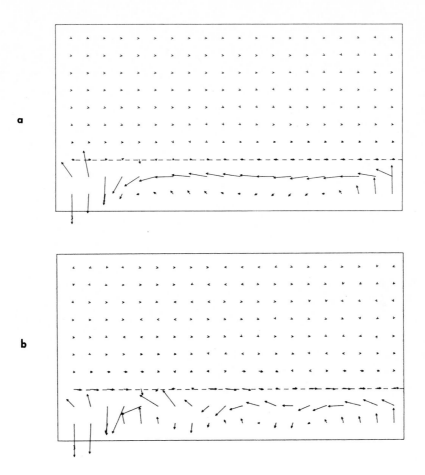

Fig. 12. Maps of potential vorticity fluxes $\langle q_o u_o \rangle$ for westerly forcing propagation in a bounded ocean.
(a) high frequency ω = .251 (3.6 days)
(b) low frequency ω = 6.28 10^{-2} (14.5 days)

Although rather similar at high frequencies again interesting differences appear at low frequencies. Besides the systematic down gradient flux found in the bounded case, it is found that the presence of the walls allows much more north south energy penetration with a prograde forcing. This is increasingly true at lower frequencies and tells us that typical atmospheric motions are able to excite some free oceanic waves through the interplay of the north south walls.

Figure 12 presents the maps of the vorticity fluxes at both high and low frequencies for westerly forcing propagation. The high frequency run reveals the familiar wall influence found earlier : large values of down gradient flux (respectively up gradient) occurs in boundary layers occurring along the western (respectively eastern wall). At low frequencies the usual zonal western intensified structure

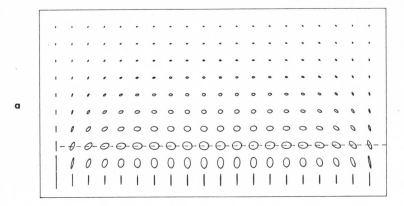

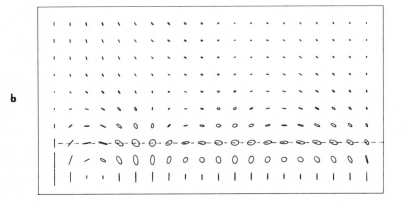

Fig. 13. Maps of variability ellipses for westerly forcing propagation in a bounded ocean.
(a) high frequency ω = .251 (3.6 days)
(b) low frequency ω = $6.28 \, 10^{-2}$ (14.5 days)

appears. For possible relevance to an analysis of current meter records we show in figure 13 the maps of the variability ellipses in this atmospheric parameter regime. Clearly simple shapes of ocean boundaries add already much structure and inhomogeneitics to these fields of variabilities and complicates seriously the task of documenting wind-current correlations and directions of "current energy" from oceanic observations.

Finally it is instructive to find if the present theory may predict valuable quantitative values for the rectified fields in the real ocean. With an ocean width a of 5000 kms, typical eddy current speed U of 5 cm/s, and a beta plane centered at 45°, the non linearity coefficient is :

$$M = \frac{U}{\beta a^2} \sim O(10^{-4})$$

The present approach with field perturbations in powers of M will be valid if M^2 times the rectification efficiency U_1^2/KE_o is much smaller than unity. A quick look at tables 1, 2, 3 shows that the present theory is correct at high frequencies but unfortunately breaks down at low frequencies just when the efficiency of the rectification process is getting strong. Dimensional periods of the order of 25 days might therefore be a maximum for an application of the present theory.

The results of table 3 allow us to evaluate the strength of the mean circulations generated by the atmospheric synoptic cyclones. Rather than evaluating the level of fluctuating currents induced by the wind stress curl, we assume a typical observed value (5 cm/s) for the background level of wavy currents. In that instance, the magnitude of the rectified currents is of the order of $2 \cdot 10^{-2}$ cm/s, being at most 0.5cm/s for the most favorable (but unlikely) retrograde low frequency case. Thus the present calculations indicate that the rectification of atmospherically forced transients in the synoptic cyclone frequency band produces mean currents which are rather too weak to be of importance in the general circulation scheme of the ocean. It is only close to the forcing region and to the western boundary that possibly stronger mean velocities could be found locally. This conclusion was also reached in a pure numerical study by Willebrand et al. (1979), who used a more exhaustive model including stratification and topography. They attributed the small amount of rectification to the stochastic nature of the forcing. In view of the present study we would rather attribute the results to the high frequency band of the atmospheric forcing : the rectification efficiency, large at low frequencies, drops off quickly at high frequencies. However it is true that more realistic modelling of the driving and higher values of the virtually unknown friction coefficient would weaken further the values of the rectification efficiencies found in the present study.

Finally the present calculations suggest that it is in the low frequency band (periods in excess of 30 days) that more important applications could be found. The observed high energy and long time scales of baroclinic oceanic mesoscale eddies suggest them as likely candidates, but it is then necessary to take non linearity and stratification fully into account.

REFERENCES

Colin de Verdière, A., 1979. Mean flow generation by topographic Rossby waves. Journal of Fluid. Mechanics, vol. 94, 1 : 39-64.
Colin de Verdière, A., 1979. Quasigeostrophic turbulence in a rotating homogeneous fluid. Submitted to : Geophysical and Astrophysical Fluid Dynamics.
Pedlosky, J., 1965. A study of the time dependent ocean circulation. Journal of the Atmospheric Sciences, vol. 22 : 267-272.
Phillips, N., 1966. Large scale eddy motions in the western Atlantic. Journal of Geophysical Research, vol. 71, n° 16 : 3883-3891.
Rhines, 1977. The dynamics of unsteady currents. The Sea, vol. 6, Marine Modeling : 189-318.
Veronis, G., 1970. Effect of fluctuating winds on ocean circulation. Deep-Sea Research, 17 : 421-434.

Whitehead, J., 1975. Mean flow driven by circulation in a beta plane. Tellus, 27 : 358-364.
Willebrand, J., Philander, G., Pacanowski, R., 1979. The oceanic response to large scale atmospheric disturbances. To appear in Journal of Physical Oceanography.

SPECTRAL STRUCTURE OF HORIZONTAL OCEANIC TURBULENCE -
SEMI-EMPIRICAL MODELS

S. PANCHEV

Department of Meteorology, Faculty of Physics, University of Sofia
(Bulgaria).

INTRODUCTION

Geophysical Fluid Dynamics (GFD) is the science of atmospheric and
oceanic motions. These motions are characterized by velocities much
smaller than the sound speed ; they have quite wide space-time spectra;
they exist on the rotating Earth and are always turbulent. It seems
to us quite reasonable to call this turbulence Geophysical Turbulence
(GT).

So far only the properties of the small scale three-dimensional
geophysical turbulence have been well investigated both theoretically
and experimentally. Now the attention is directed towards the large
scale quasi-two-dimensional (Q2-D) turbulence on the rotating Earth,
with or without topography, both in the atmosphere and in the ocean.
Among many important applications of the results from such studies we
would like to mention only the problem of predictability which, in
the modern formulation due to Lorenz (1969), Leith (1971) and others,
can be solved if we know the behaviour of the spectra of the corresponding motions. Therefore, the determination of the spectra of various vector and scalar quantities characterizing the motion, and
first of all of the kinetic energy is of primary interest.

There is no comprehensive and uncontradictory theory capable of
predicting such spectra under various conditions. Nevertheless, related scientific problems and also the practice demand for the information. In view of this situation, semi-empirical approaches, as a
temporary solution, may have great value.

The purpose of this paper is to present and discuss some of the
existing semi-empirical models for the structure of two-dimensional
turbulence and their applicability to the horizontal oceanic turbulence.

INERTIAL TRANSFER IN 2-D TURBULENCE

According to Batchelor (1965) and Kraichnan (1967), there exist two inertial subranges as shown in Fig. 1. :

$$E_1(K) \sim \eta_1^{2/3} K^{-3} \quad , \quad \varepsilon_1(K) \equiv 0 \quad ,$$

$$\eta_1(K) = \eta_1 = \text{const} > 0 \quad , \quad K \geq K_* \quad , \tag{1}$$

and

$$E_2(K) \sim |\varepsilon_2|^{2/3} K^{-5/3} \quad , \quad \eta_2(K) \equiv 0 \quad ,$$

$$\varepsilon_2(K) = -|\varepsilon_2| = \text{const} < 0 \quad , \quad K \leq K_* \quad . \tag{2}$$

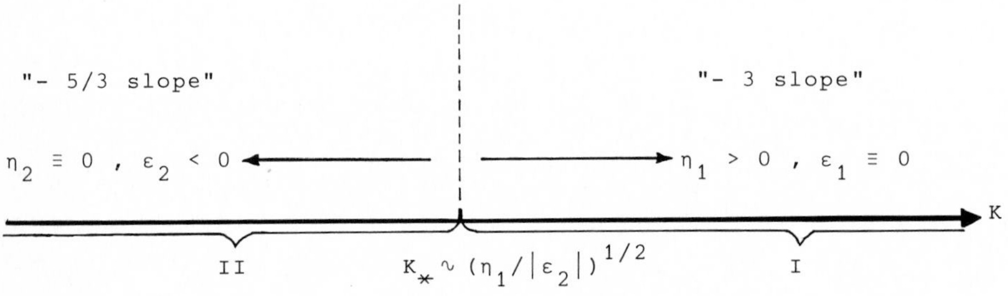

Fig. 1. Inertial subranges in 2-D turbulence.

In (1) and (2) $\varepsilon(K)$ and $\eta(K)$ are the energy and enstrophy transfer function, K_* -the wave number where energy and enstrophy are supposed to be generated, is a point of discontinuity for $\varepsilon(K)$ and $\eta(K)$, but the energy spectrum $E(K)$ is continuous, i.e.

$$E_1(K_*) = E_2(K_*) \quad . \quad \text{Hence} \quad K_* \sim (\eta_1/|\varepsilon_2|)^{1/2} \quad . \tag{3}$$

The simplest way to derive the -3 and -5/3 power laws (1) and (2) is to use Leith's (1968) diffusion approximation

$$\varepsilon(K) = -\gamma_1 K^{-1} (d/dK) K^{9/2} E^{3/2}$$

$$\eta(K) = -\gamma_2 K^3 (d/dK) K^{5/2} E^{3/2} \tag{4}$$

where $d\eta/dK = K^2 d\varepsilon/dK$ and $\gamma_1 \neq \gamma_2$. Actually, integrating (4) one obtains (1) and (2).

NON INERTIAL TRANSFER IN 2-D TURBULENCE

The simplest way to account for the existence of dissipative factors is to introduce a linear drag in the vorticity equation

$$\frac{\partial}{\partial t}\, \nabla^2\psi + J(\psi,\, \nabla^2\psi) = -R\nabla^2\psi \tag{5}$$

where $\psi(x,y,t)$ is the stream function and R the constant drag coefficient. Two spectral equations correspond to the eq. (5) - one for the energy transfer and one for the enstrophy transfer :

$$- d\varepsilon(K)'/dK = 2RE(K) \quad , \quad K < K_*$$
$$\tag{6}$$
$$- d\eta(K)/dK = 2RK^2E(K), \quad K > K_*$$

The combination between (6) and (4) yields two non linear differential equations of the second order. To simplify the solution, following Lilly (1972) we shall adopt another "local" approximation, originally proposed by Kovasznay (1948) :

$$\varepsilon(K) = \left(\frac{K^{5/3}E(K)}{\alpha}\right)^{3/2} \quad , \quad \eta(K) = \left(\frac{K^3E(K)}{\beta}\right)^{3/2} \, . \tag{7}$$

With this approximation, eqs. (6) are of the first order. After integrating, we obtain

$$E(K) = \alpha^3 R^2 K^{-5/3} \left(K^{-2/3} - K_L^{-2/3}\right)^2$$
$$\tag{8}$$
$$\varepsilon(K) = \alpha^3 R^3 \left(K^{-2/3} - K_L^{-2/3}\right)^3 \quad , \quad K_L \leq K \leq K_*$$

and

$$E(K) = \beta^3 R^2 K^{-3} \left(\frac{2}{3} \ln \frac{K_H}{K}\right)^2$$
$$\tag{9}$$
$$\eta(K) = \beta^3 R^3 \left(\frac{2}{3} \ln \frac{K_H}{K}\right)^3 \quad , \quad K_* \leq K \leq K_H$$

where K_L and K_H are low- and high cut-off wavenumbers at which $E(K_L) = E(K_H) = 0$, α and β are (possibly) universal constants. By numerical simulation of 2-D turbulence, Lilly estimated $\alpha \approx 6$, $\beta \approx 3,6$.

APPLICATION TO THE HORIZONTAL OCEANIC TURBULENCE

J. Bye (1975) pointed out that, because of the remarkable assymptotic behaviour of the solution (8), one can assume that a complete energy spectrum consists of <u>two</u> ranges of the form (8) with an energy flux to high and low wavenumbers from a single spectral energy source at the intermediate forcing wavenumber K_*, Fig. 2 (see also Fig. 1).

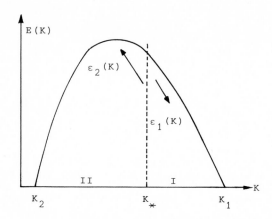

Fig. 2. The composite energy spectrum (after J. Bye).

He argues that this model is very appropriate for the description of oceanic turbulence. Actually, writing (8) once for the range I with E_1 , ε_1 , α_1 , K_1 :

$$\varepsilon_1(K) = \alpha_1^3 R^3 \left(K^{-2/3} - K_1^{-2/3}\right)^3$$

$$E_1(K) = \alpha_1^3 R^2 K^{-5/3} \left(K^{-2/3} - K_1^{-2/3}\right)^2$$

and then for the range II with E_2 , ε_2, α_2 , K_2 :

$$\varepsilon_2(K) = \alpha_2^3 R^3 \left(K^{-2/3} - K_2^{-2/3}\right)^3$$

$$E_2(K) = \alpha_2^3 R^2 \left(K^{-2/3} - K_2^{-2/3}\right)^2 . K^{-5/3}$$

and equating $E_1(K_*) = E_2(K_*)$ we obtain

$$\alpha_1^3 \left(K_*^{-2/3} - K_1^{-2/3}\right)^2 = \alpha_2^3 \left(K_*^{-2/3} - K_2^{-2/3}\right) ,$$

(10)

so that

$$\varepsilon_1(K_*)/\varepsilon_2(K_*) = (K_*^{-2/3} - K_1^{-2/3})/(K_*^{-2/3} - K_2^{-2/3}) < 0 \qquad (11)$$

since $K_2 < K_* < K_1$, or $\ell_1 < \ell_* < \ell_2$, where $\ell = 1/K$ is the length scale. For the ocean ℓ_1 and ℓ_2 are clearly determined by the effective depth and the horizontal size so that we may take $\ell_1 \sim 1$ km, $\ell_2 \sim 10^4$ km. For typical instability processes in the ocean generating eddy kinetic energy, $\ell_* \sim 10^2$ km. Thus essentially

$$\ell_1 << \ell_* << \ell_2 \qquad \text{or} \qquad K_2 << K_* << K_1 . \qquad (12)$$

Then from (11)

$$\varepsilon_1(K_*)/\varepsilon_2(K_*) \approx - (K_2/K_*)^{2/3} \sim - 0.05 \qquad (13)$$

In other words, only about 5 % of the energy is cascaded to smaller scales. The remaining 95 % go to larger scales.

It follows also that two subranges with power law dependence of the spectrum can be established.

a) High wavenumber range ($K_* \leq K << K_1$)

$$\varepsilon_1(K) = (\alpha_1 R)^3 . K^{-2} \quad , \quad E_1(K) = \alpha_1^3 R^2 . K^{-3} \qquad (14)$$

b) Low wavenumber range ($K_2 << K \leq K_*$)

$$\varepsilon_2(K) = - (\alpha_2 R)^3 K_2^{-2} = \text{const} \quad , \quad E_2(K) = \alpha_2 \varepsilon_2^{2/3} . K^{-5/3} \qquad (15)$$

Therefore, the most general properties of 2-D turbulence as established by Batchelor, Kraichnan and others can be described by the simple model due to Bye, based on the solution (8). J. Bye obtained also a number of useful relationships between the parameters of his model which allow their evaluation using experimental data.

"TWO-WAY" INERTIAL TRANSFER IN 2-D TURBULENCE

So far we considered models based on the conception of "one-way" inertial enstrophy-transfer to high wavenumbers and energy-transfer to low wavenumbers. The real geophysical flows (the atmosphere and the ocean) however are neither inviscid nor strictly two-dimensional.

48

Preserving the fundamental features of the idealized models they should display some differences.

Particularly, having in mind (13), let us suppose that the energy and enstrophy cascade is <u>nearly</u> inertial in character, but takes place in both directions, with more energy going to larger scales and more enstrophy going to smaller scales - Fig. 3. Some small parts of them however go in opposite directions. What is the consequence of this refinement ?

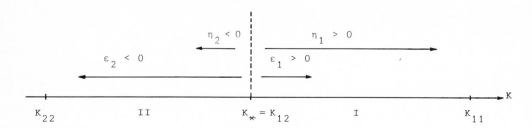

Fig. 3. "Two-way" cascade of energy and enstrophy in 2-D turbulence.

To answer this question, following Panchev (1976), we write eqs. (4) for the range I (Fig. 3.) :

$$- \gamma_1 K^3 \frac{d}{dK} (K^{5/2} E^{3/2}) = \eta_1 = const > 0 \quad ,$$

$$\tag{16}$$

$$- \gamma_1 K^{-1} \frac{d}{dK} (K^{9/2} E^{3/2}) = \epsilon_1 = const > 0 \quad .$$

Integrating we obtain

$$E_1(K) = \left(\eta_1 K^{-9/2} / 2\gamma_1 + A K^{-5/2} \right)^{2/3} \quad ,$$

$$\tag{17}$$

$$E_1(K) = (B K^{-9/2} - \epsilon_1 K^{-5/2} / 2\gamma_1)^{2/3} \quad .$$

Both expressions represent one and the same spectrum so that we must take $B = \eta_1 / 2\gamma_1$, $A = - \epsilon_1 / 2\gamma_1$. Hence

$$E_1(K) = (\eta_1 / 2\gamma_1)^{2/3} K^{-3} \left(1 - \frac{K^2}{K_{11}^2} \right)^{2/3} \quad , \quad K_{11} = (\eta_1 / \epsilon_1)^{1/2} \quad . \tag{18}$$

Similarly

$$E_2(K) = (|\epsilon_2|/2\gamma_2)^{2/3}K^{-5/3}\left(1 - \frac{K_{22}^2}{K^2}\right)^{2/3} \quad , \quad K_{22} = (\eta_2/\epsilon_2)^{1/2} \quad , \quad (19)$$

and $E(K) \equiv 0$, $K \leq K_{22}$, $K \geq K_{11}$. From the continuity of the spectrum at $K = K_*$, i.e. $E_1(K_*) = E_2(K_*)$ one obtains

$$K_*^2 = \frac{\gamma_2}{\gamma_1}\frac{\eta_1}{|\epsilon_2|}\left(1 + \frac{\gamma_1}{\gamma_2}\frac{|\eta_2|}{\eta_1}\right)\bigg/\left(1 + \frac{\gamma_2}{\gamma_1}\frac{\epsilon_1}{|\epsilon_2|}\right) \approx \eta_1/|\epsilon_2| \qquad (20)$$

since $\gamma_1 \sim \gamma_2$, $|\eta_2| \ll \eta_1$, $\epsilon_1 \ll |\epsilon_2|$.

Therefore

$$K_{22} \ll K_* \ll K_{11} \quad . \qquad (21)$$

Formally K_{11} and K_{22} are interpreted as cut-off wavenumbers. However, the obtained solutions (18) and (19) are physically justified only for $K \sim K_*$, i.e. for $K_{22} \ll K \ll K_{11}$. Therefore, (18) and (19) should be replaced by

$$E_1(K) \approx (\eta_1/2\gamma_1)^{2/3}K^{-3}\left(1 - \frac{2}{3}\frac{K^2}{K_{11}^2}\right) \approx a_1K^{-3} - b_1K^{-1} \quad , \qquad K \gtrsim K_*$$

$$(22)$$

$$E_2(K) \approx (|\epsilon_2|/2\gamma_2)^{2/3}K^{-5/3}\left(1 - \frac{2}{3}\frac{K_{22}^2}{K^2}\right) = a_2K^{-5/3} - b_2K^{-11/3}, \quad K \lesssim K_* \quad .$$

If $\epsilon_1 = \eta_2 = 0$, then $b_1 = b_2 = 0$ and we obtain (1) and (2).

INFLUENCE OF THE EARTH ROTATION AND CURVATURE

For simplicity we again assume a "one-way" cascade of energy and enstrophy. Suppose there exist additional characteristic length scales L_1 and L_2 which must be considered together with η_1 and ϵ_2 . Then, from general considerations one can write

$$E_1(K) \sim g_1(KL_1)\eta_1^{2/3}K^{-3} \quad ,$$

$$E_2(K) \sim g_2(KL_2)|\epsilon_2|^{2/3}K^{-5/3} \quad .$$

To determine the functions $g_1(x_1)$ and $g_2(x_2)$ we generalize also the diffusion approximation (4)

$$\varepsilon(K) = - \tilde{\gamma} \ h(KL) K^{\frac{3}{2}p-1} \frac{d}{dK} \left[K^{\frac{9}{2} - \frac{3}{2}p} E^{\frac{3}{2}} \right] \tag{24}$$

where $h(x)$ is another function and p is a free power.

The enstrophy conservation condition

$$\int_0^\infty \varepsilon(K) K dK = 0$$

applied to (24) yields

$$x \frac{dh}{dx} = - \frac{3}{2} ph \quad , \quad i.e. \quad h(x) \sim x^{-3p/2} \quad . \tag{25}$$

Therefore

$$\varepsilon(K) = - \tilde{\gamma} \ K^{-1} \frac{d}{dK} (K^{9/2} \phi^{3/2})$$

$$\tag{26}$$

$$\eta(K) = - \tilde{\gamma} \ K^{3} \frac{d}{dK} (K^{5/2} \phi^{3/2})$$

where

$$\phi(K) = (KL)^{-p} E(K) \quad , \quad \tilde{\gamma} \neq \gamma \quad . \tag{27}$$

Now, if there exist subranges of wavenumbers such that $\varepsilon(K) = -|\varepsilon_2|$ and $\eta(K) = \eta_1 = const$, then eqs.(7) can be solved to give

$$E_1(K) \sim \eta_1^{2/3} (KL_1)^p . K^{-3} \quad , \quad \varepsilon_1(K) \equiv 0 \quad , \quad K \geq K_* \tag{28}$$

$$E_2(K) \sim |\varepsilon_2|^{2/3} (KL_2)^p . K^{-5/3} \quad , \quad \eta_2(K) \equiv 0 \quad , \quad K \leq K_* \tag{29}$$

$$i.e. \quad E_1(K) \sim K^{-3+p} \quad , \quad E_2(K) \sim K^{-\frac{5}{3}+p} \quad .$$

What could be the nature of the scales L_1 and L_2 ? One possibility is that L_1 and L_2 are related to the curvature and the rotation of the earth, represented by two new dimensional parameters

$$f = 2\omega \sin \phi \sim 10^{-4} s^{-1} \quad , \quad \beta = df/dy \sim 10^{-11} m^{-1} s^{-1} \quad . \tag{30}$$

Five different length scales can be constructed from β , f , η_1 , $|\varepsilon_2|$:

$$D_1 = \eta_1^{1/3} \beta^{-1} \quad , \quad D_2 = |\varepsilon_2|^{1/5} \beta^{-3/5} \quad , \quad D_3 = L_* = (|\varepsilon_2| / \eta_1)^{1/2}$$

$$D_4 = |\varepsilon_2|^{1/2} f^{-3/2} \quad , \quad D_5 = f \beta^{-1} \quad (\sim 10^4 \text{km}) \quad . \tag{31}$$

The order of magnitudes of these scales for the atmosphere and the ocean can be seen in the table

	$|\varepsilon_2|$ (cm^2/s^3)	η_1 (s^{-3})	D_1 (km)	D_2 (km)	$D_3 = L_*$ (km)	D_4 (km)
Atmosphere	$10^0 - 10^1$	$10^{-15} - 10^{-16}$	10^3	10^3	10^3	10^1
Ocean	$10^{-3} - 10^{-4}$	$10^{-17} - 10^{-18}$	10^2	10^2	10^2	10^0

We conclude that D_5 is too large and D_4 is too small to be identified with L_1 and L_2 in (28), (29). The most appropriate are the scales D_1 and D_2. Consequently, the deviation of the empirical power dependences from K^{-3} and $K^{-5/3}$ can be associated with the influence of the Rossby parameter β on the eddy motion. As to D_3 $(=L_*)$, D_4 and D_5, they can be interpreted as forcing scale and cut-off scales correspondingly.

In the case of a "two-way" cascade, the solutions (28) and (29) should be corrected as indicated before (see (22)).

HORIZONTAL EDDY VISCOSITY COEFFICIENT

Having determined the kinetic energy spectrum $E(K)$, one can find also the eddy viscosity coefficient K. In the local spectral approximation (see Panchev, 1971)

$$K(\ell) = \delta \left(\frac{E(K)}{K} \right)^{1/2} \quad , \quad K = 1/\ell \quad .$$

Inserting here for instance (8) we obtain

$$K(\ell) = a \cdot \ell^{4/3} |\ell^{2/3} - \ell_L^{2/3}| \quad , \quad a = \delta \alpha_1^{3/2} R = \text{const.}$$

This simple expression can be used in the numerical models of ocean circulation with variable horizontal eddy viscosity coefficient.

SPECTRA OF SCALAR FIELDS

The temperature of the water in the upper ocean and at the surface is one of the most important scalar characteristics. We are interested here in the meso- and macroscale variability in the horizontal plane. So far there exist three different theoretical predictions for the two-dimensional temperature spectrum $E_T(K)$, based on different physical assumptions :

a) The temperature is a dynamically passive quantity in the flow. In this case a simple local type approximation for the temperature transfer function

$$\varepsilon_T(K) \sim K^{5/2} E^{1/2}(K) E_T(K) \tag{34}$$

permits to relate $E_T(K)$ with $E(K)$ in the inertial range $\varepsilon_T(K) = \varepsilon_T = $ const

$$E_T(K) \sim \varepsilon_T K^{-5/2} E^{-1/2}(K) \sim \begin{cases} \varepsilon_T \eta_1^{-1/3} (KL_1)^{-\frac{p}{2}} \cdot K^{-1} \quad , & K \geq K_* \\[3mm] \varepsilon_T |\varepsilon_2|^{-1/3} (KL_2)^{-\frac{p}{2}} \cdot K^{-5/3} \quad , & K \leq K_* \end{cases} \tag{35}$$

with $p = 0$, these predictions appeared simultaneously and independently in the papers by Saunders (1972) and Gavrilin et al (1972).

b) The temperature is not a passive quantity. In this case, following Saunders (1972) we can use the thermal wind equation

$$\vec{U}_T = \vec{U}_g - \vec{U}_g^o = \left(\frac{R}{f} \ln \frac{P_o}{P_1} \right) \vec{K} \wedge \nabla \overline{T} \,, \tag{36}$$

where $\vec{U}_g$ is the geostrophic wind and $\overline{T}$ the mean temperature in the layer (P_o, P_1). This equation implies an approximate relationship between the energy and the temperature spectra of the type

$$E(K) \sim K^2 E_T(K) \,. \tag{37}$$

In the enstrophy inertial range

$$E(K) \sim K^{-3} \quad \text{and} \quad E_T(K) \sim K^{-5} \tag{38}$$

This result contrasts strikingly with $E_T(K) \sim K^{-1}$, following from (34) and (35).

c) Charney (1971), who proposed the theory of the so-called geostrophic turbulence in the atmosphere, also stated that the predicted -3 power laws

$$E(K) \sim \eta^{2/3} K^{-3} \quad , \quad E_T(K) \sim \eta^{2/3} N^2 (\alpha g)^{-2} . K^{-3} \tag{39}$$

might apply to oceanic fields in regions of strong baroclinic excitation, e.g. in areas of upwelling. In (39), N is Brunt-Väisälä frequency, α the expansion coefficient, g the acceleration of gravity.

We see that the K^{-3} prediction lies between K^{-1} and K^{-5}. No one of them should be valid under arbitrary conditions in the atmosphere or in the ocean. Their validity must be tested experimentally.

OBSERVATIONAL EVIDENCES SUPPORTING OR NOT THE THEORIES

The experimental data concerning the horizontal meso-and macro-scale statistical structure of the velocity and temperature fields in the ocean are very scarce. Surprisingly to us, we could not find any publication confirming the -3 power law for the energy spectrum of the ocean currents. An exception is the unpublished paper of J. Bye (1975).

As to the temperature we can indicate at least four papers :
- Mac Leish (1970) presented results from the study of the sea surface temperature (SST), recorded from a remote sensing infrared thermometer. The computed spectra have a maximum in the smallest K and then decrease rapidly with increasing of K up to K = 2-5 cycle/km, remaining almost constant for larger K.
- One-dimensional spatial spectra of SST were also presented by Saunders (1972) over the range of scales 3-100 km. He found that $E_T(K) \sim K^{-P}$, $P \approx 2,2$. As he stated, this result does not fit any of the above reviewed model of geophysical turbulence.
- In contrast of the previous authors, Holladay and O'Brien (1975) performed a two-dimensional spectral analysis of the dayly horizontal SST field and found that over the range of scales 4-20 km the isotropic part of the temperature spectrum obeys a -3 power law. They do not however think that this result supports the geostrophic turbulence theory.
- Mesoscale variability of the temperature in the upper 100- meters layer in the ocean has been experimentally studied by Navrotsky (1969). Among many interesting results from the statistical treatment of the data, he published one-dimensional spatial spectra $E_T(K)$ of the

temperature over the range of scales 1,3-67,2 km for 10 horizonts. The distance between the horizonts equals 10 meters. All the spectra have maxima at scales ℓ_m = 8,5-17 km. For smaller scales ($\ell < \ell_m$) in the log-log plot the spectral curves represent straight lines with slopes between -7/3 and -4 including -3 as the most frequently observed. In other words, an universal spectral curve does not follow from these data. In a later paper by Navrotsky et al (1972) this result is confirmed and refined.

CONCLUSION

It seems to us that the predictions of the theory of two-dimensional and quasi-two-dimensional turbulence have received now substantial support from several different types of sources :

a) Direct numerical integration of the two-dimensional Navier-Stokes equations and mathematical simulation of such flows.

b) Laboratory modelling and measurements of the spectra of energy and passive scalars in two-dimensional magnetohydrodynamic flows.

c) Empirical kinetic energy spectra of large scale atmospheric motions computed from actual observations by many authors.

In contrast however, almost nothing is known about the empirical kinetic energy spectra of the large horizontal oceanic motions and very little about the temperature spectrum and the spectra of the order scalar fields of interest. Future studies in this respect will be of primary interest. The data obtained during the "POLIMODE" experiment are extremely appropriate for the purpose.

REFERENCES

Batchelor, G.K., 1965. Computation of the energy spectrum in homogeneous two-dimensional turbulence. Phys. of Fluids supp., II.
Bye, J.A.T., 1975. The two-dimensional spectrum of oceanic turbulence (manuscript obtained by personal communication).
Charney, J.G., 1971. Geostrophic turbulence. J. Atm. Sci. 28: 1087-1095.
Gavrilin, B.L., Mirabel, A.P. & Monin, A.S., 1972. On the energy spectrum of synoptic processes. Izv. Atm. Ocean Phys., 8, 5: 483-493.
Holladay, C.G. & O'Brien, J.J., 1975. Mesoscale variability of sea surface temperature. J. Phys. Oceanogr., 5, 5: 761-772.
Kovasznay, L., 1948. Spectrum of locally isotropic turbulence. J. Aeron. Sci., 15: 745-753.

Kraichnan, R.H., 1967. Inertial ranges in two-dimensional turbulence. Phys. of Fluids, 10: 1-7.

Leith, C.E., 1968. Diffusion approximation for two-dimensional turbulence. Phys. of Fluids, 11: 671-672.

Leith, C.E., 1971. Atmospheric predictability and two-dimensional turbulence. J. Atm. Sci., 28, 2: 145-161.

Lilly, D.K., 1972. Numerical simulation studies of two-dimensional turbulence. Geophys. Fluid Dyn. 3: 289-319.

Lorenz, E.N., 1969. The predictability of a flow which possesses many scales of motion. Tellus, 21: 289-307.

Mac Leish, W., 1970. Spatial spectra of ocean surface temperature. J. Geoph. Res., 75: 6872-6877.

Navrotsky, V.V., 1969. Statistical analysis of spatial fluctuations of temperature in the upper layer of ocean. Izv. Atm. Ocean Phys. Acad. Sci. USSR, 5: 94-110.

Navrotsky, V.V., Paka, V.T. & Karabasheva, E.I., 1972. Characteristics of thermal inhomogeneities at cross sections in the Atlantic ocean. Izv. Atm. Ocean Phys. Acad. Sci. USSR, 7, 3: 307-320.

Panchev, S., 1976. Inertial range spectra of the large scale geophysical turbulence. Bulg. J. of Geophys., 2, 2: 3-11.

Saunders, P.M., 1972. Space and time variability of temperature in the upper ocean. Deep Sea Res., 19: 467-480.

INTRUSIVE FINE STRUCTURE IN FRONTAL ZONES AND INDICATION OF DOUBLE
DIFFUSION

K.N. FEDOROV

Institute of Oceanology, Acad. Sci. USSR, Moscow (U.S.S.R.).

ABSTRACT

Fedorov, K.N., 1979. Intrusive fine structure in frontal zones and
    indications of double diffusion.

One of the earliest evidences of double-diffusion effects in frontal zones is reconsidered in the light of later reports. The method
of T,S-analysis is recommended for the identification of the final
outcome of double-diffusive convection. The results thus obtained together with later findings by the author in the frontal zone of an
anticyclonic Gulf Stream ring give strong evidence of the double-
diffusive convection participation in the "filling in" of cold and
fresher intrusions through mixing with the surrounding warmer and
saltier water.

In addition to a direct recording of the salt finger convection
in the ocean (Williams, 1974) observations of fine structure provide
increasing indirect evidence of the double-diffusive convection func-
tioning in the thermo-halocline. As a rule, these indications are
closely related to the intrusive interleaving across frontal zones
(Gregg, 1975 ; Williams, 1976 ; Lambert and Sturges, 1977 ; Horn,
1978 ; Gargett, 1978 ; Postmentier and Hougton, 1978 ; Joyce et al, 1978).
In this connection many questions arise concerning both the physical
nature of the phenomenon, and important methodology aspects. Without
touching now upon the role of double-diffusive convection in cross-
frontal heat and mass transfer, we shall come back to one of the
earliest evidences of the kind, as observed by the author in 1970
(Fedorov, 1971), in order to demonstrate the perspectiveness of the
T,S-analysis for the identification of the outcome of the salt finger
action. In this particular case, the results of the analysis can
hardly be questioned since they are based on multiple repeated

measurements so that the mean salinity values used for identification
purposes differ by 0,1 - 0,2 °/oo, which is one order of magnitude
more than the actual CTD-probe resolution.

Observations on the thermohaline finestructure in the upper layer
of the ocean carried out in 1970 at a hydrophysical polygon in the
Tropical Atlantic, revealed an often recurring type of temperature
inversion, at depths of 90-110 m, clearly associated with features of
the vertical salinity distribution. In the region of the polygon
(Lat. 16°30'N ; Long. 33°30'W) there is a subsurface layer of high
salinity of subtropical origin. The salinity in this layer reaches
36.9-37.2 °/oo , which exceeds that in the surface layer of the ocean
by 0.5-0.8 °/oo . The layer of high salinity is normally located immediately below the upper uniform layer, approximately from 60 to
130 m. Its thickness and the value of the maximum salinity have been
subjected to large variations in space and time. At individual points
of the polygon (grid 113 x 113 miles), at particular times, the thickness of the layer decreased to 30 m, and the size of the maximum fell
to 36,6 °/oo . The layer of high salinity occurs against a background
of an overall lowering of temperature with depth, which is sufficiently strong to compensate for the effect of the decrease in salinity
with depth below the point of the maximum, so that the hydrostatic
stability is everywhere, as a rule, positive.

In the layer of high salinity itself, a complicated structure has
been observed almost everywhere with the aid of the temperature-
salinity probe, and this has also been supported by samples, collected
with water bottles.

In particular, the distribution of the layer of high salinity in
two or three individual sub-layers (A and B in Fig. 1) with intermediate minima, has been typical.

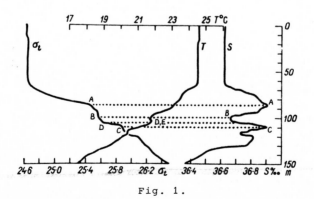

Fig. 1.

Along with these minima, which have a total thickness only of the
order of 10 m, temperature inversions like those noted above have been
revealed. These have differences of order 0.1-0.2°C (D,E) and a total
thickness of 3-6 m, and are located below the point of the intermediate minima of salinity (segment B-C in Fig. 1) at the place where the
salinity again increases with depth, forming a hydrostatically stable
"cushion". Fig. 1 shows the vertical distribution of density, from
which we may assess the hydrostatic stability of the layers. It is
seen that the segment A-B is the least stable, and the segment B-C is
the most stable in the whole structure below the uniform layer.

The subdivision of the layer of high salinity has been associated
with the occurence of a thermohaline frontal zone at the level where
these saline subtropical waters are distributed. As a result of this,
the vertical profiles of salinity in the layer containing the maximum
has often been extremely variable over a total distance of only 5-7
miles. This has also created favourable conditions for the operation
of the mechanism of lateral convection which could also produce temperature inversions, although of a somewhat different type (Stommel and
Fedorov, 1967) from those that have been considered above. These different temperature inversions are characterized only by the extreme
value of temperature against a background of increasing salinity.
The extreme value of salinity, developed somewhat lower down, is not
associated with a temperature inversion.

The advection of more saline water within a layer of increased salinity sustains its potential energy at a relatively high level. In
this case, because the vertical hydrostatic stability is close to
zero, even the most minute deviations in the advective, turbulent,
and even the diffusion regimes of heat or salt could lead to hydrostatically unstable situations, in which the excess of potential
energy passes into kinetic energy of convective motion. It has already been noted that the least stable segment is A-B (Fig. 1). The
mutually compensating vertical gradients of salinity and temperature
here, as a rule, are so large (up to 0.1 $°/_{oo}$ and 0.3°C per metre
respectively), that in fact segment A-B may be regarded as a marked
thermohaline discontinuity boundary between the layers.

Repeated soundings while drifting, with a frequency of 2 minutes,
have revealed quasi-periodic vertical fluctuations of the entire
thermohaline structure in the 80-120 m layer with an amplitude of
about 15 m and a period of 15-18 minutes, apparently associated with
internal waves (Fig. 2).

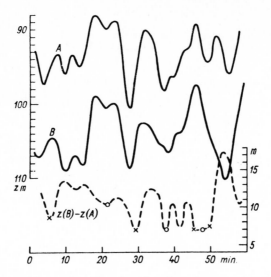

Fig. 2. Fluctuations in the depth of extreme points A and B (see
        Fig. 1.) and in the distance between them.

At least four times per hour, the hydrostatic stability of segment
A-B became negative, and a few times it was zero.   Times of instability coincided with a decrease in the distance between the upper and
lower adjacent maxima of salinity to the minimum (6-7 m).   Fig. 2.
shows the change in this distance with time, evidently associated
with an overall wave motion which was well marked by the change in
the depth of the maxima of salinity A and B.   The crosses denote
times of instability in segment A-B, and the circles times when the
stratification was uniform.

The internal waves, from time to time thinning out the intermediate layer of decreased salinity, also accentuate independently the
high vertical gradients of temperature and salinity in segment A-B.
In this case, specific situations must arise or sharpen, that are
favourable for the development of convection in the form of salt
fingers.   In Turner's experiments (Turner, 1967), minutes or fractions
of a minute have been required for its development, which is the time-
scale also characteristic of the process being considered by us.

The variety of changing situations has been graphically illustrated
on a T,S diagram (Fig. 3), where the points corresponding to the T and
S pairs of three layers with extreme values of salinity (A, B and C)
lie within three restricted regions.   Region A occupies the largest
area, which indicates the greatest variation in the T-S pairs of the
upper layer.   Four pairs, also assigned to this layer (and denoted by

black triangles), lie outside region A in a position which emphasizes
their hydrostatic instability.

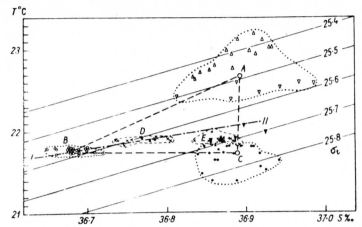

Fig. 3. Composite T-S diagram for segment ABCDE of the thermohaline
structure shown in Fig. 1.

The T-S pairs of the temperature inversion are shown on the diagram
by crosses. They lie within the triangle ABC, formed by the centres
of the three regions, approximately on a line which passes through
the group of four unstable points and the centre of region B. From
this we may conclude that the temperature inversion has been formed
from masses of water of the upper layer of high salinity (A), which
have mixed during the periods of instability with water from the
underlying layer of lower salinity (B) and rest on the hydrostatically
stable "cushion" of the lower layer of high salinity (C). In every
case, the water of the inversion layer, which occurs within the mixing
triangle ABC, is the product of mixing of the waters from layers A, B
and C.

There are strong arguments in favour of an important role of salt
fingers in the present case. One of them relates to the fact that in
all four unstable situations, the salinity of layer A, as shown in the
diagram (Fig. 3.), is practically unchanged, whereas the temperature
has fallen markedly each time. The drop in temperature in layer A is
also associated with a hydrostatic instability in segment A-B. It is
also not by chance that, on the T-S diagram (Fig. 3.), the crosses
denoting the temperature inversion form two separate groups, D and E.
The second of them is close to the field of points, C, from which it
follows that this group belongs to the inversion layer during the
process of mixing with the underlying layer C. On the other hand,

field D evidently corresponds to the temperature inversions just formed
or renewed as a result of convection. The slope of the line I-II,
which passes through the centres of groups B and D, is proportional to
the value of the ratio $\alpha\Delta T/\beta\Delta S$, where ΔT and ΔS are the increments in
temperature and salinity respectively between the centres of groups B
and D. This value gives an estimate of the ratio of the vertical buoyancy fluxes resulting from the transfer of heat and salt from layer A
through layer B, and in the present case it is equal to 0.6 , which
corresponds to the figure of 0.56, obtained by Turner (Turner, 1967)
in experiments with salt fingers. The line I-II passes somewhat above
the group of four unstable points. It is likely that the instability
observed in the four cases is a consequence of the resulting transfer
of negative buoyancy by the salt fingers.

The above temperature inversions represent partially a process of
"filling in" through mixing of the less saline intrusive lenses (B on
Fig. 1.) which penetrated into the high salinity core. An analogous
situation was observed by the author in 1978 in the frontal zone of
an antyciclonic Gulf Stream ring (Ginzburg et al, 1979). Here jet-
like intrusions of colder and fresher shelf water happened to be entrained into the saltier and warmer slope-water on the outer edge of
the ring. In this case again the "filling in" of these intrusions
occured, according to our T,S-analysis, with such contributions of
heat and salt that their ratio indicated presence of double-diffusive
effects. Such ratio for cold intrusions was earlier predicted by the
author (Fedorov, 1976).

REFERENCES

Fedorov, K.N., 1971. Sluchai konvektsii s obrazovaniem inversii tem-
    peratury v svyazi s lokalnoi neustoichivostiu v okeanicheskom
    termokline. (A case of convection with temperature inversion for-
    mation due to local instability in the oceanic thermocline).
    Doklady Acad. Sci. USSR, 198: 822-825.
Fedorov, K.N. Russian ed. 1976/English ed. 1978. Tonkaya termohalin-
    naya struktura vod okeana. (The Thermohaline Finestructure of the
    Ocean). Gidrometeoisdat, Leningrad and Pergamon Press, Cambridge,
    184-170 pp.
Gargett, A.E., 1978. Microstructure and fine structure in an upper
    ocean frontal regime. J. Geophys. Res., 83: 5123-5134.
Ginzburg, A.I., Zatsepin, A.G., Kuzmina, N.P., Sklyarov, V.E. &
    Fedorov, K.N., 1979. O tonkoi strukture frontalnoi zony tyoplogo
    koltsa Golf-strima. (On the fine structure of a frontal zone in a
    warm core Gulf Stream ring). In: Okeanologicheskie issledovania,
    33 (in press).
Gregg, M.C., 1975. Microstructure and intrusions in the California
    Current. J. Phys. Oceanogr., 5: 253-278.

Horne, E.P.W., 1978. Physical aspects of the Nova Scotian shelf break front. In: M.J. Bownan (Editor), Oceanic Fronts in Coastal Processes, Springer-Verlag, Berlin, pp. 59-68.

Joyce, T.M., Zenk, W. & Toole, J.M., 1978. The anatomy of the Antarctic Polar Front in the Drake Passage. J. Geophys. Res., 83: 6093-6113.

Lambert, R.B.Jr. & Sturges, W., 1977. A thermohaline staircase and vertical mixing in the thermocline. Deep-Sea Res., 24: 211-222.

Posmentier, E.S. & Houghton, R.W., 1978. Fine-structure instabilities induced by double diffusion in the shelf/slope water front. J. Geophys. Res., 83: 5135-5138.

Stommel, H. & Fedorov, K.N., 1967. Small-scale structure in temperature and salinity near Timor and Mindanao. Tellus, 19: 306-325.

Turner, J.S., 1967. Salt fingers across a density interface. Deep-Sea Res., 14: 599-611.

Williams, A., 1974. Salt fingers in the Mediterranean outflow. Science 185: 941-943.

Williams, G.O., 1976. Repeated profiling of microstructure lenses with a midwater float. J. Phys. Oceanogr., 6: 281-292.

RESONANT AND NON-RESONANT WAVE-WAVE INTERACTIONS FOR INTERNAL GRAVITY WAVES

I. Orlanski[*] and C. P. Cerasoli[†]

[*] Geophysical Fluid Dynamics Laboratory/NOAA, Princeton University, Princeton, New Jersey, 08540 U.S.A.

[†] Geophysical Fluid Dynamics Program, Princeton University, Princeton, New Jersey, 08540 U.S.A.
Supported by NOAA Grant No. 04-7-022-44017

ABSTRACT

A detailed study of energy transfer among two-dimensional internal gravity modes in a fully non-linear regime was performed. A number of techniques were used: They were (i) solutions of the gyroscopic equations with three and four waves. (ii) integration of a finite difference numerical model, and (iii) laboratory experiments. The solution of the four wave gyroscopic equations differed dramatically from the three wave case, and the four wave solutions were aperiodic. In both cases, the non-linear interaction time scale, T_μ , was found to be inversely proportional to the square root of the total wave energy ($T_\mu \sim E^{-1/2}$), even when the weak interaction assumption was violated. Integration of a finite difference numerical model showed that triad evolution was greatly affected when many waves other than the primary triad components could be excited. Initial condition experiments for triad evolution were performed with either a quiescent background state or a random field of waves, and the final states were similar, although the time to reach steady state was short when a background field was present. The numerical model was used to simulate surface forced, resonant modes and results were compared to laboratory experiments. Good agreement was found, not only in the initial wave evolution but also in energy level of the final states. An equilibrium state was achieved in both types of experiments, and wave-wave interactions and wave breaking were important in the energy distribution.
The numerical model was used to create a random, finite amplitude internal wave field, and a set of experiments whereby this basic state initially perturbed was performed. In these experiments energy was introduced over bands of low, medium and high wavenumbers. The results show that when the basic state energy is low and non-linear time scales are much greater than intrinsic wave periods, multiple triad interactions account for the distribution of any input energy. As the energy level increases, the high wavenumbers become saturated and localized overturning provides the dissipation mechanism. Additional energy input to low and medium wavenumbers will eventually result in an equilibrium state, whereby any extra energy input will result in very rapid, localized overturning. This equilibrium level depends on the presence of saturated high wavenumbers and once achieved, the system is very inefficient at transferring energy via wave-wave interactions while very efficient at dissipating energy via localized overturning.

† Present address: Aeronautical Research Assoc. of Princeton, 50 Washington Road, Princeton, New Jersey 08540 U.S.A.

INTRODUCTION

The important role of internal gravity waves in a geophysical system has been recognized or some time, especially in the case of very stable systems such as the ocean. The large body of evidence for the existence of oceanic internal gravity waves began with the temperature measurements of LaFond (1949) and Charnock (1965), and the long time current meter records obtained by Webster (1969) and Fofonoff and Webster (1971). The number and quality of internal gravity wave field measurements have grown dramatically since the early 1970's, and a variety of techniques has been employed to obtain measurements suitable to give wavenumber and frequency spectra in different parts of the world oceans. Concurrent with the observational advances, our theoretical knowledge of internal gravity waves has grown considerably. A lucid exposition of oceanic internal waves can be found in Thorpe's (1975) review article, and an up-to-date interpretation of the latest advances can be found in Phillips (1977).

The purpose of this study is to investigate the means by which energy is transferred among internal gravity waves. The transfer of energy in weakly inter-acting flows is characterized by resonant interactions in which the interaction time scale is much greater than the component wave periods (Müller and Olbers, 1975, McComas and Bretherton, 1977). The energy transfers occurring in a more nonlinear system, where interaction time scales can be of the same order of the wave period (strong interaction) or larger, were practically untreated. In the present study, we will discuss the dynamics of waves and the building of an energy spectrum in such a nonlinear regime. Such a regime is perhaps more charac-teristic of the natural state of the ocean with regard to internal gravity waves. Let us, therefore, summarize the highlights of the scientific works which have provided today's picture of internal waves in the ocean.

Characteristics of the energy density as a function of horizontal wavenumber have emerged from data based on towed sensor (Charnock, 1965, Katz, 1975) and moored spatial array measurements (IWEX, Müller, Olbers and Willebrand, 1978). Basically, one has that the energy spectrum is horizontally isotropic and depends upon horizontal wavenumber as the -2.5 power. Measurements of vertical structure (Hayes, 1975, Leaman and Sanford, 1975, Müller et al., 1978) show that a modal representation is appropriate for high frequency internal waves (close to the local buoyancy frequency), while low frequency waves (close to the local inertial frequency may be represented as a superposition of freely propagating upward and downward waves, where the vertical energy fluxes need not be balanced. Long, fixed point records of horizontal velocities were obtained by Foffonoff and Webster (1971) and Gould, Schmitz and Wunsch (1975), while Cairns (1975) obtained long isotherm displacement records using an instrumented buoy which "yo-yoed"

about a fixed isotherm. The spectral picture is one where energy is predominantly
contained in a band between the local inertial frequency and the local buoyancy
frequency. The energy density may be approximated by an ω^{-2} (ω is the frequency)
dependence with predominant spectral peaks at the local inertial and tidal frequencies. A less pronounced peak exists near the local buoyancy frequency and a
sharp roll-off in energy occurs beyond this frequency. Finally, the similarity
of the appropriately normalized energy spectra from various world ocean sites can
be interpreted as the existence of a degree of universality for the spectral
band representing oceanic internal gravity waves.

An empirical formulation relating wavenumber and frequency spectra for internal
waves was proposed by Garrett and Munk (1972a, 1975), and provided a consistent
framework for the large body of data taken with various techniques. Universality
and horizontal isotropy, as well as the kinematic characteristics of internal
gravity waves, were invoked to formulate their empirical relation. This formulation does not contain any description of the dynamics which give rise to the observed
spectra. The recurring spectral shape obtained from field data, universal or not,
opens the questions of what are the mechanisms for achieving this spectral shape
and what is the time scale for relaxing to this shape, since the oceans are
continually being forced in various spectral bands. The problem of energy distribution among internal waves and the description of the energy spectrum in terms of
such transfers has been studied since the last decade. Theoretical work on the
slow or weak interaction for three discrete, resonantly interacting waves was done
by Bretherton (1964) and Phillips (1966), while Davis and Acrivos (1967) were the
first to experimentally observe triad resonance, using a two-layer stratified fluid.
The time scale for such interactions was found to be inversely proportional to the
square root of the total triad energy. Typically, this time scale is much greater
than the wave periods, justifying the weak interaction assumption. Laboratory
experiments on triad resonance in linearly stratified fluids were performed for
horizontally propagating internal waves by Martin, Simmons and Wunsch (1972) and
for modes by McEwan, Mander and Smith (1972). These experiments showed that whenever a single triad is is initiated, other resonant triads will also be excited,
and that the wave-wave interaction is stronger for higher frequency waves than for
the lower frequency waves. This cascade phenomenon implies that three discrete,
interacting waves is not a good model of wave-wave interaction, even in the weak
regime, since the simultaneous excitation of other triads is continually occurring.
It is apparent from the observed oceanic spectra, that wave-wave interactions
must be treated in a continuous spectrum, and theoretical work along these lines
was done by Müller and Olbers (1975) and McComas and Bretherton(1977) using a
methodology developed by Hasselmann (1966,1967). McComas and Bretherton studied
the manner in which weakly interacting waves decay in different parts of the
spectrum, and interaction coefficients were calculated only for resonantly

interacting waves. However, their results show that the interaction time could be very rapid (a few periods). This opens the question of whether only resonantly interacting waves should be considered, or perhaps, waves which are non-resonantly forced must also be considered. More important is the question of wave breaking versus wave-wave interactions as a mechanism for limiting wave amplitudes. Phillips (1977) discusses a scenario whereby the lower modes are limited by wave breaking rather than the transfer of energy to higher modes. Two types of wave breaking can occur, (1) a shear instability, as discussed by Phillips (1966), which requires a local Richardson number to be less than 1/4, and (2) a finite amplitude gravitational instability discussed by Orlanski and Bryan (1969) and Orlanski (1972) which requires that a locally unstable stratification exists. (The criterion for a single wave is that the fluid parcel velocity exceeds the phase velocity.) Both types can occur in the ocean and their occurrence is probably enhanced due to the presence of a large number of waves. That is, a single wave may not satisfy either breaking condition, but when the many other waves are superimposed, the breaking conditions may be satisfied. The concern over which breaking mechanism is operating in the ocean may be important for the estimation of dissipation due to wave breaking, and for predicting a critical wave spectrum amplitude. However, previous attempts to enhance the importance of shear instability or gravitational instability (Garrett and Munk, 1972b, Frankignoul, 1972) show contradictory results. In fact, Eriksen (1978) recently has found that both types of breaking are possible and breaking appears equally likely at all frequencies in the internal wave range. Perhaps, some of our ideas about wave breaking limiting low modes and wave-wave interactions limiting high modes should be reconsidered. First, because wave breaking appears to be occurring at all frequencies, and second, the weak interaction assumption may not be justified when the amplitudes of the "universal" spectrum yield interaction time scales of only a few wave periods (McComas and Bretherton, 1977).

This present work is a detailed study of interacting, two-dimensional internal gravity waves, where a variety of techniques were used. In Section 2, the two-dimensional equations of motion are presented, and two approaches are used for solving these equations. The first method uses a truncated Fourier series, where either three or four resonantly interacting waves are considered. No restrictions are placed on wave amplitudes, and the solutions of this truncated series may be viewed as an extension of the much studied gyroscopic equations. The second approach is a finite difference integration of the equations of motion which allows the resolution of many waves and uses a subgrid scale parameterization scheme for modeling dissipation due to wave overturning.

Solutions to the three and four wave truncated series are presented in Section 3.1, where the effects of adding a fourth wave to the triad set is investigated and results are discussed in light of the solutions for the gyroscopic equations. In Section 3.2, the numerical model is integrated using initial conditions similar to those used in the three and four wave solutions. Two numerical experiments were performed, one where the background field was quiescent, and the other, where a random field of internal waves was present. These integrations clearly showed the effect of waves other than the main triad components on the triad evolution.

Both numerical and laboratory experiments for internal waves excited by surface forcing were performed. Section 4.1 contains numerical solutions for this forced problem using different forcing amplitudes and viscosities. Triad evolution is investigated along with determining the final state solutions for various conditions. Laboratory experiments are presented in Section 4.2, and results are discussed in light of the numerical solutions. In Section 5, the numerical model is used to study wave-wave interactions in the presence of a random, finite amplitude internal wave field. This field is established by randomly forcing the model, as described in Section 5.1. In Section 5.2, three experiments are presented where energy was introduced at a band of either low, medium or high wavenumbers. Wave breaking was an important feature of all the experiments, and its role in energy transfer processes is discussed in Section 5. A summary and conclusions are given in Section 6.

2: EQUATIONS OF MOTION: TRUNCATED SPECTRA AND TWO-DIMENSIONAL NUMERICAL MODEL

Let us consider a two-dimensional container of length L and height H filled with a linearly stratified fluid of density $\rho = \rho_o(1+\beta z)$. The two dimensionality allows the system to be described by the following equations for vorticity and density,

$$\xi_t - J(\psi,\xi) = g\theta_x + \underline{\nabla}\cdot(\nu\underline{\nabla}\xi) \tag{1}$$

$$\xi = \nabla^2\psi \tag{2}$$

$$\theta_t - J(\psi,\theta) = \beta\psi_x + \underline{\nabla}\cdot(\kappa\underline{\nabla}\theta), \tag{3}$$

where ψ is the streamfunction and $u=\psi_z$, $w=-\psi_x$. Horizontal and vertical velocity components are u and w, respectively, and θ is the density departure divided by ρ_o, $\theta = (\rho-\bar{\rho})/\rho_o$. The eddy conductivity and diffusivity are κ and ν, and their proper form will be discussed in Section 3.

Free slip boundary conditions are used for streamfunction and vorticity at the sidewalls and bottom boundary,

$$\psi = \xi = 0 \qquad\qquad x = 0, L$$

$$\psi = \xi = 0 \qquad\qquad z = 0,$$

and the adiabatic conditions for the density perturbation are used,

$$\theta_x = 0 \qquad\qquad x = 0, L$$

$$\theta_z = 0 \qquad\qquad z = 0 \ .$$

The boundary conditions at z=H are either rigid

$$\psi = \xi = 0 \qquad\qquad z = H,$$

or forced,

$$\psi = f(x, t)$$

$$\xi = f_{zz} \qquad\qquad z = H$$

$$\theta_z = 0 \ ,$$

where f(x,t) is prescribed.

2:1 Truncated Spectra Equations

Equations (1) through (3) were spectrally decomposed in Fourier series. The streamfunction is zero at all boundaries in the unforced case, and sines are the natural series for expanding ψ as follows,

$$\psi(x,z,t) = \sum_m \sum_n \phi_{mn}(t) \ \sin \frac{m\pi x}{L} \sin \frac{n\pi z}{H} \ .$$

As the main force balances for the density perturbation arise from the linearized, inviscid equations, cosine and sine series are necessary for expanding θ as follows,

$$\theta(x,z,t) = \sum_m \sum_n \theta_{mn}(t) \ \cos \frac{m\pi x}{L} \sin \frac{n\pi x}{H} \ .$$

Note that the adiabatic condition precludes such a representation in the viscous case. Our intention is to look to the inviscid case when using spectral decomposition, although this description does allow certain global dissipative forces. Equations (1) through (3) take the following form when spectrally decomposed and k=ν=0,

$$K_{mn}^2 \ \dot{\phi}_{mn} + \sum_{p=m}^{r} \sum_{q=n}^{t} K_{pq}^2 \ J(\phi_{rt}, \phi_{pq}) = \frac{m\pi}{L} \ g\theta_{mn}, \qquad\qquad (4)$$

$$\dot{\theta}_{mn} + \sum_{p=m}^{r} \sum_{q=n}^{t} J(\phi_{rt,pq}) = \frac{m\pi}{L} \ N^2 \ \phi_{mn} \ , \qquad\qquad (5)$$

where

$$K_{mn}^2 = (\frac{m\pi}{L})^2 + (\frac{n\pi}{H})^2, \quad N^2 = - g \ \bar{\rho}_z / \rho_o \ ,$$

and dot refers to time differentiation.

2:2 Two-Dimensional Numerical Model

The vorticity and density equations (1) and (3) were integrated in finite difference. A rectangular tank was modeled with 51 horizontal points and 61 vertical points, and allowed the resolution of some 200 waves. The details of the model were described by Orlanski and Ross (1973) and uses a standard leap frog, energy conserving scheme with an exact Poisson solver for the stream-function. The height and length of the tank were 81 cm and 150 cm, respectively, for comparative purposes with laboratory experiments.

The effects due to sub-grid scale motion were modeled by use of an eddy diffu-sivity and conductivity defined as follows: (Orlanski and Ross, 1973, Orlanski, Ross and Polinsky, 1974)

$$
\left.\begin{matrix}\nu\\ \\ \kappa\end{matrix}\right\} =
\begin{cases}
\nu_L\left[1 + {}^{\nu_{NL}}\!/_L \right] \\
\kappa_L\left[{}^{\nu_{NL}}\!/_L \quad \dfrac{g\theta_z}{\nu_L\kappa_L} \right]^{1/3} & \theta_z < 0 \\[2em]
\kappa_L \\
\kappa_L & \theta_z > 0
\end{cases}
\tag{6}
$$

The diffusivity and conductivity were divided into two parts, one was constant value, while the other was a non-linear part which depended on the local Rayleigh number when a locally unstable density region was present. This method was found to be very efficient for simulating the dissipation due to the breaking of inter-nal gravity waves. The following experiments used various values for the non-linear diffusivities to examine dissipative effects of wave breaking on the major triad components, while the constant part was varied to simulate the drag produced by the sidewalls in the laboratory experiments.

3: TRIAD INTERACTIONS: INITIAL CONDITIONS

Truncated Spectra

The truncation of a spectral series always involves an arbitrariness which must be justified. Typically one studies a band of low wavenumbers and para-meterizes the effects of the higher, unresolved wavenumbers; the justification being that the lower bands energy is much greater than the energy present in the higher wavenumbers. This relative energy argument is not always a valid justi-fication, as the effects of non-linear wave interactions can be distorted by such a truncation. For instance, waves in the resolved band could transfer energy to a higher band of wavenumbers, and subsequently receive energy back again. Such a process may be dynamically important even if the relative energy in the high wavenumber band is small, and the process will not be properly modeled using a truncated series. Fully aware of such processes, we will consider a truncated series with only three or four components. The solutions to this severely truncated series will be used for comparison with solutions from a more complete two-dimensional numerical integration. The most natural truncated series to consider where internal gravity waves actively transfer energy via non-linear interactions is a set of three waves in triad resonance. The case of four waves interacting in a "double triad" was also computed to study the characteristics of the solutions when an additional wave is included.

The triad resonance problem has received considerable theoretical attention (Bretherton, 1964, McGoldrick, 1965, Phillips, 1966), and results show that two time scales are present for triad resonance, a "fast" scale T, (T=2π/N) of the order of the individual wave periods and a "slow" scale, T_μ, which characterizes the time for non-linear evolution of the triad components. This evolutionary time scale is proportional to the inverse of the square root of the total triad energy, that is

$$T_\mu \alpha \; \bar{E}^{-1/2} \quad .$$

Usually T_μ is much larger than the individual wave periods because weakly interacting waves are studied. The numerically integrated solutions obtained in this work do not require this restriction.

Let us consider four internal gravity waves such that

$$\underline{k}_1 + \underline{k}_2 = \underline{k}_3$$
$$\underline{k}_1 + \underline{k}_3 = \underline{k}_4$$
$$\omega_1 \pm \omega_2 \pm \omega_3 = 0$$
$$\omega_1 \pm \omega_3 \pm \omega_4 = 0$$
$$K_i = |\underline{k}_i|$$
$$\underline{k}_i = (k_i, \gamma_i)$$
$$\omega_i = k_i N / K_i \quad ,$$

and we have chosen

$$\underline{k}_1 = (+1, +2)$$
$$\underline{k}_2 = (+1, -1)$$
$$\underline{k}_3 = (+2, +1)$$
$$\underline{k}_4 = (+3, +3),$$

corresponding to resonant modes for the particular ratio H/L used in numerical and laboratory experiments. Equations (3) and (4) reduce to the following set for the four wave, double triad case.

$$\dot{\phi}_1 + \frac{\omega_1}{k_1} \; \theta_1 = A_1 \phi_2 \phi_3 - A_1' \phi_3 \phi_4 - D_1 \phi_1$$
$$\dot{\phi}_2 + \frac{\omega_2}{k_2} \; \theta_2 = A_2 \phi_1 \phi_3 - D_2 \phi_2$$

energy as functions of time for the main components are presented in Figure 3.
The triad evolution in the quiescent experiment of Section 3.1 is similar to
that observed in the truncated spectra experiments and is shown in Figures 3(a)
and 3(b). Wave 3 decayed as waves 1 and 2 grew, and the time scale is similar to
that observed in the low energy, truncated spectra experiment. The remarkable
difference occurred following the first half-cycle. The periodicity was destroyed as the non-linear interaction time scale was continuously changing. Wave
3 never returned to its original value as it did in the truncated spectra, while
wave 2 reached a maximum value and then decayed to a fairly constant value. Wave
4 had a varying amplitude maxima and minima, while wave 1 became the most predomi-
nate wave as time evolved.

Dissipation was very small in this experiment ($\nu_L=10^{-5}cm^2/sec, \nu_{NL}=10^{-5}cm^2/sec$),
so that the decreasing sum of the four wave energy represented a loss of energy
to other, higher wavenumber components. An attempt to model this leakage of energy
to high wavenumbers was made by including an energy dependent dissipation ($D_i \neq 0$)
in equations (7) and (8). The decay of the four wave energy could be simulated,
but not the complex, aperiodic behavior observed in Figures 3(a) and 3(b).
The non-linear interaction of the main triad components with the other waves
(whether or not they are in triad resonance) was so important that the elegant,
analytic triad solution predicted the full solution for only a half period of the
long time scale, T_μ. Note that the triad energy was lower in this case,
$\bar{E}=8 \ cm^2sec^2$ versus 12.5 cm^2/sec^2, and the weakly non-linear assumption ($T>>2\pi/\omega_i$)
was valid. The dramatic differences between the two solutions underscore the fact
that neglect of the high wavenumber components was a severe restriction. The
accumulation of energy in the high wave numbers will be discussed later.

The non-linear interaction of the main triad with the higher wave-numbers
was enhanced by the addition of a random field of internal gravity waves. Streamfunction and energy versus time are presented for the main components in Figures
3(c) and 3(d), and comparison of the quiescent and non-quiescent results in
Figure 3 do not show dramatic differences. A better representation for comparing
these two experiments was the triangle energy graphs presented in Figure 4
where two vertices represented E_1 and E_3 separately and the third vertex represented total energy minus E_1 and E_3. Initial trajectory behavior in both cases
was similar to that observed in the truncated spectra. The quiescent experiment
(Figure 4(a)) showed a trajectory which underwent a large excursion from the
initial straight line during the second half period. After that, the trajectories become straight lines with slopes similar to the value given by theory,
although they drift toward the upper vertex. This result was consistent with the
fact that the main triad was losing energy to higher wavenumber components.
The random field case was greatly different than the previous case. After

78

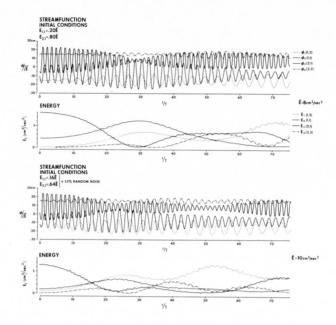

Fig. 3. Streamfunction and energy of the four primary waves versus time from numerical integrations of the equations of motion (T=2π/N)

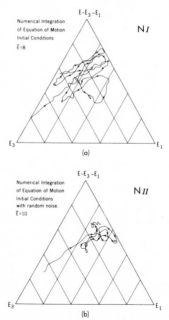

Fig. 4. Energy space triangle graphs from the numerical integration shown in Figure 3.

undergoing the first half non-linear oscillation, the system never returned to its original state and was not periodic. The presence of a background field was very efficient at removing energy from the main triad components and in destroying the system's periodicity (further discussions will be presented in Section 5). Both cases showed trajectory motion toward the upper vertex, which represented energy in either wave 2 and 4, or energy in wave 2 and 4 plus all the other waves. The motion toward the upper vertex represented energy flow into the higher wavenumber components, as the presence of waves 2 and 4 resulted in a meandering of the trajectories (Figure 2) without a net movement toward the upper vertex. The departure of these results from the single and double triad solutions demonstrated how drastically different system evolution was when a full spectrum was present. To complement the previous experiments, a set of experiments where a resonance mode was excited by surface forcing were conducted, and simulated laboratory experiments which will be described in Section 4.

4: TRIAD INTERACTION: SURFACE FORCING

A variation of the evolution of the single triad experiment (Section 3) is the case of a resonantly growing wave excited by surface forcing. This type of experiment can be performed using numerical simulation, or in an experimental laboratory set-up, and both types of experiments will be discussed in this section. The forcing was chosen to resonantly excite a mode which grew until the loss of energy via non-linear interactions became important. The conditions for the various experiments are summarized in Table II.

4:1 Two Dimensional Numerical Simulation

The numerical experiments with forcing use the following conditions at the upper surface,

$$h_{Int} = -a_o \sin(_3t) \qquad\qquad \ell < |x|$$
$$h_{Int} = a_1 \sin(_3t) + H(t-t_o)a_2\sin(_1t) \qquad -\ell < x < 0$$
$$h_{Int} = a_1 \sin(_3t) - H(t-t_o)a_2\sin(_1t) \qquad 0 < x < \ell$$

where

$$H(t-t_o) = \begin{cases} 0, & t < t_o \\ 1, & t > t_o. \end{cases}$$

The analytic solution for single wave forcing ($a_2=0$) was given by Orlanski (1972), who found that when a resonance condition was satisfied, the forced wave grew linearly in time until non-linear effects became important. At this point, growth of the triad components was observed and the case of continual forcing raises some questions. At what time do the triad members begin to grow and how effective are they at draining energy from the primary wave? The triad members might grow along with the primary forced wave until non-linear interactions with

other, higher wavenumber, triads took place. Alternately, the triad interaction
might be so effective at removing energy from the primary wave that the removal
either equalizes or overcomes energy flux due to forcing. McEwan, Mander and
Smith (1972) attempted to answer these questions, but considered only a highly
viscous, steady state case case where the balance was among viscous dissipation,
forcing and non-linear terms.

Results from numerical experiment NIII ($\nu_L=10^{-5} cm^2 sec$, $\nu_{NL}=1 cm^2/sec$)
are presented in Figure 5, where streamfunction contours are given at various

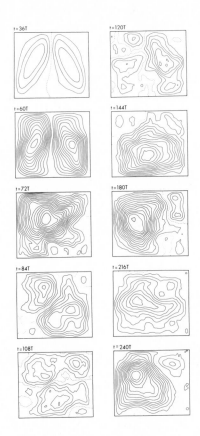

Fig. 5. Streamfunction contours at various times from the forced, resonant wave
numerical integration.

times. The force (2,1) mode dominates the field at t=36T and the tilting of the
closed streamlines is due to the presence of the (4,2) mode. This was a consequence
of the spatial distribution of the forcing which excited the (4,2) mode as well as

(2,1). Very quickly, by t=60T, the (2,1) mode is being distorted by the presence
of the triad instability components. During the period from t=72T to 120T, the
flow pattern is one where the contributions from the (2,1) and (1,1) modes are
outstanding. Analyses similar to that used for the initial condition experiments
were done and results are shown in Figure 6. The upper portion of the figure
shows the streamfunction for the four main waves as a function of time after
approximately 30 forcing periods. (The behavior during the first 30 periods

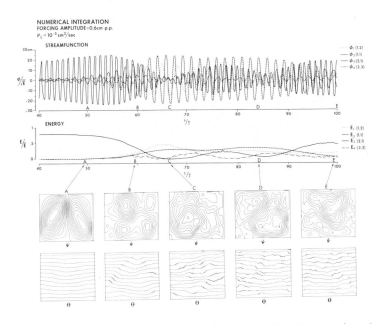

Fig. 6. Streamfunction and energy versus time from the integration shown in Figure
5 are presented in the upper portion. Streamfunction and density perturbation
contours at various times are shown in the lower portion, and the stippling in the
plot shows regions of probable wave breaking.

agreed with the theoretical results of Orlanski (1972). The middle portion of
Figure 6 shows the energy of the four components and oscillations with a time
scale of order the non-linear interaction time are present along with much longer
period oscillations. Streamfunction contours and isopycnics are presented at
various times in the lower portion of Figure 6, where the letters A thru E are
used to mark points in time. Figure 6a shows the (2,1) mode dominating, and the
isopycnics do not exhibit large deviations from ambient as the time chosen was
one where the streamfunction is large and the perturbation density field is

small. One half period later, this situation would be reversed. Figure 6, points B through E again show the evolution from the (2,1) mode dominating to a period when (2,1), (1,1) and (1,2) are present, and finally the (1,1) dominates. Regions of low static stability are marked with stipling in the isopycnic plots and wavebreaking can be expected in these regions. Inspection of the computer analyses shows indeed that these are areas where the non-linear diffusivity is activated due to unstable density gradients.

4:2 Laboratory Experiments

The discussion of the numerical solution depends on the waves which are well resolved as well as the manner in which turbulent processes are parameterized, as previously discussed. No matter how sound a scheme appears, close verification with laboratory experiments is desirable, and the experiments described in this section are geared to such verification. Details of the techniques and apparatus in Delisi and Orlanski (1975), and will not be discussed here. Streak photographs and spectra are presented in Figure 7 for a parametrically resonant experiment. The exposure time for the photographs was equal to one period of the primary wave, and the upper photograph shows the (3,1) mode clearly at t=90T. The spectrum was obtained from probe data at (.98L, .25H) during the interval 40 to 100T and the $\omega_{3,1}$ peak is seen to be dominant. The middle photograph was taken at t=130T, while the data interval for the spectrum was from 160 to 220T. The photograph shows the presence of unclosed streaks indicating the presence of $0.5\omega_{3,1}$ oscillations, and the tilting of the primary mode suggests the presence of the (1,1) mode. The spectrum shows clearly the $0.5\omega_{3,1}$ peak along with the larger peak at $\omega_{3,1}$. The lower picture was taken at t=240T and the spectra interval was from 220 to 280T. The streak photograph shows a field dominated by the (1,1) mode and shows an indication of the (2,2) mode by the presence of wavenumber 2 in the vertical. The spectrum shows that the energy at frequency $0.5\omega_{3,1}$ is now greater than at $\omega_{3,1}$, and the spectrum remained stationary from this time on.

5: WAVE INTERACTIONS IN A CONTINUOUS SPECTRUM, TWO-DIMENSIONAL NUMERICAL SIMULATION.

The results described in Section 3.1 - 3.2 concerning first, the difference between triad evolution in a quiescent versus random background field, second, the cascade of energy to high wavenumbers, and third, the lack of an energy cycle even for ideal conditions, make it imperative to treat the energy transfer problem in a finite amplitude spectrum. For a more complete discussion of this section, see Orlanski and Cerasoli 1980). To this end, numerical experiments were performed where energy was given to a band of wavenumbers in a finite amplitude field of internal gravity waves, and the energy transfer from either the modes or bands to the rest of the spectrum was investigated. Three different

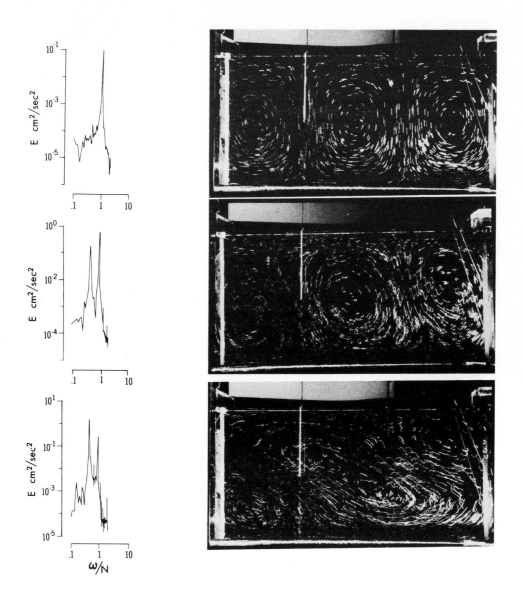

Fig. 7. Frequency spectra and streak photographs from a parametric resonance experiment. The degradation of the initial (3,1) mode along with the growth of the triad partners can be seen.

experiments will be discussed in this section.   In these three experiments energy
was introduced in low, medium and high wavenumber bands in the three experiments.
A description of the background random field will be presented before discussing
the energy transfer experiments.

5:1 Random Internal Wave Field

   The numerical model discussed in Section 4 was integrated for a long duration
(1050 T), where random body forcing was introduced in space and time to the
vorticity and density fields. This forcing resulted in the build-up of a random
field of internal gravity waves at all resolvable space[1] and time scales. The
spectral energy level increased monotonically in time until the combined effects
of the small amount of random noise plus the existing finite amplitude waves
produced localized overturning, which in turn, increased the overall eddy
diffusivity and caused a reduction in the spectral energy growth. From that point
on, it became more and more difficult to increase the overall energy level, implying that a saturation level was reached for the model. We note that this does
not mean all portions of the spectrum are saturated and this question will be
addressed in the subsequent series of experiments.

   Frequency spectra for the horizontal and vertical kinetic energies
($\dfrac{|u_\omega|^2}{2}$, $\dfrac{|w_\omega|^2}{2}$ and potential energy ($\dfrac{g^2|\rho_\omega|^2}{2\,N^2}$) were computed from time

series taken at different positions in the tank. The positions were: C1(0.20 L,
0.16H), C2(0.33L, 0.38H), C3(0.60L, 0.93H) and C4(0.81L, 0.50H), where the origin of
the (x,z) coordinates is at the lower left hand corner of the tank. The spectra
were calculated from data blocks of duration 500 seconds (2000 time steps or
approximately 60 T) with a total of 33 blocks ($\sim 7x10^4$ time steps). The average
spectra and the dispersion due to the individual spectra for the horizontal and
vertical kinetic energies and the potential energy at position C2 are presented in
Figure 8.   Note that the vertical scale represents the energy (not the spectral
energy density) at each frequency increment; the more standard representation of
spectral energy density versus frequency on log-log scales will be presented later.
The horizontal kinetic energy is shown in Figure 8(a) and the slope is negative
from low frequencies to N and falls with a steep slope beyond N. At first sight,
the dispersion appears large, although one must remember that early data blocks
enter into the dispersion when the energy level is still low, and the dispersion

---

[1]The model resolves approximately 750 waves with a total of 3000 grid points, and
approximately 150 waves are very well resolved (up to approximately (10, 15)).

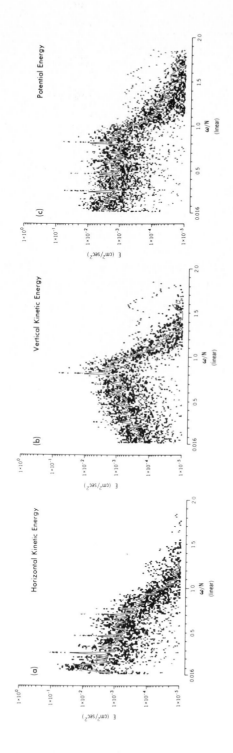

Fig. 8. The average spectra and the dispersion due to individual spectra for the horizontal and vertical kinetic energies $\dfrac{|u_\omega|^2}{2}$, $\dfrac{|w_\omega|^2}{2}$ and the potential energy $\dfrac{g^2}{2N^2}|\rho_\omega|^2$ are presented for the finite amplitude, random internal wave field. The probe position is (0.33L, 0.38H).

for the final ten data blocks is very small. Small peaks can be observed superimposed on the background spectrum, but it is surprising to find that the field is
not dominated by discrete modes favored by the tank geometry. The average vertical
kinetic energy is shown in Figure 8(b) and the energy level increases with
frequency until a maximum is reached near N and then the energy abruptly falls off;
a result consistent with the spectral shape for internal gravity waves. The average
potential energy spectrum is given in Figure 8(c) and shows a shallow decreasing
of energy as frequency increases up to N, and then falls rapidly. Inspection of
the spectra shows that the sum of the horizontal and vertical kinetic energies will
yield a spectral shape similar to the potential energy spectral shape, as might be
expected from the equipartition of energy as applied to internal gravity waves.

This system of internal gravity waves is a fully nonlinear one, and the degree
of wave breaking is substantial. In fact, energy contained in frequencies above N
arises from either forced nonlinear oscillations or turbulence due to wavebreaking.
Even so, the relationships between the various spectral components and the spectral
shapes could be predicted by simple linear theory applied to internal waves
randomly forced. The lack of discrete modes dominating the solution can be clearly
observed in Figure 9, where the average potential energy spectra from the four

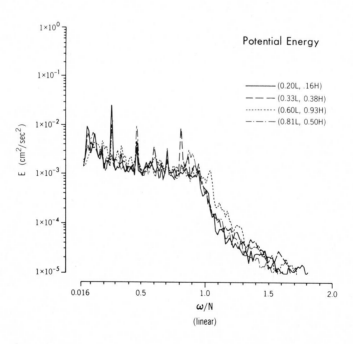

Fig. 9. The average potential energy spectra at four probe positions ((0.20L)

positions, C1 through C4, are presented. The similarity is striking, although some small peaks due to modes appear and are a function of position.

A measure of the stationary of the random wave field energy can be seen in the wavenumber spectra corresponding to the last 1000 seconds of the experiment, which is presented in Figure 10. These wavenumber spectra were computed at every time step and then averaged over a block 250 seconds ($\sim$ 30T) in duration. The spectra represent the energy at each wavenumber interval, as opposed to a spectral density. The spectra are seen to be red in nature and two regions of different slope can be discerned. Wavenumbers below 0.6 cm^{-1} are very well resolved by the model, and the energy level is low at the poorly resolved wavenumbers above 1.0 cm^{-1}.

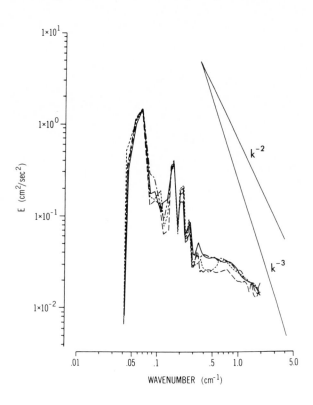

Fig. 10. Wavenumber spectra for the random internal wave field energy are presented for four consecutive time intervals following the build-up of the random wave field.

5:2 Band Random Experiments

It is unquestionably realized that determining the energy transfer for internal waves in a realistic system requires the investigation of energy input over a band of wavenumbers rather than at discrete modes (Sec. 3-4), although the mode

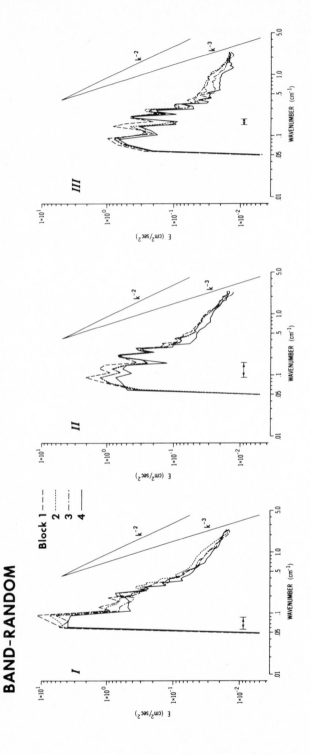

Fig. 11. Total energy wavenumber spectra from the Band-Random experiments for four consecutive time intervals. The marked intervals show the range of wavenumbers at which energy was introduced.

experiments were useful as a basis for understanding the more complex band experiments. Three final numerical experiments were conducted where energy was placed
as follows: Case I, low wavenumber band, .043 to .090 cm^{-1}, Case II, intermediate
wavenumber band, .090 to .158 cm^{-1}, and Case III, high wavenumber band, .158 to
.204 cm^{-1}. The initial amount of energy placed in the wavenumber bands was
approximately inversely proportional to wavenumber. This is because the high
wavenumbers cannot sustain the same energy as the low wavenumbers due to wavebreaking. This initial energy is not an important factor in determining the
final behavior and shape of the spectra.

Wavenumber spectra for the total energy are presented in Figure 11 for the
Band-Random experiments. The marked intervals in the lower portion of the figures
show the band of wavenumbers at which energy was introduced. In Case I, a transfer
of energy from low intermediate wavenumbers occurs within the first block (250
seconds), and energy fills the high wavenumbers by the second block, while the
energy at low and intermediate wavenumbers decreases. A loss of total energy occurs
over the entire spectrum due to dissipation resulting from wavebreaking. Case II
shows some initial transfer of energy from intermediate to low wavenumbers, and the
high wavenumbers are now being fed energy directly and more rapidly than in Case I.
Following block 1, the overall energy level decreases due to wavebreaking, and,
again, some minor transfer of energy to the intermediate wavenumbers occurs. In
Case III, where energy is placed at high wavenumbers, a small degree of wave-
wave transfer occurs in the first block and then energy dissipates at all wavenumbers due to wavebreaking. The spectral representation is not ideal for observing
energy transfers, but does show the characteristic slopes at various times during
the experiment.

A better representation for observing energy transfers in wavenumber space is
the wave band analysis shown in Figure 12, where the cumulative energy in each
wave band is given as a function of time. Case I is presented in Figure 12(a).
The energy initially contained in bands 1 and 2 is rapidly ($\sim$ 10 T) distributed
to other bands containing higher wavenumbers. This cascade continues until
approximately 30 T and most of the input energy is lost to either diffusion
directly or has been transferred to high wavenumbers. A very small degree of
energy transfer back to band 1 occurs after 30 T, and the system is now predominately losing energy via wavebreaking. Case II, shown in Figure 12(b), shows a
rapid (less than 10 T) energy loss of the intermediate waveband 3 to bands 2 and
6. Subsequent times show band 2 losing energy to high wavenumbers with the total
energy (band 6) decaying due to breaking. One should note that the dissipation
rate is band, or wavenumber, selective where band 1 decreases slowly, while bands 4,
5 and 6 decrease much more rapidly with respect to time. Case III is shown in
Figure 12(c), where the energy input is at high wavenumbers and energy is transferred

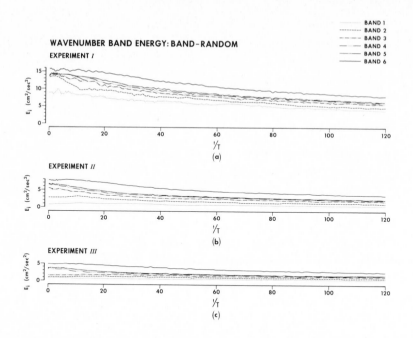

Fig. 12. Energy within six wavenumber bands for the three Band-Random experiments.
Band 1, $0.043 \leq k \leq 0.150$; Band 2, $0.150 \leq k \leq 0.226$; Band 3, $0.226 \leq k \leq 0.300$;
Band 4, $0.300 \leq k \leq 0.376$; Band 5, $0.376 \leq k \leq 0.400$; Band 6, the rest.

to higher wavenumbers very rapidly with only a small amount of energy transfer
to lower bands via wave-wave interactions. The common feature for all three
experiments is the rapidity of the energy transfers, the pumping of energy to
the high wavenumbers, and the diffusion due to wave breaking.

The wave-wave interactions occur on a relatively short time scale of order 10 T
in all three experiments (questioning the validity of any weak interaction assumptions), and this time scale depends on the band energy levels. Diffusive time
scales, τ_i, can be computed for the overall dissipation due to localized wavebreaking, and they are as follows: Case I, $\tau_I = 124$ T, Case II, $\tau_{II} = 104$ T, and
Case III, $\tau_{III} = 168$ T, where τ_i depends on the energy levels of the various bands.
The eddy diffusitivities for the gravest mode from the three cases can be estimated
and are 0.50, 0.76 and 0.47 cm^2/sec, respectively; these diffusivities agree
remarkably well with the estimated oceanic values.

A more descriptive representation of this complicated, interacting system of
internal gravity waves, where both wave-wave interactions and dissipation due to
wavebreaking are simultaneously occurring, is given in Figures 13(a-c). Case I of
the Band-Random experiment is presented in Figure 13(a), while Cases II and III
are shown in 13(b) and (c), respectively. Case I is shown for a duration of 960

seconds, while Cases II and III are shown for 480 seconds, as by this time, diffusive processes outweigh wave-wave interactions. The graphs depict the amount of energy in each band, in contrast to the cumulative band energy shown in Figure 12. The previous representation showed the energy contained in a band of wavenumbers from K_o to K_i, while the present representation shows the energy contained in a band of wavenumbers from K_i to K_{i+1}. The vertical scale is in energy units (cm^2/sec^2), and the block height (stippled and drawn with a full line) represents the energy contained in that band at the given time. The diagonal lines in the upper portion of the block represent an increase in band energy relative to the previous time, whereas the areas marked with parallel lines represents energy loss relative to the previous time. The seventh band represents energy loss over all wavenumbers relative to the previous time, and this loss is due to wavebreaking. Such a representation provides a clearer picture of energy transfers between bands due to wave-wave interactions and energy losses due to wave breaking.

Case I, Figure 13(a) shows the t=0 background spectra plus the additional energy placed in the lower bands 1 and 2. After a time of 60 seconds (less than 8 T), band 2 losses a large amount of its energy, while band 1 losses a smaller amount, and this loss goes to bands 3 to 6 in differing proportions with a small amount of wave-breaking dissipation occurring. The time scale for various wave-wave interactions can be observed in this representation, where band 1 gradually loses energy until t=360 seconds and then begins to regain energy, whereas band 2 loses energy until t=180 seconds and begins to regain energy. Band 3 receives most of its energy within 120 seconds, and loses energy after t=180 seconds. Band 6 reaches a maximum amplitude due to wave-wave transfers and by t=180 seconds, wave breaking becomes significant and is a substantial percentage of the energy transfers due to wave-wave interactions. What cannot be concluded, and we are not implying, is that breaking is predominantly taking place in band 6. As a matter of fact, band 6 gains energy during the times 180 and 240 seconds when breaking and dissipation are intense. Wave-wave interactions and breaking continue throughout the experiment as the system slowly reaches an equilibrium state.

Case II is presented in Figure 13(b) and during the first 120 seconds the cascade and de-cascade characteristic of wave-wave interactions is occurring within the low and intermediate wave bands. A considerable depletion of the input energy band occurs in a short period, followed by a period when wave breaking becomes an important factor in the energy loss for each band. Again, the system slowly approaches an equilibrium state. Case III is presented in Figure 13(c), and the initial energy is predominantly in band 4. Similar to the previous case, most of the wave-wave interactions occur in the first 120 seconds. It is interesting to note the transfers at different times, at t=60 seconds, band 4 transfers energy to bands 5 and 6, while at t=120 seconds, bands 2 and 3 are gaining energy. Again,

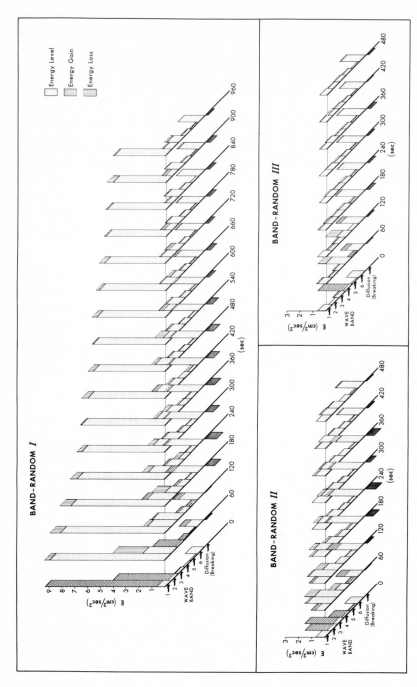

Fig. 13. Waveband analysis for the Band-Random experiments. The height of the vertical blocks represents the energy in each of six wavebands at the given time. The diagonal lines denote an increase of energy relative to the previous time, and the parallel lines denote a loss of energy. The seventh block labeled diffusion represents energy loss over all wavenumbers due to wavebreaking.

energy loss due to wave breaking represents a significant percentage of the overall energy transfer.

We conclude from these results that wave bands 4, 5, and 6 become saturated by time 480 seconds, and inspection of the three experiments shows that the amplitude of these bands remains similar. We can also conclude that the equilibrium amplitude for wave band 3 is given in Cases I and II. Results for wave band 1 and 2 are inconclusive, but it appears that the amplitudes present in Case I represent approximate equilibrium values. In fact, one may argue that Case I represents an equilibrium spectrum.

It is not implied that further energy input to any band will produce instantaneous breaking, but what is implied is that further energy input may be rapidly transferred by wave-wave interactions to higher wavenumbers where breaking becomes likely. Such a cascade of energy, however, is not necessary for breaking, as the high wavenumbers are near saturation and are superimposed on the lower wavenumbers in physical space. This superposition means that any additional energy input to the low wavenumber modes can result directly in wave breaking. This implies that the amplitude of a single wave is a sufficient criterion for wavebreaking, but not necessary. In fact, the collective contribution of all waves at a single position in physical space determines the conditions for breaking in a full spectrum. This point is relevant to the discussions of Frankignoul (1972) and Garrett and Munk (1972b) where the criterion for a single wave or band of waves was considered without treating the background contributions arising from low-frequency, low wavenumber oscillations. In the case of overturning, the criterion developed by Orlanski and Bryan (1969) involved the comparison of an exchange velocity (the time derivative of the density divided by the horizontal derivative of the density) to the advection velocity. The exchange velocity equals the phase velocity only for a single wave and, in this case, overturning occurs when the advection velocity exceeds the phase velocity. The criterion of advection velocity exceeding exchange velocity, on the other hand, can be used in the full spectrum case provided that the quantities represent contributions from all waves. In the case of shear instability, the Richardson number criterion discussed by Phillips (1966) must account for the collective values of wave shear and density gradient. A related conclusion which can also be drawn from the numerical experiments is that although the final amplitude for each wave individually was small, the collective nature of the wave field yields a system in equilibrium. The small amplitude for individual waves in this saturated spectrum resulted in linear relationships for the kinematical wave properties, but the wave field became very nonlinear when additional energy was introduced.

Throughout this paper, we have purposely avoided comparing our two-dimensional results for internal gravity waves with the three-dimensional oceanic counterparts.

This was because of the model's strong assumptions, that is, two-dimensionality and non-rotation. One may argue that for frequencies far from the inertial frequency, or close to the buoyancy frequency, the rotationless approximation is justified. The justification for two-dimensional is harder to support. It was unclear whether a two-dimensional equilibrium spectrum existed, and if so, what shape it took; this alone justified the study of a two-dimensional simulation. Also, the study of wave-wave interactions requires a large number of waves, and an even larger number if wavebreaking is included. At least fifty waves in each direction should be resolved, making any three-dimensional model prohibitively large. The interplay between wave-wave interactions and wave-breaking is present in two-dimensional systems. Perhaps this interplay does not have the intensity present in three dimensions, or precisely the same role, but we believe that the physics will be similar and accurately modeled with sufficient resolution. The

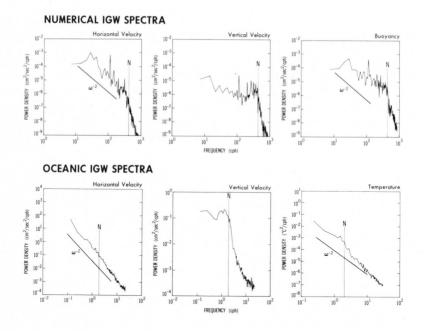

Fig. 14. The upper portion shows spectral densities for horizontal and vertical velocities and the buoyancy taken from the numerically generated random internal wave field. The lower portion shows similar spectral densities taken from oceanic observations by Eriksen (1978).

two-dimensionsal model will lack interactions found in a three-dimensional simulation; if anything, this should underestimate wavebreaking phenomena as it is well known that three-dimensional systems cascade energy to high wavenumbers more readily than two-dimensional systems. With the preceeding in mind, we present some recent data on oceanic power spectra (Eriksen, 1978) in the lower portion of Figure 14, where the frequency range is well separated from the inertial frequency. Power spectra from the two-dimensional model are shown in the upper portion of Figure 14, and the two sets have some striking similarities. The slopes for frequencies below N are approximately equal, and the rise in the vertical velocity spectra near N is similar. Oceanic spectra fall off less rapidly for frequencies above N that for the model. Some of the energy at frequencies above N is probably due to microstructure, while turbulence is the prime candidate for the majority of energy above N. If so, the oceanic spectra and the numerically derived spectra share a similar phenomenological description. Eriksen (1978) observed sporadic wave breaking at all frequencies, and no clear distinction of overturning (Orlanski and Bryan, 1969) or shear instability (Phillips, 1966) could be made. Our numerical simulation is one where localized, sporadic patches of overturning (as shown in the lower portion of Figure 6) account for dissipation.

6: SUMMARY AND CONCLUSIONS

6:1 Summary

Numerical computations were performed concerning wave-wave interactions for a broad range of conditions from gyroscopic equation solutions at low and high energies to the most general solutions of a two-dimensional numerical model which included a finite amplitude spectrum of waves. Let us then summarize the highlights from the various experiments. (i) The classical triad gyroscopic solutions agreed well with the less restrictive, numerical computations of the truncated spectra. In particular, the relationship between the non-linear evolutionary time scale, T_μ, and energy, $T_\mu \ \bar{E}^{-1/2}$, was found to hold even when the condition $\omega \gg 1/T_\mu$ was violated. Study of the double triad, four wave resonance problem indicated that the presence of additional waves alters the triad evolution significantly with respect to the simple triad solutions, and destroys the long time periodicity. A two-dimensional numerical simulation capable of resolving a large number of waves showed that the evolution of the primary triad components differed dramatically from the simple analytical solutions after only one half a non-linear period, T_μ. This difference was due to the presence of other waves with which the primary triad components interacted (both resonantly and non-resonantly). Two numerical experiments concerning triad evolution were performed, one where the background state was quiescent and the other where a random background field of internal waves was present. The final states for both experiments were very similar indicating that the initially quiescent background so becomes filled with internal waves due to wave-wave interactions, however, the random case achieves its final state in a shorter time. (ii) Numerical simulations of surface forced, resonantly growing waves were performed to compare with laboratory results, and a close agreement was found between the two. The numerical solution of the forced triad showed similar behavior to that of the initial condition experiments. Both showed that the time scale for wave-wave interactions was very short (∿20 buoyancy periods) and that a significant amount of wavebreaking occurred at all stages of evolution. Frequency spectra from the simulation agreed well with the spectra taken from laboratory experiments. Differences due to the manner in which sidewall drag was parameterized using a body diffusivity were not found to be significant. However, neither sidewall drag nor diffusivity appeared to be a factor in determining the final, forced wave amplitude. Results indicated that a level was achieved, and was controlled by instantaneous wavebreaking. Once the system reaches the equilibrium level, wave-wave interactions and wave breaking are equally important for the distribution of energy. (iii) A finite amplitude, random internal wave field was generated by a long time integration of the two-dimensional model with random body forcing. The wavenumber and frequency spectra were relatively smooth and red in nature, and a set of experiments was performed using this background spectrum. This set of experiments labelled Band-Random, energy was introduced over a band of wave-numbers and its main results are shown in Fig. 13(a), (b) and (c), where the time evolution of wavebands is shown along with the dissipative energy loss. The background spectrum appears to be at a saturation level for the high wavenumbers, but not for the low ones. The second experiment introduced energy at medium wavenumbers, and both decascading of energy to lower wavenumbers and cascading to high wavenumbers accounted for the loss of energy from medium wavenumbers during the early times. By around 15T (180 seconds), wavebreaking becomes an important feature, as seen in Figure 11(b), where block 7 grows in amplitude. The system reaches an equilibrium from medium wavenumbers upward, and the lowest wavenumbers do not grow even though they are far from saturated. This point will be taken up later. In the third experiment, energy was introduced at high, although not the highest, wavenumbers and energy was rapidly transferred to higher wavenumbers where dissipation occurred. Dissipation rates for the three Band-Random experiments were calculated for the gravest mode found to be of order 1 cm^2 sec.

6:2 CONCLUSIONS

The main conclusions of the final set of experiments, along with much of the previous work, can best be described if one assumes a sequence of steps whereby the energy level is gradually increased. Multiple triad interactions will result

in a filling of the energy spectrum when energy is introduced in a particular band
of wavenumbers. For bands where the energy level is high enough to result in non-
linear time scales of only a few intrinsic periods, wave-wave interactions
(resonant and non-resonant) provide the mechanism for filling the spectrum. The
energy transfer becomes more and more rapid with increasing energy, and no universal spectrum appears to result from these processes. As the energy input
increases, energy will accumulate in high wavenumbers until localized instabilities
(overturning) occurs. From that point on, these high wavenumbers will remain at
a saturation level, such that any additional energy input at the saturated band,
either directly or via wave-wave interactions, will result in localized mixing.
On the other hand, additional energy input at bands other than the saturated band
will result in an increase of low and medium waveband energy (via wave-wave
interactions) until an equilibrium level is achieved. The equilibrium level of
any particular band will depend on the high wavenumber bands being saturated.
For instance, any energy above the equilibrium at low wavenumbers will produce
localized mixing in physical space almost instantaneously. This does not mean
that the low wavenumbers are saturated, as their energy levels can be much lower
than a saturation level. What takes place at or near an equilibrium level is that
the contributions from high and low wavenumbers result in localized regions in the
physical space where the criterion for instability is almost met. In fact, this
superposition effect means that low and medium wavenumbers are far from meeting
any breaking criterion when taken individually, yet cannot tolerate any additional
input energy when in the presence of a saturated band of high wavenumbers. It is
clear that the sporadic, localized patches of turbulence in physical space are
by no means localized in wavenumber space. Even though a band of saturated high
wavenumbers is necessary to produce the localized mixing, a direct cascade from
low to high wavenumbers is not necessary. Consider the case of overturning,
where the criterion is that the total vertical density gradient must reverse sign
at some point in space. Suppose that a low and high wavenumber are superimposed,
such that the criterion for gravitational instability is nearly met. Energy
input at the low wavenumber will produce a region in space where this criterion
is met and results in overturning without an energy cascade from low to high
wavenumber. This statement must be qualified, as overturning is a manifestation
of very high wavenumbers and one may view the process as a cascade. However,
overturning is such a localized feature that the wavenumber representation is
inappropriate, and we prefer to assume that instantaneous mixing is taking place.
Once an equilibrium amplitude is reached, the system is very inefficient at
transferring energy from band to band via wave-wave interactions, but on the
other hand, is very efficient at dissipating energy by localized overturning.
Well defined criteria for overturning or mixing are possible in physical space,

98

but no criteria can be given involving wavenumber space because of the complex superimposing of many waves. It is for this reason that no simple argument can be put forward to explain the shape of the equilibrium energy spectrum as a function of wavenumber. It is apparent from this study that any attempts to understand equilibrium spectra must address the question of mixing and dissipation with an appropriate closure scheme.

ACKNOWLEDGEMENTS

The authors wish to thank Dr. P. Ripa and Dr. G. Philander for their comments and Mr. L. Polinsky for programming and analysis work. We also thank Ms. J. Kennedy and Mrs. B. Williams for typing the manuscript, and Mr. P. Tunison for scientific illustrations. One of the authors (C.P.C.) was supported by a NOAA Grant No. 04-7-022-44017 and the experimental laboratory facilities were supported by a National Science Foundation, Grant No. ATM77-19955.

REFERENCES

Bender, C. M. and Orszag, S. A., 1977: Advanced Mathematical Methods for Scientists and Engineers, International Series in Pure and Applied Mathematics. McGraw Hill.

Bretherton, F. P., 1964: Resonant interaction between waves. The case of discrete oscillations. J. Fluid Mech., 20, 457-479.

Cairns, J. L., 1975: Internal wave measurements from a midwater float. J. Geophys. Res., 80, 299-306.

Charnock, H., 1965: A preliminary study of the directional spectrum of short period internal waves. Proc. 2nd U.S. Navy Sym. Mil. Oceanogr., 175-178.

Davis, R. E. and Acrivos, A., 1967: Solitary internal waves in deep water. J. Fluid Mech., 29, 593-608.

Delisi, D. P. and Orlanski, I., 1975: On the role of density jumps in the reflexion and breaking of internal gravity waves. J. Fluid Mech., 69, 445-464.

Eriksen, C. C., 1978: Measurements and models of fine-structure, internal gravity waves and wave breaking in the deep ocean. J. Geophys. Res., 83, 2989-3009.

Foffonoff, N.P. and Webster, F., 1971: Current measurements in the Western Atlantic. Phil. Trans. Roy. Soc., A, 270, 423-36.

Frankignoul, C. J. 1972: Stability of finite amplitude internal waves in a shear flow. Geophys. Fluid Dyn. 4(2)91-99.

Garrett, C.J.R., and Munk, W. H., 1972a: Space-time scales of internal waves. Geophys. Fluid Dyn., 2, 225-264.

Garrett, C.J.R. and Munk, W. H., 1972b: Oceanic mixing by breaking internal waves. Deep-Sea Res., 19, 823-32.

Garrett, C.J.R. and Munk, W. H., 1975: Space-time scales of internal waves: A progress report. J. Geophysical Res., Vol. 80, 291-297.

Gould, W. J., Schmitz, W. J. and Wunsch, C., 1974: Preliminary field results for a mid-ocean dynamics experiment. (MODE-Ø). Deep-Sea Res., 21, 911-32.

Hasselmann, K., 1966: Feymann diagrams and interaction rules of wave-wave scattering processes. Rev. Geophysics, 4, 1-32.

Hasselmann, K. 1967: Nonlinear interactions treated by the methods of theoretical physics (with application to the generation of waves by wind). Proc. Roy. Soc. London, A299, 77-100.

Hayes, S. P., 1975: Preliminary measurements of the time-lagged coherence of vertical temperature profiles. J. Geophys. Res., 80, 307-11.

Katz, E. J., 1975: Tow spectra from MODE. J. Geophys. Res., 80, 1163-67.

LaFond, E. C., 1949: The use of bathythermographs to determine ocean currents. Trans. Amer. Geophys. Un., 30, 231-7.

Leaman, K. D. and Sanford, T. B., 1975: Vertical energy propagation of inertial waves: a vector spectral analysis of velocity profiles. J. Geophys. Res., 80, 1975-78.

Lorenz, E. N., 1963: Deterministic nonperiodic flow. U. of Atmos. Sci., 20, 130-141.

Martin, S., Simmons, W. F., and Wunsch, C., 1972: The excitation of resonant triads by single internal waves. J. Fluid Mech., 53, 17-44.

McComas, D. H. and Bretherton, F. P., 1977: Resonant interactions of oceanic internal waves. J. Geophys. Res., 82, 1397-1412.

McGoldrick, L. F., 1965: Resonant interactions among capillary-gravity waves. J. Fluid Mech., 21, 305-331.

Muller, P. And Oblers, D. J., 1975: On the dynamics of internal waves in the deep ocean. J. Geophys. Res., 80, 3848-3859.

Müller, P., Olbers, D. J., and Willebrand, S., 1978: The IWEX spectrum. J. Geophysical Res., 83, 479-500.

Orlanski, I., and Bryan, K., 1969: Formation of the thermocline step structure by large amplitude internal gravity waves. J. Geophys. Res., 74, 6975-83.

Orlanski, I., 1971: Energy spectrum of small-scale internal gravity waves. J. Geophys. Res., Vol. 76, 5829-5835.

Orlanski, I., 1972: On the breaking of standing internal gravity waves. J. Fluid Mech., 54, 577-98.

Orlanski, I. and Ross, B. B., 1973: Numerical simulation of generation and breaking of internal gravity waves. J. Geophys. Res., 78, 8806-8826.

Orlanski, I., Ross, B. B., and Polinsky, L. J., 1974: Diurnal variation of the planetary boundary layer in a mesoscale model. J. Atmos. Sci., 31, 965-989.

Orlanski, I. and Cerasoli, C. P., 1980: Energy transfer among internal gravity modes: weak and strong interactions. (submitted for publication to J. Fluid Mech.)

Phillips, O. M., 1966: The Dynamics of the Upper Ocean. Cambridge University Press.

Phillips, O. M., 1977: The Dynamics of the Upper Ocean. Cambridge University Press, 2nd. Edition.

Ripa, P., 1978: Non-linear interactions among ocean waves. (submitted to J. Fluid Mech.)

Thorpe, S. A., 1975: The excitation, dissipation and interaction of internal waves in the deep ocean. J. Geophys. Res., 80, 328-38.

Webster, F., 1969: Vertical profiles of horizontal ocean currents. Deep-Sea Res., 16, 85-98.

Wunsch, C., 1975: Deep ocean internal waves. J. Geophysical Res., 80, 339-343.

KINETIC ENERGY DISSIPATION OBSERVED IN THE UPPER OCEAN

T. M. DILLON and D. R. CALDWELL

School of Oceanography, Oregon State University, Corvallis, Oregon 97331 (USA)

ABSTRACT ONLY

   Observations of vertical temperature microstructure at Ocean Station P
(50°N, 145°W) were made with a freely falling tethered microstructure recorder.
Two storms were observed during the MILE experiment, and during the second
milder storm temperature gradient fluctuations were resolved to the smallest
dissipative scales in more than 90% of the data. The temperature gradient
profiles were broken into blocks approximately 60 cm in vertical extent and
spectrally analysed with a fast Fourier transform.

   A cut-off wavenumber (the wavenumber at which the spectrum falls to approximately 10% of its peak value) was found for each spectrum, and a Batchelor
scale (Batchelor, 1959) was determined from the cut-off wavenumber; a detailed
description of this method of calculation of the Batchelor scale has been described in a previous work (Dillon and Caldwell, 1980). After non-dimensional-
izing each spectrum by the varience and the Batchelor scale, the spectra were
grouped into three classes according to Cox number, $\overline{(dT/dz)}^2/\overline{(dT/dz)}^2$, which is
a rough measure of the relative strength of turbulent mixing to stratification.
The non-dimensional spectra within each group were then ensemble averaged to
determine a characteristic spectral shape or form for three classes of Cox
number. For low Cox number (less than 500), the ensemble average spectrum did
not agree well with the one-dimensional Batchelor spectrum (Gibson and Schwartz,
1963) in the linear range, but for intermediate Cox number (between 500 and
2500) the agreement was better, and for large Cox number (greater than 2500) the
agreement was remarkably close. The approach to the Batchelor spectrum with
increasing Cox number is attributed to an increasing separation between the
finestructure and microstructure wavenumber ranges (Gregg, 1977). For small Cox
number, the turbulence is not sufficiently intense to mix the fluid at larger
length scales, and the finestructure can dominate at wavenumbers usually thought
to be in the microstructure range. For large Cox number the fluid is well mixed
to larger length scales and the linear range of the spectrum agrees well with
the Batchelor prediction.

   Dissipation rate profiles were estimated from the Batchelor scale under
conditions of both low (0-6 m s^{-1}) and high (6-15 m s^{-1}) wind speed. During low

winds, the average dissipation profile exhibited a great deal of structure associated with the mean stratification in the mixed layer-thermocline transition zone where turbulent stirring was relatively intense but not strong enough to completely overwhelm the stratification. The highest dissipation during low winds ($\sim$5 x 10^{-4} cm^2 s^{-1}) was seen at the bottom of the transition zone. In contrast, the high wind speed dissipation profiles were more uniform in the vertical, revealed little correlation with temperature structures above the seasonal thermocline, and had an average value of 2 x 10^{-3} cm s^{-1}, about 30 times as large as during low winds. Within the seasonal thermocline, the distribution was patchy in both space and time, but even here the dissipation was much larger during high winds than low. Details of the dissipation profiles may be found in Dillon and Caldwell (1980).

REFERENCES

Batchelor, G. K., 1959. Small-scale variation of convected quantities like temperature in a turbulent fluid. J. Fluid Mech., 5:113-133.

Dillon, T. M. and Caldwell, D. R., 1980. The Batchelor spectrum and dissipation in the upper ocean. J. Geophys. Res. (in press).

Gibson, C. H. and Schwartz, W. H., 1963. The universal equilibrium spectra of turbulent velocity and scaler fields. J. Fluid Mech., 16:365-384.

Gregg, M. C., 1977. Variations in the intensity of small-scale mixing in the main thermocline. J. Phys. Oceanogr., 6:528-555.

OBSERVATIONS OF AIR-SEA INTERACTION PARAMETERS IN THE OPEN OCEAN

M. REVAULT D'ALLONNES[1] and G. CAULLIEZ[1]

[1]Laboratoire d'Océanographie Physique du Muséum National d'Histoire Naturelle de PARIS
43-45, rue Cuvier, 75005 PARIS, FRANCE.

ABSTRACT

This paper describes an experiment conducted during October 1976 in the Mediterranean Sea, on board the Bouée-Laboratoire BORHA II moored on three points at 42°N,
4°45'E (fig. 1). These results have been obtained with the support of the "Action
Thématique Programmée" set up by the "Centre National de la Recherche Scientifique"in
the field of Physical Oceanography. They are shown here as an empirical and experimental evaluation of the relative parts of energy the wind gives to the sea for waves and
for "mean" current.

I. METHOD

The physical problem considered here is that of the penetration of a turbulent
boundary layer under unstationary conditions in a preturbulent and stratified medium,
while taking into account the growth of waves and the entrainment. The equations
governing that problem are know, but remain unsolvable in the general case, owing
to their non-linearity and to the large number of factors entering the problem, as
turbulent variables. Thus, the mathematical and numerical approach to this problem
reduces to the problem of "closure" of the equations.

Together with the dead-ends inherent in this mathematical formulation -the only
one known in the field considered- stand numerous experimental difficulties : difficulty of marine measurements in the open sea, number of the parameters to be taken
into account, range of the scales affected by these interactions, technological problems associated with this last question when choosing the probes, etc... Working
from the Bouée-Laboratoire BORHA II as a platform, the following experiment has been
carried out. The basic idea of this experiment is premised on the fact that the
effects of an abrupt increase of the wind velocity at one fixed observational site
are related, through a relation which has to be precised, to the effects of a stationary wind with variable fetch. This approximate analogy involves in particular the
assumption that the air and water flows have statistically a one-dimensional character. It follows that we work only short periods of time so that it is possible to neglect the effects of the earth's rotation,which could give a "two-dimensional" character to

the problem in the horizontal plane, and that we look for macroscopic conditions propitious to the analogy.

The fundamental interest of the idea lies in the fact that if we are able to identify and to follow with time the penetration in the water of the wind effects, we are thus able to give a lower limit to the superficial water layer set in motion, then to calculate the time evolution of "mean" potential and kinetic energies given to this layer. It should be noticed that we are not working with absolute quantities, but with the difference between the values at some time, and at the origin of the experiment. The latter corresponds to the end of a "residual" initial state. This last statement entails important experimental advantages.

In short, we have been looking (i.e. waiting) for the following sequence of events: 1°) the wind has a low and constant value during enough time so that the initial state of interactions can be said to be "residual" ; 2°) the wind then sharply rises until it reaches the limiting value authorized by the measurement array, and remains constant (in velocity and direction) until the time scales considered approach the limits of validity for the hypotheses outlined above.

Owing to the fact that it is not possible to know what is to become of any macroscopic situation, many recordings were stopped for having not been in accordance with the two characteristics. However, during an experiment, a nearly ideal situation appeared, on October 26, 1976 between 2 and 11 p.m. The results presented hereafter concern only this 9 hours-observation.

II. EXPERIMENTAL ARRANGEMENT

The experimental array is summarized in figures 2 and 3. We will thus only recall its principle. In the atmosphere, the array of probes provides measurement of the wind stress by means of the three classical methods : profile method (6 cup anemometers), flux method and inertio-dissipative method (X hot wires anemometers fixed on a low inertia vane). The recording of the vertical temperature gradient allows to check the hypotheses for the air boundary layer, that is to say the constant flux hypothesis.

Under the circumstances, one obtains (progressing from large to small time scales): 1°) through the profile method

$$u_* = (\frac{\tau}{\rho_a})^{1/2} = \frac{\kappa \, \overline{U(z)}}{Ln \, z/z_0}$$

provided that the flow is stationary, one-dimensioned, without horizontal pressure gradient, and that the Monin Obukhov stability length L remains much larger than the height $z = a$ of the measurement, i.e. $L > 10a$ here;

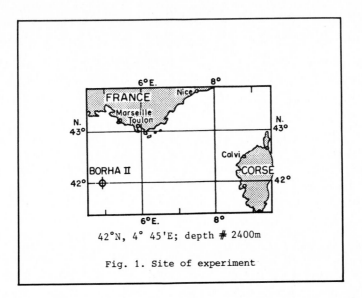

42°N, 4° 45'E; depth # 2400m

Fig. 1. Site of experiment

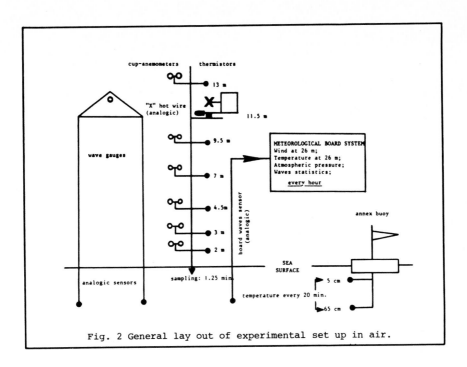

Fig. 2 General lay out of experimental set up in air.

106

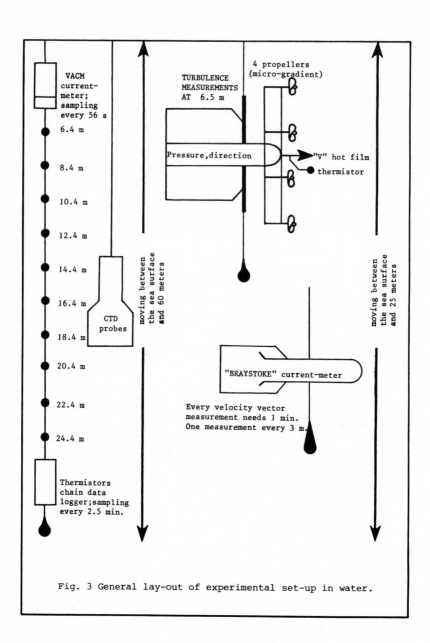

Fig. 3 General lay-out of experimental set-up in water.

2°) through the flux method

$$u_*^2 = - \overline{u'w'} = \int_0^\infty C_{u'w'}(n) \, dn = \int_{n=0}^{n=\infty} n \, C_{u'w'}(n) \, d(Ln \ n),$$

$C_{u'w'}$ being the cospectrum of horizontal (u') and vertical (w') velocity fluctuations at the height a . The last expression of u_* includes in the calculation all the structures contributing to the vertical flux -towards high and low frequencies as well- and thus avoids the "problem of the mean".

3°) through inertio-dissipative method

$$u_* = (\kappa a \varepsilon)^{1/3} = \kappa^{1/3} a^{1/3} \left(\frac{E_1(k_1)}{\alpha} \right)^{1/2} k_1^{5/6}$$

$$= \frac{\sqrt{3}}{2} \kappa^{1/3} a^{1/3} \left(\frac{E_2(k_1)}{\alpha} \right)^{1/2} k_1^{5/6} .$$

This relation is only correct in the inertial range, that is when the physical conditions for three-dimension local isotropy are verified. These highly restrictive hypotheses are checked through their consequences over the velocity spectra on the basis of the four classical criterions :

a) enough separation between the production maximum and the dissipation maximum (high Reynolds number) ;

b) nullity of the cospectrum in the inertial range ;

c) spectra slope in -5/3 within this range ;

d) spectra ratio E_2/E_1 equal to 4/3 within this range.

Furthermore it should be noted that the transformation from the frequency (n) domain to the wave number (k_1) domain involves the use of a space-time relation in the spectral plane. We will see later on that the use of Taylor's hypothesis $k_1 = \frac{2\pi \ n}{\overline{U}}$ brings about some difficulties in the context of our measurements.

The "X" wire anemometer provides velocity measurements used to obtain the wind stress through the two last methods. The probe is fixed over the vertical axis of a low-inertia vane, whose transfer function has been thoroughly adjusted and calibrated in a wind tunnel in order that its frequency coincides with the spectral limit separating production and inertial ranges (1,2). Using the direction fluctuation of the vane, we obtain the tree components of the velocity fluctuations which contribute to the flux, while the "mean" velocity component perpendicular to the wire plane remains negligible compared with the fluctuation in the inertial range.

This experimental realization is thus very similar to the classical conditions encountered in a wind-water facility. The interface movements are followed with two vertical capacitive gauges, the distance of which can be changed remotely, and which determine a vertical plane free to turn. Thus, for example by the method of space-time correlations, we can obtain the dispersion relation of waves and the statistical characteristics of the superficial stirring (4,9).

In the water, the penetration of the effects of mixing and entrainment due to the wind is followed with the arrangement shown in figure 3. A VACM current-meter fixed at a depth of 5 meters, a chain of 10 thermistors between 6 and 24 meters, CTD between 0 and 60 meters, vertical profile of mean horizontal velocity with a curent-meter profiling between 2 to 20 meters. The time and length resolution is shown in this figure. The most original part of this array is a "fish" previously described (11, 12), wich measures among others, in the water, vertical gradient of velocity over small space (30 cm) and time (0,1 sec) scales, and horizontal and vertical turbulent fluctuations of velocity up to a frequency of 180 Hz (double "V" hot film probes whose non-dimensional behaviour has been extended specially for ocean studies, 5,10). Temperature fluctuations are measured close to the hot film with a time resolution of 0,05 sec, for both scientific (vertical thermal flux) and technical reasons (hot film behaviour in a non-isothermal flow). This "fish" is deployed at a fixed depth of 6 meters.

The data from all these probes are recorded either on digital or analog tapes. These last recording -essentially turbulence or high frequency signals- are done with 21 FM channels after treating each signal. The details concerning the electronics, synchronisation, filtering, amplification, calibration of the registration and digital acquisition systems and the various technical precautions during each step of the signal processing are described in 11 and 12.

III. DESCRIPTION OF MACROSCOPIC RESULTS

The conditions we were looking for happened during an experiment on the 28th October 1976 between 12 and 22 p.m. Figure 4 shows, for that date, the evolution in time of the wind velocity at 26 meters, the friction velocity u_* and the wave field total energy $\overline{\eta^2}$ determined by integrating the spectra. We notice in this figure :

a) that the data obtained over friction velocity by means of the mean wind profiles are corrected from the distorsion due to the presence of the buoy in the air flow. Two methods are used : first an analytical method (12), and second an experimental method based on a wind tunnel simulation of the flow structure around the buoy (13) ;

b) that between 4,30 and 6,30 p.m. the wind velocity and wind stress remain nearly constant, while the waves grow linearly with time.

The time evolution of waves spectra in linear coordinates (fig. 5) shown a narrowing around a dominant wave,wich grows with time without changing its frequency ($\pm$ 0,17 Hz). This selective amplification (fig. 6) occurs near the equilibrium state, with energy saturation as described by PHILLIPS(spectral slope close to -5).

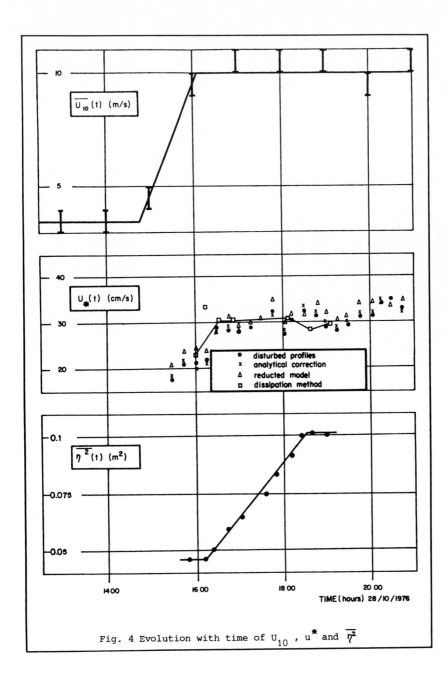

Fig. 4 Evolution with time of U_{10} , u^* and $\overline{\eta^2}$

110

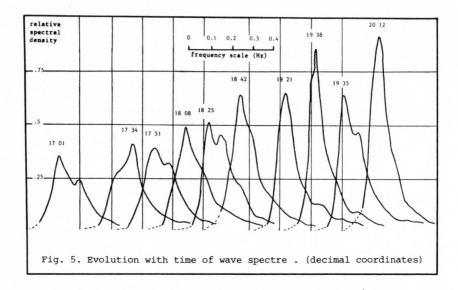

Fig. 5. Evolution with time of wave spectre . (decimal coordinates)

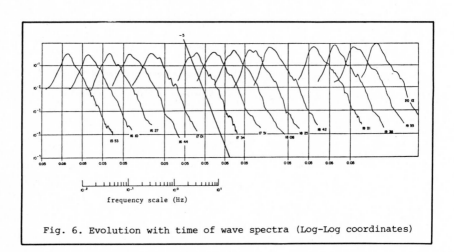

frequency scale (Hz)

Fig. 6. Evolution with time of wave spectra (Log-Log coordinates)

These remarks lead to the choice of an hypothesis of a local generation of waves. If the waves were generated away with a certain spectrum, the classical dispersion relation $c_o = \dfrac{g}{2\pi\, n_o}$ suggests that we should observe, at the Bouée-Laboratoire, a different evolution with time of the spectra. The high frequency waves would propagate less quickly than the low frequency waves, and thus the evolution of the spectra would be characterized by a slipping towards higher frequencies. These arguments are premised on the validity of the linear theory.

The vertical profiles of mean current (fig. 7) show the acceleration of the upper layers by the wind, the deeper layers accelerating more slowly. This entrained layer thickens proportionally to the square root of time (fig. 8), which is in agreement with the results of KATO and PHILLIPS (7) for Richardson numbers smaller than 60. Then the superficial layers show an homogeneous velocity after the passage along the vertical of a zone of active mixing. The continuous thermal profiling operated at 10. 02 p.m. confirms the homogeneity of the layer situated above 20 meters, while the thermal profiles given by the thermistors (fig. 9) confirms the existence of a vertical mixing starting from an initial gradient of 0. 02°C per meter.

A difficulty appears with regards to these temperature measurements which might be thought of as more useful than velocity measurements in order to determine the lower limit of the layer affected by the gust of wind. It is interesting for this to look at the same data drawn as time series of temperature measurements for each depth. Figure 10 shows a number of particularities of the vertical temperature field :

a) the vertical mixing observed begins, at the depth of the probes, by a warming up. This may be interpreted as resulting from local interactions and vertical convection ; the required heat being taken from the layer situated above the first probe (6. 4m). A superficial cooling of 0. 7°C observed on the data of the little buoy confirms this hypothesis ;

b) the warming (time of "arrival" at each depth of the first effects of mixing) happens later at greater depths. The interpretation of this observation is evident ;

c) the same happens of course for the end of the warming, from which time, and all depths, the upper layer is homogeneous. It should be noticed that the heating is more rapid at deeper stations corresponding to an increase of the temperature gradient ("steepening" of the vertical thermal front) ;

d) the amplitude of the warming gets smaller as the depth increases ; this can be interpreted in terms of global conservation of heat quantity ;

c) the precceding observations are hidden, beneath 20.4 meters, by a thermal oscillation whose amplitude is of the same order as that of the warming at the same depths and whose period is around one hour. It has not yet been possible to determine whether this is a direct consequence of a wave propagating at a lower depth around 35 meters along the main thermocline.

112

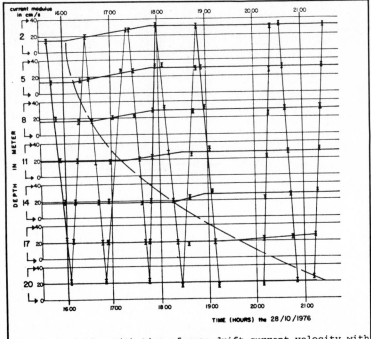

Fig. 7. Variation with time of mean drift current velocity with respect to depth.

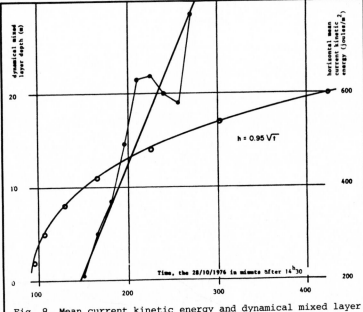

$$h = 0.95 \sqrt{t}$$

Fig. 8. Mean current kinetic energy and dynamical mixed layer depth versus time.

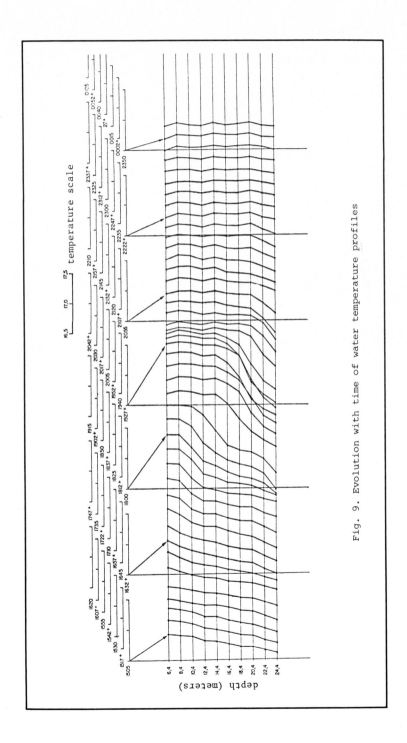

Fig. 9. Evolution with time of water temperature profiles

114

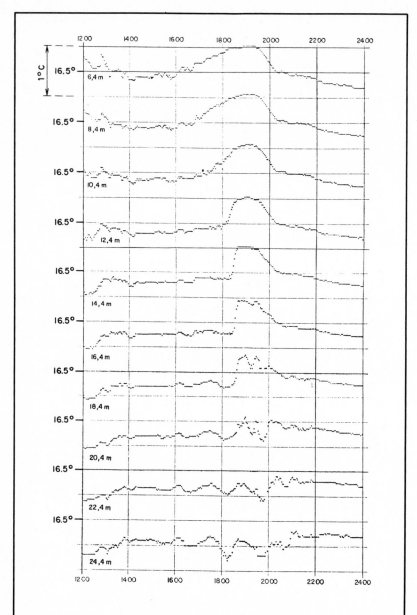

Fig.10. Evolution with time of water temperature at each level
of measurements (same data as Fig. 9).

Analyses are still underway concerning the origin of this wave, in order to "remove" its effects from those of the mixing and thus to try to establish the time evolution of the Richardson number of the active zone of convection.

IV. AN ATTEMPT TO ESTIMATE THE PARTITION OF WIND ENERGY

The preceeding macroscopic results allow the empirical determination of the rate of variation of the kinetic energy of the mean current per unit area, i. e.

$$E_c(t) = 1/2 \ \rho_w \int_0^{h(t)} (\overline{U}^2 + \overline{V}^2)(z,t) \ dz,$$

and the rate of variation of the total energy of waves per unit area, i. e.

$$E_v(t) = 1/2 \ \rho_w \ g \ \overline{\eta^2} = 1/2 \ \rho_w \int_o^\infty n \ S(n,t) \ d(Ln \ n),$$

where $S(n,t)$ is the frequency spectrum of waves at the time t.

Within these approximations,

$$\begin{cases} dE_w/dt &= 0.019 \ J/m^2/sec \\ \\ dE_c/dt &= 0.113 \ J/m^2/sec \equiv 6 \ dE_w/dt \ . \end{cases}$$

On the basis of the following hypotheses it is possible to relate the rate of change of the wave energy to the total wind stress, starting from the analysis proposed by DEARDORFF (3). Assuming that (GIOVANANGELI and Al., 5) :

a) the wave field is strictly two-dimensional,

b) the total energy is entirely associated with the dominant wave,

c) the only source of energy is the work of pressure fluctuations,

d) the various dissipative mechanisms can be neglected,

the general equation of radiative transfer

$$\frac{\delta F}{\delta t}(n,\theta,\underline{x},t) + c_g \ \nabla F(n,\theta,\underline{x},t) = \sum_i S_i(n,\theta,\underline{x},t)$$

then becomes

$$\frac{\overline{\delta\eta^2}}{\delta t} = \frac{c_o}{g \ \rho_w} \overline{\tilde{p}_\eta \frac{\delta\tilde\eta}{\delta x}} = \chi_v \ \rho_a \ u_*^2$$

when writing the form drag ($\overline{\tilde{p}_\eta \frac{\delta\tilde\eta}{\delta x}}$) as a certain fraction χ_v of the total wind stress ($\tau = \rho_a u_*^2$).

One obtains in the present situation $\chi_v \equiv 0.05 = 5\%$, comparable to the values given by :

HASSELMANN and al., 1973, 6 : ....$\chi_v \equiv 3\%$ to 10%

TOBA, 1972, 15 : ....$\chi_v = 0.15 \exp (-1.9 c_o/\overline{U}_{10})$,
which would give here $\chi_v \equiv 4\%$.

The same results given in terms of drag coefficient lead to $c_v \equiv 6.10^{-5}$ for $c_{10} = 1.2.10^{-3}$.

V. AIR AND WATER TURBULENCE

In the air, 8 meters above the surface, the application of the inertio-dissipative method for determining u_* presented the following difficulty : the frequency spectra (fig. 11) show large zones (of about 2 decades) where the slope is equal to -5/3, but the ratio between the horizontal and vertical spectra in the same zone does not reach to the theoretical value of 4/3 it should have in the inertial range. It should be recalled that this is the most strict criterion of isotropy of the second order in the spectral plane.

A review of the experimental and analysis procedures did not reveal any experimental fault which might explain this result. Thus, to explain this difficulty we are brought to postulate the following two hypotheses :

a) a significant departure from Taylor's hypothesis, larger than that given by LUMLEY's theory (8). This point is considered further in 12.

b) a week departure of the air turbulence from local isotropy, in the presence of waves. Similar results have been obtained by SCHMITT and Al. (14), who, after a detailed review of the existing litterature, note that the theoretical ratio is rarely obtained in the marine environment (where it is closer to 1 than to 1.33), while it is often reached when experiments are done over land.

These two possibilities are still being examined, but in water, at 6 meters depth, the spectra of horizontal and vertical velocity fluctuations (fig. 12) show a large zone where the slope is equal to -2 and where the ratio of the two spectra is close to its theoretical value in the inertial range (where the slope should be -5/3). We outline here a theory which could lead to a complete view of a locally homogeneous and isotropic turbulence, by means of the spate-time relation in the spectral plane. The details of this work may be found in reference 12.

The basic idea consists in assuming that there exists a statistical transport of small turbulent structures by the larger structures, and thus that the "convection velocity" (u_c) depends upon the arbitrary separation one introduces between the "small" structures transported and the bigger ones which contribute to that transport. Such an effect must be more evident when the turbulent transport is large

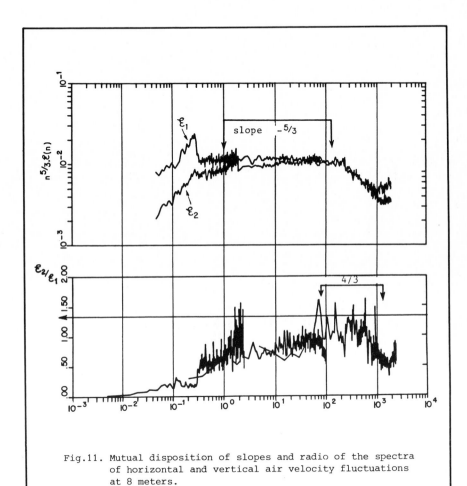

Fig.11. Mutual disposition of slopes and radio of the spectra
of horizontal and vertical air velocity fluctuations
at 8 meters.

118

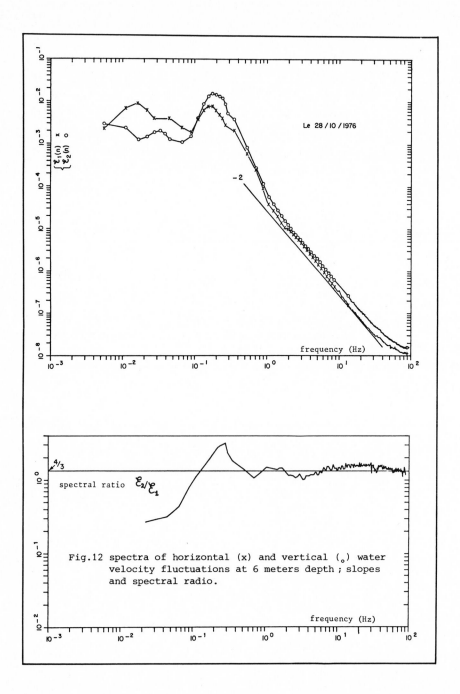

Fig.12 spectra of horizontal (x) and vertical ($_o$) water
velocity fluctuations at 6 meters depth ; slopes
and spectral radio.

with respect to the mean flow, that is to say when the turbulent intensity is large. This is exactly the case considered here where the mean velocity remains small compared to the orbital velocities due to waves which are not dissociated from the turbulence spectra.

If there exists in this case an inertial range (in which we know the dynamics are entirely characterized by the local rate of energy transfer ε), the scale-dependent variations (for exemple the wave number k) of the convection velocity in that zone can be written :

$$\frac{\partial U_c}{\partial k} = \frac{\partial U_c}{\partial k}(k,\varepsilon)$$

Dimensional analysis leads to :

$$\frac{\partial U_c}{\partial k}(k,\varepsilon) = \gamma\,\varepsilon^{1/3}\,k^{-4/3} \quad,$$

where γ is a "non-dimensional constant close to unity". Upon integration this becomes :

$$U_c(k,\varepsilon) = U_o + \gamma'\,\varepsilon^{1/3}\,k^{-1/3} \quad,$$

where U_o is an unknown convection velocity, characteristic of the global transport of the inertial structures by the larger, anisotropic, structures and by the mean flow.

In the ficticious case where such a global transport should be null, one should have :

$$U_c(k,\ ) = \gamma'\,\varepsilon^{1/3}\,k^{-1/3} \quad.$$

Using the two classical relations :

$$\left\{ \begin{array}{l} k = 2\pi\,\dfrac{n}{U_c} \quad (1) \text{ and} \\[2em] E(n) = 2\pi\,\{E(k)/U_c(k,\varepsilon)\} \quad (2) \quad, \end{array} \right.$$

the expression of the wave number spectrum in the inertial range :

$$E(k) = \alpha\,\varepsilon^{2/3}\,k^{-5/3}$$

leads to a frequency spectrum like :

$$E(n) = \beta\,\varepsilon\,n^{-2} \quad.$$

One notices that the relation :

$$E_2(k_1) = \frac{1}{2} \{ E_1(k_1) - k_1 \frac{\partial}{\partial k_1} E_1(k_1) \} ,$$

which leads to a ratio $E_2/E_1 = 4/3$ with a -5/3 power law in wave number, also leads to the same ratio with a -2 power law in frequency.

One also notices that the exponent of the dissipation rate is changed when writing in frequency the turbulence spectra, as well as the "situation" of the inertial range due to the change in the exponent of the variable n .

Thus we see that a theory such as that roughly sketched here is capable of explaining some of the features of the observations which have been described and, in particular, departures from the classical Kolgomorov constants.

In order to confirm (or cancel) this way of considering the spacetime relation problem (wave number-frequency), works are actually going on at the Laboratoire d'Océanographie Physique du Muséum National d' Histoire Naturelle in Paris.

REFERENCES

1- Larsen, S.E., and Bush, N.E., 1974. Hot-wire measurements in the atmosphere. Part 1 : calibration and response characteristics. DISA Information, 16:15-36.
2- Larsen, S.E. and Bush, N.E., 1976. Hot-wire measurements in the atmosphere. Part 2 : a field experiment in the surface layer. DISA Information, 20:5-21.
3- Deardorff, J.W., 1967. Aerodynamic theory of wave growth with constant wave steepness. Journal of the Ocean. Soc. of Japan, 23,6:278-297.
4- Giovanangeli, J.P., Revault d'Allonnes, M. and Ramamonjiarisoa, A., 1978. Open sea simultaneous observations of air and water moyiond during active air-sea interactions, in : turbulent fluxes through the sea surface, wave dynamics and prediction, ed. by Favre, A. and Hasselman, K., NATO Conferences Series, Plenum Publishing Corporation, New-York.
5- Giovanangeli, J.P., 1980 (in press). Non dimensional heat transfer law for a slamped hot-film in water flow. DISA Information, 25.
6- Hasselman, K., Barnett, Y.P., Carlson, H., Cartwright, D.E., Enke, K., Ewing, J.A., Gienapp, H., Hasselmann, D.E., Krusman, P., Meerburg, A., Multer, P., Olbergs, D.J., Richter, K., Sell, W., Walden, H., 1973. Measurements of wind wave growth and swell ducay during the Joint North Sea Project (JONSWAP). Deutsches Hydrographishes Institut, Hambourg.
7- Kato, H. and Phillips, O.M., 1969. On the penetration of turbulent layer into a stratified fluid. Journal of Fluid Mech., 37,4:643-655.
8- Lumley, J.L., 1965. Interpretation of time spectra measured in high intensity shear flows. The Physics of fluids, 8,8:1056-1062.
9- Ramamonjiarisoa, A., 1974. Contriburion à l'étude de la structure statistique et des mécanismes de génération des vagues de vent. Thèse de Doctorat d'Etat, Université d'Aix-Marseille, 160.
10-Resch, F.J., 1973. Use of dual-sensor hot film probe in water flow. DISA Information, 14:5-11.
11-Revault d'Allonnes, M., 1976. Quelques éléments sur la turbulence en mer. Société Hydrotechnique de France, 14ème journée de l'hydraulique, question 2, rapport 1:1-9.
12-Revault d'Allonnes, M., 1978. Contribution à l'étude de la micro-turbulence naturelle au voisinage de l'interface air-mer. Thèse de Doctorat d'Etat, Université de Paris VI, 289.
13-Ruimy, D., 1977. Etude en similitude de la déformation du profil moyen de vitesse du vent au voisinage de la Bouée-Laboratoire BORHA II du CNEXO. Journal de Recherche Océanographique, 11,3:31-39.
14-Schmitt, K.F., Friehe, C.A., Gibson, C.H., 1977. Sea surface stress measurements. Report contract NOAA 03-7-042-35113, Dept. of App. Mech. and Eng. Sciences, University of California, San Diego.
15-Toba, Y., 1972. Local balance in the air-sea boundary processes. I- On the growth process of wind waves. Journal of Ocean. Soc. of Japan, 28,3.

DISSIPATION IN THE MIXED LAYER NEAR EMERALD BASIN

N.S. OAKEY and J.A. ELLIOTT

Bedford Institute of Oceanography, Dartmouth, Nova Scotia, Canada

ABSTRACT

Velocity microstructure measurements were made during a 10-day mixed layer experiment on the Scotian Shelf during late September 1976 using the vertical profiling microstructure instrument OCTUPROBE II. The velocity information obtained with thrust probes has been used to determine the dissipation, ε. Our results show a strong correlation between ε and the wind speed, $\overline{U}_{10}^{-3}$.

INTRODUCTION

During a ten-day experiment from 27 September to 7 October, 1976, scientists from the Bedford Institute of Oceanography conducted a mixed layer study near Emerald Basin on the Scotian Shelf. The part of the experiment which will be the topic of this paper is the examination of an extensive time series of vertical profiles of velocity gradient microstructure. These profiles through the mixed layer and to depths of 100 m were obtained under a variety of wind conditions and it will be shown that the observed levels of dissipation correlate well with the atmospheric forcing.

The dissipation term, ε, of the turbulent energy balance equation represents the rate of viscous dissipation of velocity fluctuations. For isotropic turbulence this can be estimated (Osborn, 1978) from

$$\varepsilon = \frac{15}{2} \nu \left(\frac{\partial u}{\partial z}\right)^2 \qquad (1)$$

where ν is the kinematic viscosity and u is one component of the velocity. To make a comparison to the surface energy input we calculate the net dissipation (per unit area) as follows

$$\varepsilon_I = \int_{-h}^{o} \varepsilon(z)\,dz \qquad (2)$$

where h is the mixed layer depth. The rate of working by the wind on the sea
surface can be estimated from

$$E = \tau \overline{U}_{10} = \rho_a C_{10} \overline{U}_{10}^3 \tag{3}$$

where τ is the surface stress; $\overline{U}_{10}$ is the wind speed at 10 meters; C_{10}, the
drag coefficient and ρ_a, the density of air. The comparison of the net dissipation, ε_I, and E forms one of the major conclusions of this paper.

THE EXPERIMENT

   The site of the experiment near Emerald Basin (43°55'N, 62°40'W) off the coast
of Nova Scotia was in an area where the water depth was 200 m. This area, with
a well-defined mixed layer approximately 20 m deep, was selected because it was
generally free of strong fronts, was horizontally homogeneous, and shallow enough
that the ship could be anchored relatively easily. A mooring was placed at the
center of the experimental area with an Aanderaa current meter at 10 m depth in
the mixed layer and one at 45 m depth in the pycnocline with two thermistor chains
(thermistors at 3 m spacing) spanning the interval from 15 to 75 m. The time
series from these instruments and CTD profiles indicated that although the area
was not dominated by strong intrusive features there was horizontal variability in
salinity $\sim$ 0.3 °/$_{\circ\circ}$ at scales of $\sim$ 10 km and a mean current of $\sim$ 25 cm/sec.
   The microstructure measurements were obtained while the vessel, the CSS Dawson,
was at anchor, using the profiling instrument, OCTUPROBE (Oakey, 1977). This
instrument is approximately 2 m long with a leading sting supporting a variety of
sensors. The instrument drops vertically and falls freely at approximately 0.5 m/s
to the desired depth but is tethered by a light line used for recovery. Using
this "tethered free-fall" mode of operation, many profiles closely spaced in time
can be obtained.
   Velocity gradients relative to the instrument are measured using a pair of
mutually-perpendicular thrust probes similar to those described by Siddon (1965)
and by Osborn and Crawford (1979) and fabricated at the Bedford Institute (Oakey,
1977). Temperature gradient microstructure is measured using thin film platinum
thermometers (DISA 55R41). Several other slow-response sensors measure parameters
such as conductivity, pressure, temperature, and instrument tilts and accelerations.
Data are recorded internally as analogue multiplexed FM on a miniature tape recorder
which has a capacity of 25 minutes, sufficient for five 100 meter profiles.

DATA SET

   An example of the microstructure data is shown in Figure 1 for Station 88. Five
vertical profiles of $\partial u/\partial t$ are shown from near the surface to approximately

50 meters. A similar set exists for the perpendicular component ∂v/∂t. Because OCTUPROBE oscillates slowly, these data have been high-pass filtered for display at 0.2 Hz (∿ 0.4 cycles/m). The data are converted from time derivative to gradient using the pressure measurement. The 22 m depth which is near the bottom of the mixed layer is used as a reference and is indicated by a dashed line. Because of measurement difficulties near the surface, and in particular the draft of the ship, the top few meters of data (typically 5 m) cannot be used and a closed dot indicates the point below which we arbitrarily consider the data reliable. In the lower traces are shown the corresponding temperature gradient records and the temperature profile.

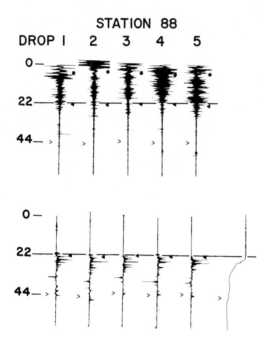

Fig. 1. Successive profiles at 5-minute intervals of velocity gradient (above) and temperature gradient (below) to a depth of 50 m for Station 88. The temperature profile is shown below at the right. Full-scale velocity gradient is 0.5 (m/s)/m. The depth of the mixed layer is indicated by ◀ , the depth below which we have analyzed the data by ● and 44 m depth by >. The 22 m depth used as a reference is marked by a dashed line.

A summary of the data set for the whole experiment is shown in Figures 2A, 2B, 2C, presented in similar format to that of Figure 1. This time series shows only one of the five profiles for each station. These data were recorded at 4 to 6 hour intervals under a range of wind speeds from < 5 to 15 m/s. The wind speeds are indicated below the velocity gradient records. The data in these records have not been high-pass filtered and the low-frequency instrument motion can be seen.

EMERALD BASIN EXPERIMENT

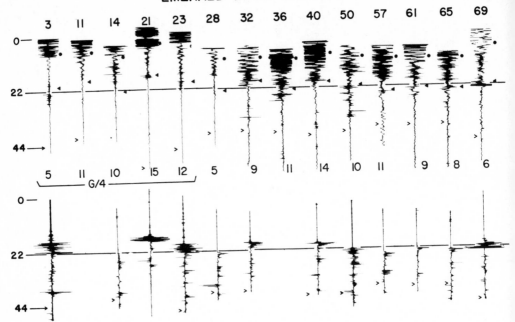

Fig. 2A

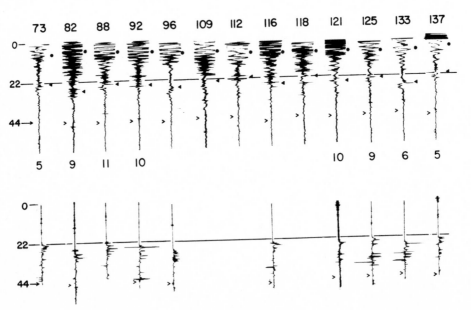

Fig. 2B

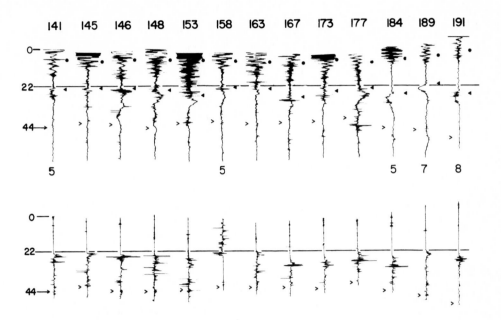

Fig. 2C

Fig. 2A, 2B, 2C. The summary time series for the experiment is shown in these
three pictures. The first profile of each station (of 5 or more profiles) is
presented. The format is similar to Fig. 1 except for the inclusion of the wind
speed in m/s below the velocity profiles. The wind speed remained near 10 m/s
from 92 to 121 and near 5 m/s from Station 141 to 184. The region marked G/4
was recorded at a sensitivity 1/4 that of the other stations.

Variations in the length of each record are a result of differences in the instrument drop speed from series to series. The microstructure intensity varies
considerably from one station to the next but is in general larger for higher
wind speeds.

DATA ANALYSIS

For each station, the first vertical profile of velocity microstructure was
digitized and analyzed using Fast Fourier Transform techniques to obtain spectra.
The data were digitized at 1000 Hz and divided into blocks of 1024 samples corresponding to approximately 0.5 m of the profile. The profiles were divided into two
parts, the lower mixed layer corresponding to the bottom 10 m of the mixed layer,
and the upper mixed layer (the rest of the data which were generally good to within
5 m of the surface). This was done to investigate the depth dependence of ε.
For these two groupings, spectra of velocity shear were determined. They were
corrected at this stage for the frequency response of the electronics and the
thrust probes. The scale response of thrust probes has never been adequately
determined experimentally. From geometrical arguments (Osborn and Crawford,

1979) related to averaging a sine wave of wavelength λ over an interval L, one obtains a minus 3 db or half power point at $\lambda = 2.25L$ where L is the length over which there is a lift force on the sensor. For our thrust probes this gives a cut-off of ~ 2 cm which is the value used in this analysis. An example of a corrected spectrum is shown in Figure 3 for Station 23. The data for both components of shear are shown as (x and ●) corrected for $\lambda_c = 2$ cm and for comparison the same data corrected for $\lambda_c = 1$ cm (▲). The solid curve is the empirical universal isotropic turbulence curve obtained by Nasmyth (1970). In those cases where the spectrum was less well defined than in Figure 3 because of lower signal or higher noise at high frequency, the universal curve was used to determine where to terminate the integration so as not to include noise. From the spectral variance, the dissipation, ε, was determined from Equation (1). A comparison of the measured values of ε for the two perpendicular components of velocity shear are shown in Figure 4. The straight line corresponds to a ratio of 1 (or horizontal isotropy).

There are a few points differing from "isotropy" by a factor of 2 but the average ratio between the determinations of ε is 0.95 ± 0.27. The self consistency of the analysis technique should not contribute errors greater than 25% so there are many examples of velocity microstructure which are not horizontally isotropic for the averaging scheme used.

Because of the variability from profile to profile even at the same station, one velocity component for each of the total of five or six profiles at each station was analyzed to provide a better average of ε. Because of the time-consuming effort of digitizing and fully analyzing these data profiles (more than 200) using the FFT method, a spectrum analyzer system (Ubiquitous UA14A) was used to process the bulk of the data. Comparisons of analyses of the same data with the FFT method and spectrum analyzer method were made for the first profile for each station. The average ratio of the results determined by the different methods was approximately unity with an RMS scatter of $\sim 50\%$ and none differing by more than a factor of 2. While this is a large discrepancy, it was felt necessary to make this compromise so that all of the data could be included. This comparatively large analysis error is, however, much smaller than the variations in ε by as much as an order of magnitude even at the same station separated by only a few minutes as can be seen in Station 88 (Figure 1).

RESULTS AND DISCUSSIONS

Figure 5 summarizes the results of the experiment as a plot against time of wind speed, net dissipation, ε_I, and the volume dissipation, ε, for the upper and lower mixed layer. Wind speeds ranging from 3 to 15 m/sec were strongest near the beginning of the experiment with three periods greater than 10 m/sec.

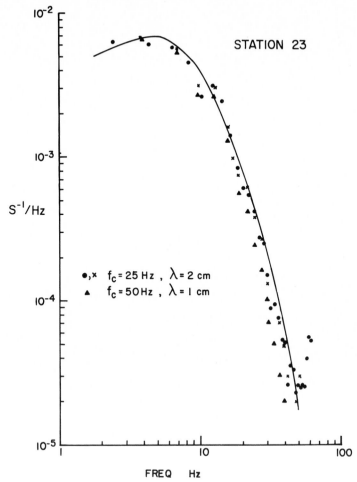

Fig. 3. The velocity shear spectra obtained for Station 23 is shown for the two perpendicular velocity sensors (x,•) corrected for a sensor cut-off, λ_c=2 cm, and the same data corrected for λ_c=1 cm (▲). The data for λ_c=2 cm agree with the isotropic turbulence curve of Nasmyth (1970), shown as a solid line. For this station with a drop speed of 36 cm/s, λ_c= 2 cm occurs at 18 Hz.

During the last half of the experiment the winds were uniformly low. The correlation between higher values of wind speed and higher values of ε_I is evident in this figure. The dissipation per unit volume has been examined only for two levels, the upper and lower half of the mixed layer. This was sufficient to determine any large depth dependence. The fact that the measured dissipation is nonstationary in space and time makes it very difficult to attempt a smaller vertical average. Figure 1 is an example of five profiles measured over a time of 20 minutes and horizontal scales of a few hundred meters. Each profile has a depth dependence $\varepsilon(z)$ which is quite different and for this reason only two large intervals were selected. Using this division we find higher levels of

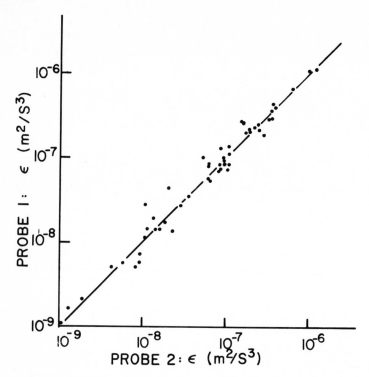

Fig. 4. The dissipations, ε, determined for two perpendicular sensors and averaged over 10 m in depth are plotted. The ratio of $\varepsilon_1/\varepsilon_2=1$ ("isotropy") is represented by the solid line.

dissipation in the upper mixed layer than the lower. This difference would probably be even more marked if we could measure in the top 5 m where dissipation by wave breaking is almost certainly the highest. Although not pronounced, there is a tendency for the dissipation to be proportionately higher in the upper mixed layer than in the lower half during lower wind speed periods and little dependence on depth during periods of higher winds.

The correlation between wind speed and net dissipation, ε_I, is shown in Figure 6. There is for any determination of ε_I a large variability associated with the ensemble average of approximately five profiles. The variability at one station is shown by a line connecting the maximum and minimum values for the station. They are often different by an order of magnitude. A significant improvement in the estimate is achieved by taking the average of 5 profiles. While the scatter is large for different determinations at the same wind speed, there is a strong correlation between high values of ε_I and high wind speeds. There are large possible errors associated with the values of $\overline{U}_{10}^3$ also since even a 10% error in the wind speed estimate yields a 30% error in $\overline{U}_{10}^3$. As an example, data from Figure 6 give $\varepsilon_I = 8 \times 10^{-3}$ W/m^2 at a wind speed of $\overline{U}_{10}^3 = 10^3$. Using Equation (3) with

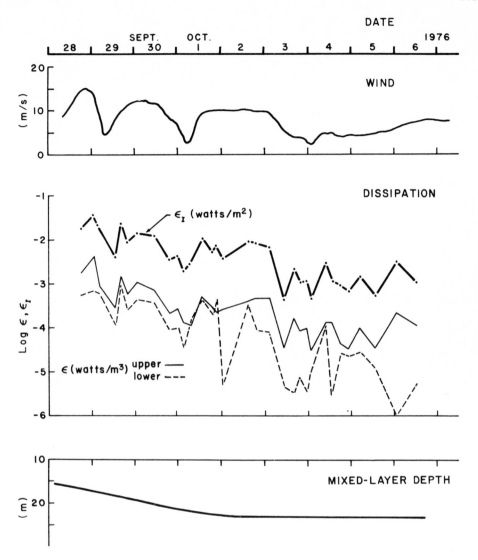

Fig. 5.   The time series of the wind speed during the experiment is shown at the top of the diagram.   Directly below is the time series of the net dissipation, ϵ_I watts/m^2, and the dissipation per unit volume, ϵ watts/m^3, for the upper and lower parts of the mixed layer.   The mixed layer depth is shown at the bottom of the figure.

ρ_a = 1.2 kg/m^3 and C_{10} = 1.5x10^{-3} the energy input is 0.18 W/m^2.   Thus, approximately 0.5% of the energy flux from the wind appears as energy dissipated in the mixed layer below 5 meters.

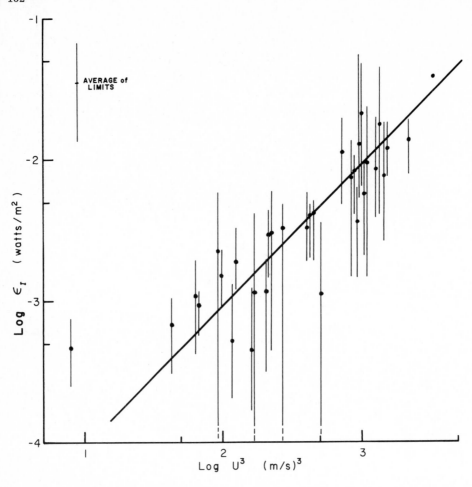

Fig. 6. The net dissipation, ε_I, integrated over the mixed layer is plotted versus $\overline{U}_{10}^3$. The dot is the average of several measurements at the same station (usually five) and the bar joins the largest and smallest value obtained. The average of limits represents the mean deviation of the largest and smallest measurement from the average.

CONCLUSIONS

The results lead us to the following conclusions. The spectra of turbulent velocity fluctuations in the mixed layer during our experiment have a shape which is similar to the isotropic turbulence spectrum obtained by Nasmyth when averaged over 10 m in the vertical. The flow is sufficiently anisotropic and intermittent in space and time that we observe significant differences between the variances from a 10 m average of one sensor and: (1) a similar average for a perpendicular sensor; (2) a profile obtained several minutes later; and (3) a profile obtained at a similar mean wind speed.   There is some variation of ε with depth, that is,

ε tends to be smaller for the lower half of the mixed layer as compared to the

upper half.   Finally, there is a strong correlation between the dissipation rate

in the mixed layer and energy input from the wind.

REFERENCES

Nasmyth, P.W., 1970. Oceanic turbulence, Ph.D. Thesis. Institute of Oceanography,
    University of British Columbia.
Oakey, N.S., 1977. Octuprobe III: An instrument to measure oceanic turbulence
    and microstructure. Bedford Institute Report Series BI-R-77-3, 52 pp.
Osborn, T.R., Measurements of energy dissipation adjacent to an island. J.G.R.
    83, 2939.
Osborn, T.R. and Crawford, W.R., 1979. Air sea interaction instruments and
    methods. Plenum Publishing Co., In Press 1979.
Siddon, T.E., 1965. A turbulence probe using aerodynamic life. Tech Note 88,
    Institute for Aerospace Studies, University of Toronto, 14 pp.

ZERO CROSSINGS OF TEMPERATURE MICROSTRUCTURE

MICHAEL C. GREGG

University of Washington, Seattle, (U.S.A.)

ABSTRACT

The vertical distribution of temperature microstructure has been examined by locating zero crossings in the temperature gradient. Since these crossings are associated with gradient features that are a few centimeters thick, the diffusive time scales imply that the temperature structures are associated with corresponding velocity fluctuations.

By using the zero crossings as an indicator function it is possible to determine the fractions of the profile that are occupied by active microstructure, the thickness of individual events, and their relationship to the finestructure.

INTRODUCTION

Mixing processes in the oceanic thermocline have been found to be very intermittent in space and time. To determine what processes are responsible for small-scale mixing in the thermocline it is necessary, but undoubtedly not sufficient, to determine the spatial scales of the mixing events. This task is complicated by the great variability that has been found at the smallest scales, termed microstructure, so that it is difficult to distinguish variations of intensity within single events from fluctuations due to different events.

Microstructure profiles from two depth ranges, 200 - 400 m and 800 - 1200 m , in the Central North Pacific have been used in a study of the distribution of the microstructure. The spectral content and average intensity of these records was considered previously (Gregg, 1977).

ZERO CROSSINGS

The rate of entropy generation is the fundamental thermodynamic quantity used to describe dissipative processes (de Groot and Mazur, 1969) and has a thermal component given by the variance of the gradients, i.e. $(\nabla T)^2$. Since, in a stratified profile, small-scale turbulence will produce large local increases in the rate of entropy generation, the root-mean-square gradient can be used to indicate the presence of major dissipative events.

Expanded-scale plots of high resolution profiles suggest a direct connection between elevated rms gradients and regions with repeated zero crossings of the microscale gradients (Fig. 1). However, it is also apparent that there are sections containing many lower-amplitude zero crossings that produce only moderate rms gradients because the regions have relatively weak mean gradients (or fine-structure). However, these structures also suggest mixing events. Due to the low noise levels and wide dynamic range of the data, the occurrence of zero crossings is much less sensitive to variations in the background finestructure than are the rms gradients and hence better indicators of mixing events.

The data records were searched with an algorithm that located the zero crossings by defining a crossing where the gradient record crossed the zero axis by more than a specified threshold. The number of crossings in successive 0.5 m intervals were then recorded. An example from a fresh water lake (Fig. 2) shows the gradient excluded by the threshold in shading. The threshold level was chosen to be sufficiently above the noise level to avoid spurious indications. In this example there were six crossings in a 0.7 m-thick section.

From examples such as Fig. 2 it is apparent that the individual gradient features in sections containing zero crossings are no more than a few centi-meters thick. By using the e^{-1} thickness of the negative gradients, it was determined that 95% of the features were less than 5 cm thick. If these gradients had been formed as infinitely thin interfaces and then permitted to diffuse without further velocity stirring, they would obtain an e^{-1} thickness of 5 cm in 13 minutes.

In the fresh water data the larger-amplitude negative gradients can be shown to be density inversions as well. Due to the difficulty in computing salinity at the cm scale, a similar direct demonstration is not possible at the present with the oceanic records. However, the short diffusive lifetime (less than the stability period N^{-1}) and the similarity to the fresh water structures are strong, if indirect, evidence that the cm scale negative temperature gradients are also density instabilities. If in fact the temperature inversions were density-stabilized by compensating salinity gradients, they would be diffusively unstable and would soon decay by double-diffusive convection.

The presence of patches of cm scale density inversions implies that correspond-ing vertical velocity fluctuations must exist in order to produce and maintain the structures. A strong argument was advanced by Stewart (1959) that turbulence in a stratified fluid is completely suppressed when the vertical velocity fluctuations are damped by the stratification. Therefore, the patches of zero crossings should also be those of small-scale turbulence. Stewart's discussion treated turbulence as three-dimensional small-scale random velocity fluctuations, with no restriction that it follow the "universal" form for fully-developed homogeneous turbulence.

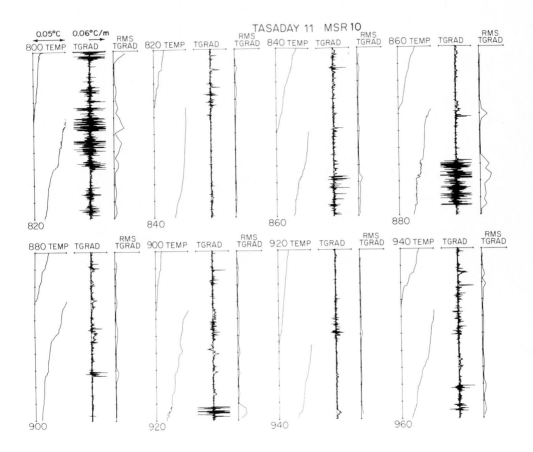

Fig. 1   An expanded -scale plot of the microscale temperature gradients shows
that the high rms gradients occur where there are numerous cm-scale
positive and negative gradients.   Other regions with lower amplitude
zero crossings are found in sections with weak mean gradients.   The
depth span of each panel is 20 m.

138

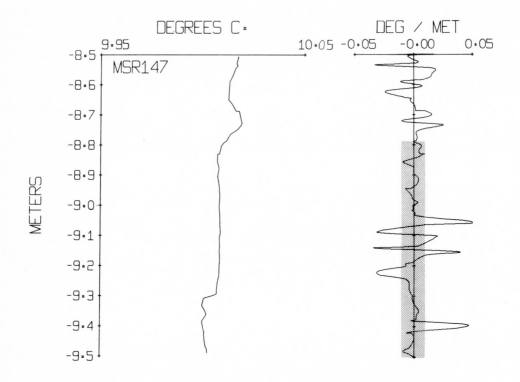

Fig. 2   Example of zero crossing algorithm in data from a fresh water lake.
the threshold gradient level (0.01 °C/m) is shown by the shaded band.
Six zero crossings are indicated for the 0.7 m - thick shaded depth
span.

DISTRIBUTION OF THE ZERO CROSSINGS

Taking the existence of one or more zero crossings as in a 0.5 m-thick section of the profile as an indication of "mixing activity" the average patch distributions in Table 1 were obtained.

TABLE 1

Average statistics for the distributions of the zero crossings.

Length of Contiguous Sections (m)	0.5-1.0	1.5-3.0	3.5-5.0	5.5-7.5	8.0-11	Net Volume Fraction
200-400 m	0.104	0.091	0.033	0	1.6	0.245
800-1200 m	0.063	0.060	0.024	0.026	0	0.172

The net volume fractions varied from 7 to 36% in the individual records and were 17 to 25% in the ensemble averages over the shallow and deep data. These valves are surprisingly high, especially since a conservative threshold level (0.01° C/m) was used. The actual volume fractions may be twice those in Table 1 and are evidence for frequent mixing events.

Regions of contiguous zero crossings varied from 0.5 to 10 m in thickness. Again, these lengths tend to be underestimates due to the threshold level; inspection of the records reveals that in some cases contiguous patches are separated by only one 0.5 m - thick interval in which zero crossings lower than the threshold were present. Such instances generally are found in sections with very low mean gradients. In one example a patch actually appeared to 30 m thick.

The occurrence of the zero crossings with respect to the finestructure was also examined. The conditional probability distributions of the 0.5 m mean gradients, given that a zero crossing occurred, were found to be nearly identical with the unconditional distribtuion of the gradients. This demonstrates that the patches of mixing occur randomly with respect to the background finestructure. By comparison, the occurrence of large or small rms gradients is strongly biased by the finestructure. Two examples of patches extending across several finestructure regions are given in Fg. 3 and 4. In both cases the modulation of the amplitudes of the individual gradients by the finestructure is apparent.

140

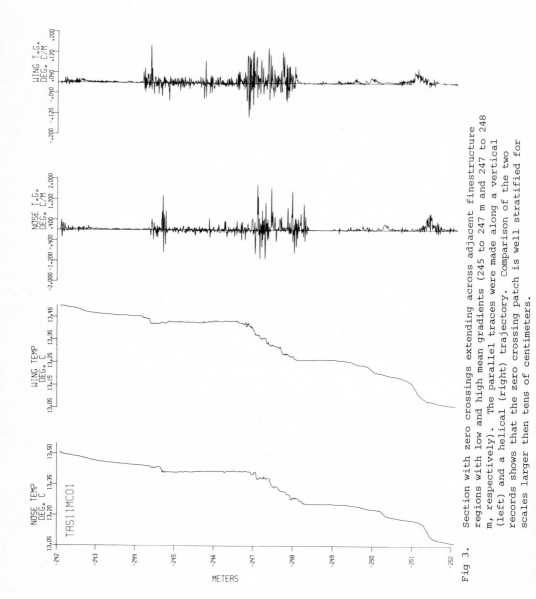

Fig 3. Section with zero crossings extending across adjacent finestructure regions with low and high mean gradients (245 to 247 m and 247 to 248 m, respectively). The parallel traces were made along a vertical (left) and a helical (right) trajectory. Comparison of the two records shows that the zero crossing patch is well stratified for scales larger then tens of centimeters.

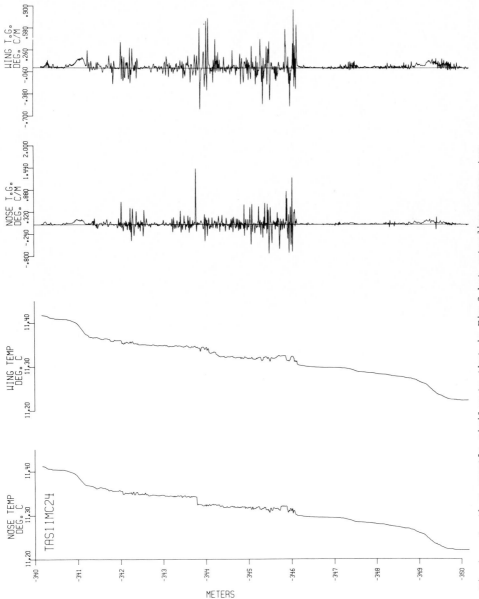

Fig 4. Another example similar to that in Fig 3 but extending across two
sections with average gradients that are separated by a sharp step.

142

DISCUSSION

The random occurence of the zero crossings with respect to the finestructure is contrary to the patterns expected if either salt fingering or Kelvin-Helmholtz instabilities on high-gradient regions were the dominant mixing process. (The upper interval of 200-400 m was diffusively unstable but the deeper section was not).

The lack of overturning signatures with scales corresponding to those of the thicker patches is also notable in the records. For example, in Fig. 3 and 4, the maximum overturning scales based onthe temperature inversions are much less than the thickness of the corresponding patches. This is expected if the profiles were made after the main overturning event (Thorpe, 1973, Koopand Browand, 1979). However, it is almost uniformly the case in the data that the maximum displacements are appreciably less than the patch thicknesses, suggesting that the mixing occurs as many contiguous events with relatively short overturning scales.

ACKNOWLEDGEMENTS

This work was supported by the Office of Naval Research of the U.S. Navy, Department of Oceanography, University of Washington. CONTRIBUTION No. 1102.

REFERENCES

de Groot, S.R. and P. Mazur, 1969. Non-Equilibrium Thermodynamics, North-Holland Publishing Co., 510 pp.

Gregg, M.C., 1977. Variations in the intensity of small-scale mixing in the main thermocline. J. Phys. Ocean. 1:436-454.

Koop, C.G. and F.K. Browand, 1979. Instability and turbulence in a stratified fluid with shear. Submitted to J. Fluid Mech.

Stewart, R.W. 1959. The problem of diffusion in a stratified fluid. Advances in Geophysics 6:303-311.

Thorpe, S.A., 1973. Turbulence in stably stratified fluids: A review of laboratory experiments. Bundy layer meteor. 5:95-119.

DISSIPATION MEASUREMENTS OF OCEANIC TURBULENCE

T.R. OSBORN[1]

[1]Department of Oceanography, University of British Columbia, Vancouver, B.C., Canada. V6T 1W5

ABSTRACT

Work done at the University of British Columbia on direct measurements of the rate of turbulent energy dissipation is described. The relationship between dissipation and mass diffusion is discussed.

INTRODUCTION

This paper describes work at the University of British Columbia on measuring small-scale velocity fluctuations in the ocean. We use these measurements to estimate the local rate of turbulent energy dissipation (denoted by ε) as a function of depth. First, I will briefly describe the instrumentation, showing that we get a direct estimate of ε which is independent of any assumptions about the shape of the spectrum of the velocity fluctuations. Second, we will examine some measurements from the ocean and see that the measured dissipation rates vary from 4×10^{-6} cm^2/sec^3 (the noise level of the instrument) to greater than 10^{-2} cm^2/sec^3. Third, we will look at profiles of dissipation and see that there are thick patches of relatively high dissipation with vertical extents ranging from 10m to 45m. Fourth, a possible explanation for these patches will be offered. It is suggested that they are maintained by the Reynolds stress working against the local mean shear, rather than by the decay of large single events, such as Kelvin-Helmholtz billows. Throughout the paper there will be references to various work that has been done with our instrumentation. The bibliography contains these references, although some are still in press or in preparation. Thus this paper outlines the work that has been done and indicates what work is in progress to give a survey of available results and data under analysis.

144

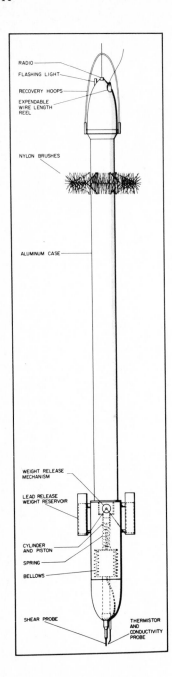

Figure 1. Schematic drawing of the free-fall instrument, Camel. It is approximately 3m in length with the airfoil probes at the lower end.

INSTRUMENTATION

We use a free-fall instrument to carry the probes. It is approximately 3m long and is described in detail by Crawford (1976) and Osborn (1977) and is shown in Figure 1. The instrument falls freely through the ocean measuring temperature, its gradient, pressure and two perpendicular components of the small-scale velocity with resolution on the 1 - 40 cm scales using an airfoil probe. The data are telemetered to the surface from the instrument using standard FM telemetry and a Sippican expendable wire link. No recording is necessary inside the instrument and the data can be monitored on the ship in real time. The probe is an airfoil of revolution (Figure 2). The transverse force on the tip is due primarily to the potential flow and is essentially linear in the cross stream velocity. Considerable literature is available on this probe; Osborn and Crawford (1979) should be consulted as an introduction to that literature. The velocity signal is differentiated inside the instrument before telemetry and used to estimate the mean square shear in the dissipation range. Dissipations are estimated over vertical intervals ranging from 2 to 5m. Spectra of the shears are calculated for each channel of the probe over these intervals. Examination of the spectra indicates the frequency bandwidth over which we need to integrate in order to calculate the total variance. These spectra are then integrated and the variance converted to dissipation rate using the formula

$$\varepsilon = 7.5\nu \left[\frac{\overline{(du/dz)^2} + \overline{(dv/dz)^2}}{2} \right]$$

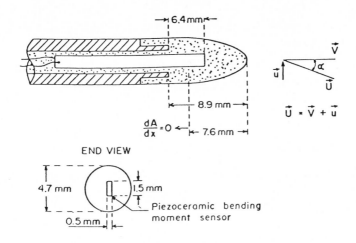

Figure 2. Schematic drawing of the airfoil probe, showing the soft epoxy tip set on the end of a stainless steel tube.

where u, v are the horizontal velocity components, z the vertical coordinate and ν the kinematic viscosity. Thus the dissipation rate is not determined from the spectral shape of the data, although spectra are calculated for individual stretches of data in order to ascertain what frequency range contains signal and what portion contains noise. The formula used is valid for isotropic turbulence, but the requirement here is that the turbulence be isotropic in a dissipation range rather than in the cascade range which is required when fitting a -5/3 spectral shape. It should be noted that the coefficient 7.5 is incorrect if the turbulence is not isotropic, but is probably in error by less than 50%. This accuracy is quite satisfactory for a study of the changes of turbulence intensity with depth, as well as for many other applications. The question of isotropy in oceanic turbulence is in itself of great interest; it can be examined using these probes which measure cross-stream velocity fluctuations in conjunction with heated anemometry which normally measures the downstream component. Such work is being done by Dr. Ann Gargett at the Ocean Mixing Group, Institute of Ocean Sciences, Sidney, B.C., Canada, from a Pisces submersible using two of our airfoil probes [from U.B.C.] and the heated sensors developed by Grant, Stewart and Moilliet (1961); these data are presently under analysis.

DATA

Measurements with our free-fall instrument have been taken on four major oceanographic cruises. First, measurements were taken in the Equatorial Atlantic to

examine the dissipation associated with the Atlantic Equatorial Undercurrent.
Second, data were taken adjacent to the island of Santa Maria in the Azores. Third,
data were collected during the Fine and Microstructure Experiment in relatively open
ocean, in the Gulf Stream and near Bermuda. Four, data were collected in the
Equatorial Pacific to look at the dissipation rate and energetics associated with
the Pacific Equatorial Undercurrent. Figure 3 shows a sample of the results from
the Atlantic Equatorial Undercurrent. This figure is taken from Crawford and Osborn
(1979a) which discusses the measurements. Plotted are salinity, temperature, σ_T
and relative velocity from the surface to 300m. Dissipation is plotted on the right
hand side of the figure. Each bar represents an ε-value over approximately 2m
intervals. The scale is logarithmic, ranging from 10^{-6} cm^2/sec^3 to 10^{-2} cm^2/sec^3.
Large values of dissipation are seen in the high shear region on the upper side of
the core of the Equatorial Undercurrent. Relatively low dissipations are seen in
the core region and the dissipations again increase when one gets into the shear
region below the core. This picture is consistent for all the drops through the
Atlantic Undercurrent with average values on the order of 3 x 10^{-3} cm^2/sec^3 in the
upper shear zone, 4 x 10^{-5} cm^2/sec^3 through the core and 2 x 10^{-4} cm^2/sec^3 in the
shear region below the core of the Undercurrent. These data can be used to examine
the energetics of the Equatorial Undercurrent. Crawford and Osborn (1979b) discuss
the energy balance for the turbulent kinetic energy as well as the mean kinetic
energy. A simplified version of the argument is as follows: in the turbulent kinetic
energy equation the dissipation is balanced by the Reynolds stress acting against
the mean shear. One can then turn to the mean kinetic energy equation and equate

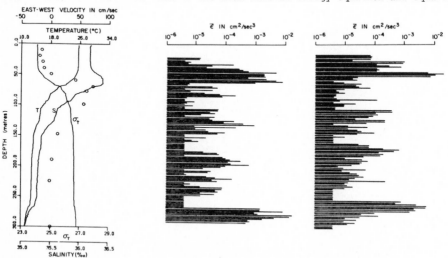

Figure 3. Data from the Atlantic Equatorial Undercurrent from Crawford and Osborn
(1979a). Shown on the left-hand side are salinity, temperature, σ_T and relative
velocity from the surface to 300m. That data was collected by John Bruce and Eli
Katz of Woods Hole Oceanographic Institution. On the right-hand side, two profiles
of dissipation with the averaging interval of approximately 2m. The two drops were
taken about one hour apart.

the turbulent production term, which also appears in that equation, to the dissipation. The mean kinetic energy equation is integrated from the surface to the shallowest zero velocity point. This layer corresponds to the South Equatorial Current. The equation is also integrated from the zero velocity level to the level of the maximum velocity. This zone corresponds to the upper portion of the Equatorial Undercurrent. A balance in the South Equatorial Current is found between the work done by the wind stress against the integrated dissipation, and the energy put into the system by the water flowing uphill, to the west, against the pressure gradient. For the upper half of the Atlantic Equatorial Undercurrent the dissipation balances the energy extracted by the water flowing eastward down the pressure gradient. This examination of the energetics indicates that the values of dissipation we measured in the Atlantic Equatorial Currents are in line with the basic energetics of the currents, whereas the much higher dissipation range reported by Belyaev et al. (1975) (2.7×10^{-2} cm^2/sec^3 to 3.9×10^{-1} cm^2/sec^3) would appear to be much larger than can be maintained by the sources of energy.

Osborn (1978) reports on data collected 16 km from the island of Santa Maria in the Azores during a one-week period in March, 1975. Figure 4 shows temperature and dissipation averaged over 5m intervals plotted as a function of depth. (Note: ε in watts/m^3 x 10 = ε in cm^2/sec^3.) Several features are apparent in this data set. First, there is a well-mixed portion of the upper layer in which the dissipations are higher than in the rest of the upper layer. Second, there is a very thick patch of relatively high dissipations in the thermocline.

The original data for drop 6 temperature, temperature gradient and the two shears

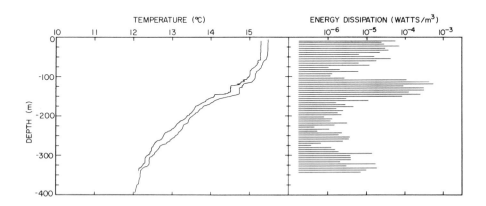

Figure 4. From Osborn (1978). Temperature and dissipation versus depth with the dissipation averaged over approximately 5m intervals. Left temperature trace is from U.B.C. instrument; right temperature trace from CTD.

148

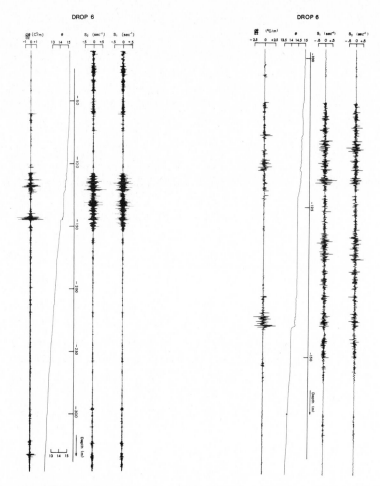

Figure 5. The original data for the profile shown in Figure 4. Temperature, temperature gradient and two shears versus depth.
Figure 6. The expanded portion of Figure 5 showing the high dissipation region in
the seasonal thermocline.

versus depth are shown in Figure 5. Visible are the well-mixed part of the upper
layer and the very active continuous patch of turbulence in the thermocline. This
intensely active region is expanded in Figure 6 showing the feature marked at the
top and bottom by temperature gradient fluctuations which do not exist in the middle
of the feature due to its isothermal nature.

Data taken the day following drop 6 are located 80 km further east from the island.
Here a very thin diurnal mixed layer was found on top of a much thicker well-mixed
layer (Figure 7). The largest values of epsilon at depth are associated with relatively isothermal temperature features and there is no apparent high dissipation
patch in the thermocline. Figure 8 is an expanded view of the data for Figure 7
for the deepest high dissipation region. Note how the turbulence responsible for

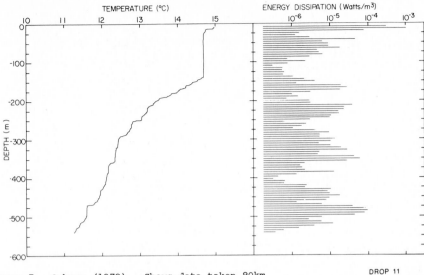

Figure 7. Osborn (1978). Shows data taken 80km
from the island of Santa Maria.

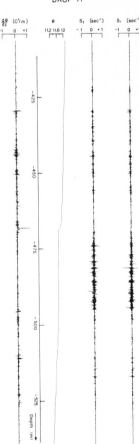

DROP 11

the dissipation is apparently continuous throughout
the homogeneous layer. The next day profiles were
taken close to the island. Again very high dissipations in the thermocline were found, as well as
relatively low dissipations at depth which are
much lower than those at similar depths seen the
previous day well away from the island (Figure 9).
Also notice the decrease in dissipation with depth
from the surface until the instrument reaches the
high dissipation feature in the thermocline.

Figure 10(a) shows 50m averages of the dissipation data for three profiles 80km from the
island, as Figure 10(b) for the three profiles
the following day adjacent to the island. Except
for the diurnal mixed layer (in Figure 7), the
upper layer away from the island is not exception-
ally strong in dissipation. Near the island one
sees the decrease in dissipation with depth, the
variation in one hour intervals between drops 13,
14 and 15, the relatively high dissipation in the
themocline and the low epsilon values below
the seasonal thermocline. It is somewhat

Figure 8. An expanded view of the data in Figure 7,
showing the high dissipation region at depth.

150

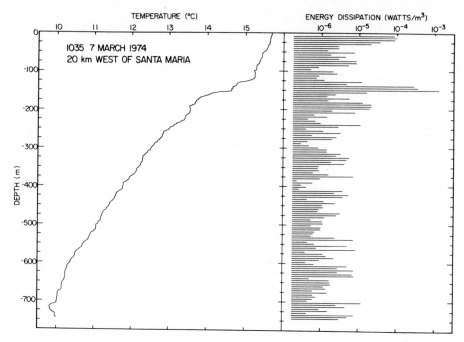

Figure 9. Osborn (1978). Taken adjacent to the island the day following Figure 7 and two days after Figure 6.

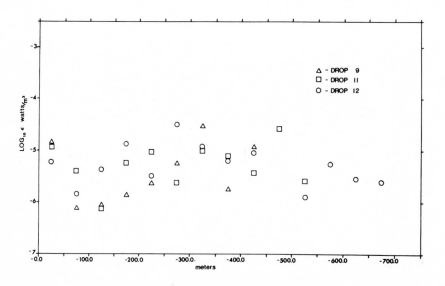

Figure 10(a). 50m averages of dissipation for the three profiles 80km from the island. There are approximately 3 hours between drops 9 and 11 and 2 hours between drops 11 and 12.

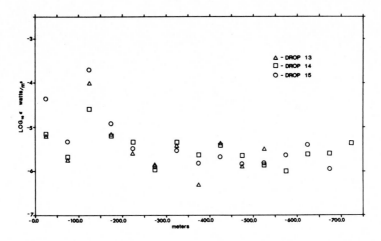

Figure 10(b). 50m averages for the dissipation for three profiles adjacent to the island of Santa Maria. These three measurements are separated approximately 1 hour in time.

surprising that the dissipations below the thermocline are lower near the island than they are away from the island, in view of the location of this site amidst a triangle of sea mounts rising to within 500m of the surface. Epsilon versus percent occurrence for the data is plotted in Figure 11 in three different formats: the first is for all the data, the second is for all the data below 150m and the third is for the last three drops below 150m. This latter is expected to be a more uniform sample, and should perhaps approach the straight line associated with a log normal distribution. Also plotted is the percentage of total epsilon versus dissipation rate. Perhaps the most interesting aspect of this is that it allows one to calculate percentage of dissipa-

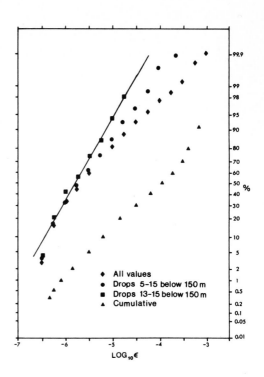

Figure 11. Statistics of the data; ε in units of watts/m^3.

tions above a certain rate. For example, about 80% of the epsilons are less than 10^{-4} cm^2/sec^3 (10^{-5} watts/m^3) but these account for less than 20% of the total dissipation.

The third major cruise was during the Fine and Microstructure Experiment. Results are available in three different publications. First, there is a paper on surface mixing layers in the Sargasso Sea (Gargett, Sanford and Osborn, 1980) where the energetics for the upper layer are calculated and compared to dissipation rates in several different profiles. Second, there is a manuscript report which contains dissipation profiles from all the data collected during the cruise (Gargett and Osborn, 1979). Calibrations and other pertinent data are included in this report, as well as a discussion of each individual profile. Third, a paper is in preparation (Gargett and Osborn, 1980) describing the data from the cruise and presenting some averaged statistics from the data.

For comparison with other data, Figure 12 shows one profile that was taken within the 2000m contour off Bermuda. Plotted are temperature, dissipation and root mean square temperature gradient over the interval from 0 to 725 db. The temperature gradient variance is not calibrated in this figure and we will not discuss it here. Notice the relatively thick regions of dissipation throughout the record.

One final set of data is available, collected during January and February 1979 in the eastern Equatorial Pacific. Preliminary results, including estimates of dissipation rates, will be reported by Crawford and Osborn in early 1980 and it is hoped that some of these results will be available for the IUGG meeting in Australia in December 1979. Preliminary examination of the data does indicate that the dissipation rates do not exceed those found in the Atlantic Equatorial Undercurrent. This result is distinctly different from that of Williams and Gibson (1974) who report a rather high upper bound to the dissipation rate by fitting the universal form to a very noisy temperature spectra.

DISCUSSION

I would now like to consider the sources of the turbulent energy found in the ocean and see if we can derive a relationship between the dissipation and the mass flux. One popular explanation for the source of small-scale turbulent velocity fluctuation in the ocean is the gravitational collapse of Kelvin-Helmholt billows. Starting with the pictures of Woods (1968) and the laboratory work of Thorpe (1973), much evidence has accumulated that these are a suitable source of energy for the turbulence. Oceanic measurements by Gregg (1977), as well as the results of Thorpe and Woods, all indicate the oceanic scale to be about a metre. Thus the thin ($\sim$1m) turbulence patches in the velocity shear data are likely to be associated with the gravitational collapse of Kelvin-Helmholtz instability.

SHEAR INSTABILITY GENERATION OF THE TURBULENCE IN THE OCEAN (FIELD EVIDENCES)

V.S. BELYAEV

P.P. Shirshov Institute of Oceanology, Academy of Sciences, Moscow (U.S.S.R.).

ABSTRACT

Field data confirming shear instability of ocean current velocity fields are presented. Using data of repeated soundings made with time intervals of 110 s in a layer of 20 m thickness which incorporates an upper part of the seasonal thermocline, estimates of local gradient Richardson numbers Ri are obtained. It is shown that, if internal gravity wave motions are superimposed on stable, on the average, stratified flow with a shear, then the stability criterion $Ri > 1/4$ is violated in 30 % of cases in agreement with theoretical predictions (Bretherton, 1969). The data obtained at a seasonal thermocline in the Indian Ocean using a free-falling microstructure probe show a layer of 20 m thickness with enhanced level of small-scale velocity fluctuations. The estimates of the Richardson number for layers of 4 m thickness testify to the possibility of the generation of turbulence therein by shear instability of currents.

INTRODUCTION

The efficiency of some possible mechanisms for the generation of ocean turbulence has been discussed by Monin (1977), but untill now the predominant mechanism had remained open to question. The origin of turbulent layers in the main density stratified body of the ocean was, in most cases, related either to internal waves instability (Phillips, 1966 ; Woods, 1968 ; Woods and Wiley, 1972 ; Belyaev et al, 1975) or to double-diffusive phenomena (Gargett, 1976). In doing so many conclusions were drawn on the basis of indirect considerations. In the present paper new ocean turbulence data and accompanying local background conditions are presented, and the possibility of formation of turbulent patches by shear instability of the velocity field in the ocean is estimated.

FIELD MEASUREMENTS IN THE PACIFIC OCEAN

Soundings from a drifting vessel were carried out with a microstructure probe developed by the Experimental Design Office of Oceanological Technique (branch of the P.P. Shirshov Institute of Oceanology, Academy of Sciences, USSR). In a series of soundings at Station
N° 7546 in the Pacific (60-th cruise of R/V "Vityaz" ; 13 September
1976 ; location : 28°40'N, 155°10'E) the probe was displaced up and
down within a layer of 20 m thickness. Measurements of the temperature profiles, the current velocity and the small-scale current velocity fluctuations u' were performed with probe submerging speed of
0.3 m/s, and with horizontally directed sensors. The time interval
between successive soundings was 110 s. During the course of the measurements, the mean horizontal component of the relative velocity of
the probe changed within the range 0.3 - 0.4 m/s, so that the angle of
inclination of the probe trajectory on the horizontal plane was 37-45°,
on the average. This corresponds to the procedure used in other investigations of the ocean microstructure (e.g. Gargett, 1976).

Fig. 1. shows the mean temperature profile $\overline{T}(z)$ for the whole
series of measurements, and the r.m.s. scatter of temperature values
at fixed depth z in the thermocline is 0.5 - 0.7°C. So according to
a bathymetrical series of measurements, the main contribution to the
variation of water density with depth, is due to temperature in the
present case. The vertical profile of the squared Brunt-Väisälä fre-
quency $\overline{N}^2 = (g/\overline{\rho}) \, \partial\overline{\rho}/\partial z$ (Fig. 1) was calculated from the distribution
$\overline{T}(z)$, for the mean salinity value in the layer. Here and below, the
estimates of derivatives were obtained using finite differences with
depth step of 0.5 m. The maximum value, $\overline{N}^2_{max} = 0.093 \text{ s}^{-2}$, corresponds
to the period of the free internal waves 3.4 min. The profile of the
mean velocity $\overline{V}(z)$ over the whole period of observations is relatively
smooth, and absolute values of the velocity vertical gradient $\partial\overline{V}/\partial z$
did not exceed 0.08 s^{-1} (Fig. 1). The distribution of the logarithm
of the gradient Richardson number $\overline{Ri} = \overline{N}^2/(\partial\overline{V}/\partial z)^2$ in the thermocline
is shown in Fig. 1 by separate points. The vertical line on this fi-
gure corresponds to the critical value of the Richardson number
$Ri_{cr} = 1/4$ for stationary parallel flow. All the values $\overline{Ri}$ obtained
are more than 1/4, so the mean flow is dynamically stable (Miles and
Howard, 1964).

A space-time section of the temperature field, according to the
data of 74 soundings, is shown in Fig. 2. This figure shows the location of the lower boundary of the upper mixed layer (curve 1) and the

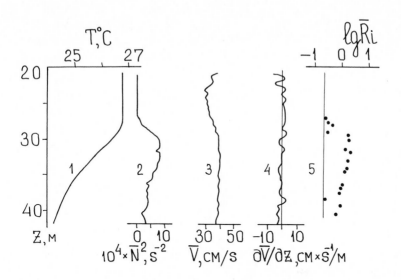

Fig. 1. Profiles of mean temperature (1), squared Brunt-Väisälä frequency (2), velocity (3), velocity vertical gradient (4) and gradient Richardson number (5).

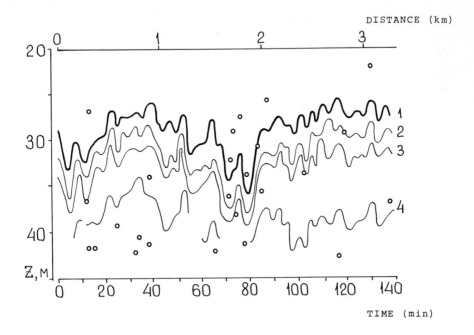

Fig. 2. Space-time section of temperature field : 1 - lower boundary of the upper mixed layer ; 2,3 and 4 - isotherms 26.5°C, 25.5°C and 24.5°C, correspondingly.

individual isotherms (curves 2 to 4). The root mean square scatter
of the isotherm deviations from horizontal position is 2.2 - 2.5 m.
It is natural to relate the variations of the vertical location of
the seasonal thermocline to internal wave propagation in a stratified
ocean. According to Fig. 2., internal waves occur here with at least
two essentially different scales, with periods of $\sim$90 and $\sim$8 min.

Standard deviations of $T(z-h)$ (here h is the thickness of the upper
mixed layer) relatively to the mean value $\tilde{T}(z-h)$ with respect to ten
successive profiles do not exceed 0.4°C, and is, as a rule, of the
order of 0.1°C. This means that, for an observer located on the
lower boundary of the upper mixed layer, the deformation of the vertical temperature profile is relatively small. For all profiles
$\tilde{T}(z-h)$, the thickness of the layer of maximum temperature gradients
ranges from 2 to 4 m. Corresponding profiles of $\tilde{N}^2(z-h)$ were computed
for a mean salinity value in the measured layer, and the minimum free
internal wave period, here 2 min.

The values of the velocity vertical gradients $\partial\tilde{V}/\partial z$ were computed
using $\tilde{V}(z-h)$, averaged with respect to ten successive profiles. Inhomogeneities with vertical scales 2-3 m can be clearly seen on some
particular profiles of $\partial\tilde{V}/\partial z$, especially in the lower part of the
upper mixed layer. The absolute values of the velocity vertical gradients reach 0.16 - 0.18 s^{-1} in some cases.

The estimates of the gradient Richardson number $\tilde{R}i$ were obtained
using the profiles of $\tilde{N}^2(z-h)$ and $\partial V(z-h)/\partial z$. The histogram of $lg\tilde{R}i$
is shown in Fig. 3 where f is the frequency in the bandwidth 0.2.
Apparently, for the internal wave shear motions, the stability crite-
rion for steady parallel flow can be valid if max $|\partial\tilde{V}/\partial z| \gg \omega$ where
ω is the circular frequency of the internal wave (Phillips, 1966).
The vertical profiles of $\tilde{N}^2(z-h)$ and $\partial\tilde{V}(z-h)/\partial z$ are the results of
averaging of ten profiles which correspond to random phases of high
frequency internal waves and to, approximately, the same phase of low
frequency internal wave with $\omega = 0.0012$ s^{-1}. Minimum velocity gradients used to estimate $\tilde{R}i$ are at least one order of magnitude larger
than the above value of ω. The most probable value of $\tilde{R}i$ is 1. This
means that, in the case considered, the internal wave motion in stratified shear flow is on the average hydrodynamically stable. But a
substantial fraction (30 %) of the values of $\tilde{R}i$ is smaller than 1/4,
and one can expect a manifestation of internal wave shear instability.

The estimation of the probability of satisfying the condition
$Ri < Ri_{cr}$ in some localized region of the internal wave field has been
obtained by Bretherton (1969). Using a linear superposition of high

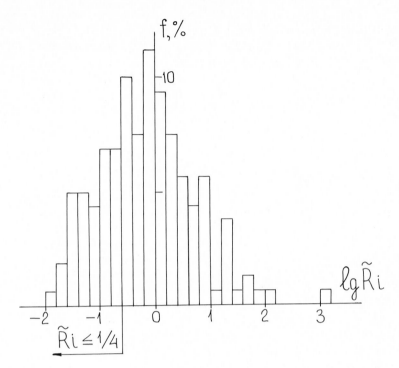

Fig. 3. Histogram of the logarithm of the gradient Richardson number.

number of independent spectral components and assuming a relation between velocity and density gradients which is valid for a flow with
a small velocity gradient (Phillips, 1966), it can be shown that the
probability that Ri < Ri_{cr} depends only on the parameter
$\sigma^2 = \overline{(\partial u/\partial z)^2}/\overline{N}^2$. The probability that Ri < Ri_{cr} in some localized
space-time domain is

$$P = \frac{1}{2\pi\sigma^2} \int_{-\infty}^{\infty} \exp\left(-\frac{x^2}{2\sigma^2}\right) \left[\int_{1-Ri_{cr}x^2}^{\infty} \exp\left(-\frac{y^2}{2\sigma^2}\right) dy \right] dx \quad .$$

Transforming to polar coordinates and performing one integration we
then obtain for $Ri_{cr} = 1/4$

$$P = \frac{1}{\pi} \int_{-\pi/2}^{\pi/2} \exp\left(-\frac{1}{2\sigma^2\cos^4(\phi/2 - \pi/4)}\right) d\phi \quad .$$

For the particular conditions of the field experiment in the seasonal
thermocline, $\sigma^2 = 0.89$ and P = 21 % in agreement with the above experimental estimate P = 30 % .

Because of the variability of the flow velocity it came out that for a fixed frequency band of gauges the window of space scales was variable. Therefore the values of the u' variance obtained by averaging over layers of 0.3 m thickness with a narrow velocity band (from 0.4 to 0.5 m/s) were selected for the analysis. Under the circumstances, the values of sample variances $\overline{(u')^2}$ are randomly distributed at space-time section of the temperature field. For the upper isothermal layer as well as for the seasonal thermocline the $\overline{(u')^2}$ distribution histograms are asymmetric. The location of the highest $\overline{(u')^2}$- values, which exceed the sum of the mean value and r.m.s. value of corresponding empirical distributions, is shown in Fig. 2 by circles. There is a certain tendency to their localization in some layer in the thermocline and in some region above, because of low-frequency internal wave in the upper isothermal layer. Insufficient space-time discreteness of $\overline{(u')^2}$- values does not permit a clear distinction of particular regions of enhanced small-scale current velocity fluctuation levels.

The superposition of an internal wave motion and a mean stable shear flow can, according to the experimental data, lead to the violation of the stability criterion in some localized space regions. In this case the shear instability is manifested as an enhanced level of small-scale velocity fluctuations in the corresponding localized space regions.

FIELD MEASUREMENTS IN THE INDIAN OCEAN

Interesting results have been obtained during the course of 22-th cruise of R/V "Dmitry Mendeleev" when sounding with a free-falling microstructure probe at a polygon in the Indian Ocean. The sounding speed was 3.4 m/s. As an example, Fig. 4 shows simultaneous realizations of the temperature signal T, the small-scale current velocity fluctuations u', the depth D and the r.m.s. current velocity fluctuation rms u' (average over 0.3 s period by 55D35 "DISA" module) in the upper 250 m layer (Station N° 1833 ; 11 March 1979 ; location : 8°32'S, 104°51'E). In the left part of Fig. 4 every-second time marks are indicated. The value of rms u' in the near surface layer is ∿1 cm/s. The sharp increase of the small-scale current velocity fluctuation level occurs in the depth range 60-80 m approximately (quasi-isothermal layer in seasonal thermocline) ; the rms u'-values reach 4 cm/s. The vertical distribution of small-scale current velocity fluctuations at Station N° 1833 is typical of the whole polygon and testifies to the presence of an intensive source of small-scale

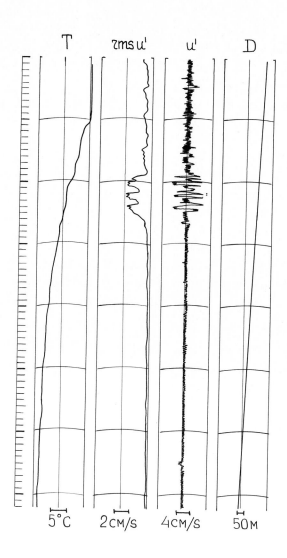

Fig. 4. Realizations of signals of the temperature T, root mean square current velocity fluctuations rms u', current velocity fluctuations u' and depth D.

turbulence in the upper part of seasonal thermocline.

In accordance with the data measured at the polygon (Station N° 1836 ; 11 March 1979 ; location : 8°46'S, 104°47'E) in the depth range from 20 to 80 m the mean current direction continuously varies with depth, approximately 2-3°/m, and the value of $\partial U/\partial z = \overline{(\partial u/\partial z)^2 + (\partial v/\partial z)^2}$, where u and v are the orthogonal horizontal current velocity components, varies in the range from 0.01 to 0.06 s^{-1}. Estimates of $\overline{Ri}$ are shown in Fig. 5. (Density and velocity profiles were averaged over layers of 4 m thickness). In the layer between 50 and 70 m, $\overline{Ri} \sim 1$ and at the depth of 30 m, the Richardson number is smaller than the critical value $Ri_{cr} = 1/4$. Thus, the most probable mechanism for the turbulence generation in the upper part of the seasonal thermocline in the ocean region investigated is shear instability of mesoscale current velocity field.

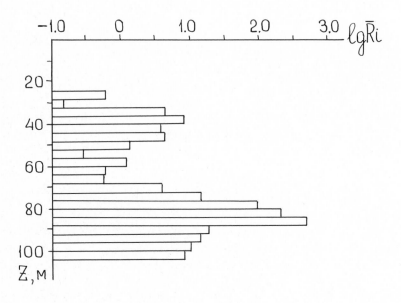

Fig. 5. Gradient Richardson number distribution with respect to depth.

CONCLUSIONS

The field data presented above undoubtedly testify to the possibility of small-scale turbulence generation in stably ocean stratified layers due to velocity field shear instability. Further investigations will permit to estimate the efficiency of this mechanism for turbulence generation under different hydrometeorological conditions.

ACKNOWLEDGEMENTS

I should like to thank Prof. R.V. Ozmidov for helpful discussions.
Thanks are also extended to V.I. Fedorov and his group from the
Experimental Design Office of the Oceanological Technique for providing the field measurements, and to Dr. A.N. Gezentsvey for taking
part in the field data processing.

REFERENCES

Belyaev, V.S., Lozovatsky, I.D. and Ozmidov, R.V., 1975. O svyazi pa-
    rametrov melkomasshtabnoi turbulentnosti s localnimi usloviyami
    stratificatsii v oceane. Izv. Akad. Nauk SSSR Fiz. Atmos. Oceana,
    11: 718-725 (in Russian, with English abstract).
Bretherton, F.P., 1969. Waves and turbulence in stably stratified
    fluids. Radio Sci., 4: 1279-1287.
Gargett, A.E., 1976. An investigation of the occurence of oceanic
    turbulence with respect to fine structure. J. Phys. Oceanogr.,
    6: 139-156.
Miles, J.W. and Howard, L.N., 1964. Note on a heterogeneous shear
    flow. J. Fluid Mech., 20: 311-313.
Monin, A.S., 1970. Osnovnii osobennosti morskoi turbulentnosti.
    Oceanologiya, 10: 240-248 (in Russian, with English abstract).
Phillips, O.M., 1966. The dynamics of the upper ocean. Cambridge
    University Press, London.
Woods, J.D., 1968. Wave-induced shear in summer thermocline. J.
    Fluid Mech., 32: 791-800.
Woods, J.D. and Wiley, R.L., 1972. Billow turbulence and ocean micro-
    structure. Deep-Sea Res., 19: 87-121.

SMALL-SCALE OCEANIC TURBULENCE

R.V. OZMIDOV

P.P. Shirshov Institute of Oceanology, Academy of Sciences,
Moscow (U.S.S.R.).

In recent years extensive studies of small-scale oceanic turbulence and hydrophysical fields fine structure have been performed at the PP. Shirshov Institute of Oceanology (Academy of Sciences, U.S.S.R.) (Ozmidov, 1973 ; 1974a). For these investigations, first of all, it was necessary to create new, rather sensitive, gauges, to develop the procedure of their use in the ocean, and also to solve the problems of bulk statistical processing of the field information obtained using vessel and shore installed computers.

The complex of the model measurement system consists of :

- towing devices supplied with a set of temperature, current velocity and electrical conductivity sensors. The transmission band of the fluctuation sensor is up to several hundreds Hz, its space resolution is portions of centimetres, its maximum depth of submergence is up to 250 m for a vessel speed of 6 knots (Fig. 1.);

- Probes equipped with the same sensor for vertical sounding from drifting vessel, submerged up to 2 km depth using a cable ;

- general purpose probes for vertical sounding as well as for towing with small speed, using a cable ;

- free-falling probe (sliding along a rope or cable) for measurements of fine structure with autonomous registration of signals on a magnetograph, situated in the probe's body, or with signal transmission through a transformer gripping rope ;

- completely autonomous devices moving in water along a prescribed trajectory till definite depth and coming to the surface after throwing down a ballast ;

- radiometeric systems situated in a buoy, equipped with sensors for temperature measurements at 10 horizons at depth up to 1 km. The transmitter "audibility" distance is not less than 10 miles, the autonomy of the system is about 5 days, the interrogation time for all horizons ranges from 1 to 30 seconds.

For registration of the instruments information, analog and digital tape recorders and punchers were used. In doing so, necessary signal transformations (amplification, filtration, discretization and so on) were performed. The important property of the system is the visualization of the registered phenomena using pen recorders and oscilloscopes.

The processing system was based on vessel installed computers, because the character of the processes investigated required statistical calculations (at least involving spectral analysis).

The investigations of the fine structure of hydrophysical fields of small-scale oceanic turbulence and hydrological processes caused by this turbulence were performed using the above mentioned measurement system during the course of special expeditions of the research vessels of the P.P. Shirshov Institute of Oceanology (Academy of Sciences, U.S.S.R.) in the Atlantic, Pacific and Indian Oceans (Ozmidov, 1973 ; 1974 b). The measurements were performed in polygons located in typical ocean regions with respect to the mean hydrometeorological conditions. The polygons were located in the North Atlantic, in equatorial zones of the Atlantic, Pacific and Indian oceans, in moderate latitude regions of these oceans as well as in Antarctic regions. The measurements were carried out from the ocean surface till depths of 1500-2000 m in strong current zones (Gulf Stream, equatorial currents), in dynamically calm regions of the oceans, in regions of various water density stratification, in winter and summer conditions, in calm and heavy weather conditions.

Together with the measurements of the fine structure of hydrophysical fields and turbulence, the mean vertical temperature, salinity and density profiles as well as the wind speed, wave properties and so on were determined in the polygons, using routine gauges. Buoy stations supplied with current meters and photothermographs were also installed in the polygons.

Data processing was carried out using digital computers as well as analog devices supplied with special programs for computation of correlation, structural and spectral functions, cross statistical characteristics of signal pairs, rates of kinetic and heat energy dissipation, current variances, probability distribution densities and so on. Spectral characteristics of measured turbulence realizations were presented, as a rule, in the three forms : spectral density, energy distribution and dissipative spectra.

Arrays of velocity, temperature and electrical conductivity spectra, gathered in the unified graphs, have clearly shown the possible

variations of the levels and forms of turbulence spectra under various
conditions in the ocean (Fig. 2.). The temperature, salinity, water
density and current velocity distribution patterns for every polygon
were compared with corresponding set of turbulence spectra (Belyaev
et al, 1974a). For the whole array of spectral characteristics, the
distribution laws were obtained which show parameters (variances,
third and fourth moments) of possible scattering of spectral function
values for fixed wavenumber values (Belyaev et al, 1974b). Root mean
square values of hydrophysical fields fluctuations, internal turbulence scale, buoyancy scales were also obtained for various hydrometeorological conditions.

Fig. 1. Towing device for investigation of oceanic turbulence.

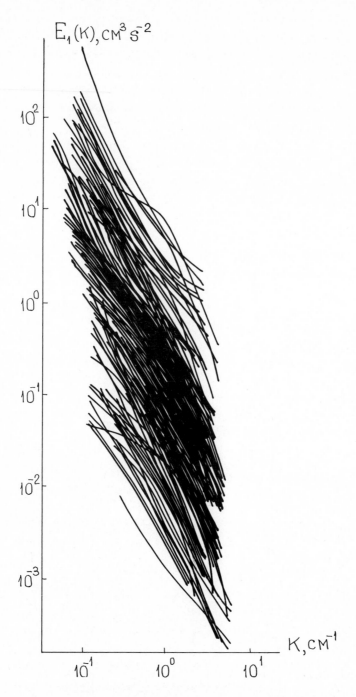

Fig. 2. Family of current velocity fluctuation spectral density curves obtained at a number of polygons. Scattering of spectral density values reaches 4-5 orders for fixed wave-number values, but the mean value is the most probable.

Using available records of velocity and temperature fluctuations
the computations of the rate of turbulent energy dissipation and the
rate of temperature nonuniformity smoothing in the ocean were performed. The fluctuations of the rate of turbulent energy dissipation in
the ocean were also investigated. Drawing the spectra of these fluctuations, one can obtain the universal constant in Kolmogorov's third
hypothesis. This constant appears to be 0.56 ± 0.11 (Belyaev et al,
1973). For the fluctuation velocity field in the ocean the computation of higher order structural functions was carried out. The de-
pendencies obtained well correspond, to a certain extent, to available
atmospheric turbulence data and to the theoretical model of intermittent turbulence (Vasilenko et al, 1975).

Using the data of measurements by complex towing lines supplied
with sensors, measured turbulent fluctuations and mean temperature
and electrical conductivity, the simultaneous computations of turbulence statistical characteristics and local background conditions were
carried out. Using the data of background sensors the local gradients
of the fields in zone of fluctuation measurements were determined.
This gives a possibility to compare variances of velocity and electrical conductivity (temperature) with values of parameters defining generation and decay of turbulence (Belyaev et al, 1975a,b).

The analysis of the processed small-scale oceanic turbulence data
and the corresponding background conditions leads to new concepts.
In particular, the analysis of small-scale turbulence spectral characteristics has led to the revision of earlier existing notions of oceanic
turbulence as a well developed flow at very high Reynolds number. It
turned out that, in most cases, the turbulence in the ocean has no
clear-cut ranges of universal similarity, - a consequence of the small
values of the Reynolds number - and, hence, the turbulence in the ocean
is not sufficiently developed. For this reason a model of oceanic turbulence has been suggested in which the turbulence is defined by local
background conditions with Reynolds numbers based on vertical dimensions of uniform layers in the fine structure density field and on
current velocity differences in such layers (Ozmidov and Belyaev,
1972). According to this model the "conjunction" of viscous and buoyancy subranges in turbulence spectra can often occur without clear-
cut "5/3 power law" ranges. Also, it turned out that, in some cases,
there can exist inertial, viscous and buoyancy subranges in the same
wavenumber range from 0.1 to 10 cm^{-1} (Belyaev et al, 1976). The bounds
separating these ranges as well the parameters of turbulence itself
are essentially dependent on local velocity and density gradients.

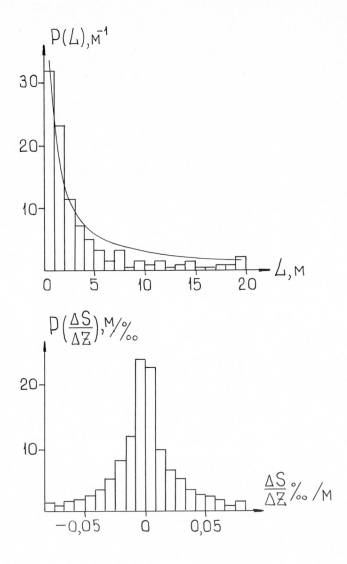

Fig. 3. Layer thickness probability distributions for fine structure of salinity field (a) and probability distributions of gradient salinity values (b) (accordingly to the data obtained in the Antarctic region).

Intensity of turbulent fluctuations of hydrophysical fields in the ocean in the whole layer investigated (to a depth of 2000 m) was found highly space and time variable. Often a vertical displacement of the gauge led to sharp changes of intensity and spectral composition of the fluctuating signals. The same picture often took place during towing a turbulimeter behind a vessel at some fixed depth. The existence of relatively thin layers with increased level of turbulence at various depths in the ocean (including very large depths), which have limited dimensions in the horizontal plane (so called turbulent "patches"), testifies of the existence of "internal" energy sources for small-scale fluctuations. One of such mechanisms for the generation of intermittent turbulence in the ocean body is the shear instability caused by the existence of internal waves. In most cases, the internal wave fields themselves are nonhomogeneous and nonisotropic, but they have a complicated structure. One often can observe wave fields of train form. Internal wave spectra are usually continuous and without clear-cut peaks, that testifies of the multimode nature of the field and the complicated and various mechanisms of wave generation.

The fine structure, which occurs practically throughout the oceans (Korchashkin and Ozmidov, 1975), is manifested in an alternation of layers with practically uniform properties and with large gradients of these properties between the layers ; the thickness of these layers may vary from tens of metres to centimetres. It turned out that the layer thickness probability distribution is close to the logarithmic-normal distribution. Fig. 3. shows the distribution laws for the thicknesses of the layers and for the gradients of the properties in these layers.

Further investigation of fine structure and turbulence in the ocean must develop in two ways. The first one is a more detailed study of the relations between the parameters of turbulence and the local properties of the fine structure and internal waves in the immediate vicinity of the point of turbulence measurement. These investigations must define more precisely our knowledge of the mechanisms for the generation and decay of turbulence and of the relationship between turbulence, internal waves and fine structures. The second one is the acquisition of bulk information on hydrophysical fields fine structures and turbulence for the sake of the determination of the probability distribution laws of their main parameters and for finding relations between the parameters and the mean typical hydrometeorological conditions for a given polygon.

REFERENCES

Belyaev, V.S., Lubimtsev, M.M. and Ozmidov, R.V., 1973. O skorosti
dissipatsii turbulentnoi energii i skorosti vyravnivaniya tempera-
turnykh noednorodnostei v okeane. Izv. Akad. Nauk SSR Fiz. Atmos.
Oceans, 9: 1179-1185 (in Russian, with English abstract).
Belyaev, V.S., Monin, A.S. and Ozmidov, R.V., 1974a. Tysyacha spectrov
okeanskoi turbulentnosti. Dokl. AN SSSR, 217: 1053-1056 (in
Russian).
Belyaev, V.S., Ozmidov, R.V. and Pyjevich, M.L., 1974b. Empiricheskie
zakony raspredeleniya znachenii odnomernykh spectralnykh plotnos-
tei pulsatsii skorosti i elektroprovodnosti v okeane. Okeanologiya,
14: 802-805 (in Russian, with English abstract).
Belyaev, V.S., Lozovatskii, I.D. and Ozmidov, R.V., 1975a. O svyazi
parametrov melkomsshtabnoi turbulentnosti s lokalnymi usloviyami
stratifikatsii v okeane. Izv.Akad. Nauk USSR Fiz. Atmos. Oceana,
11: 718-725 (in Russian, with English abstract).
Belyaev, V.S., Lozovatskii, I.D. and Ozmidov, R.V., 1975b. Issledovanie svyazi kharakteristik fluctuatsyi electroprovodnosti vody s
osobennostyami verticalnykh profilei temperatury v okeane. Izv.
Akad. Nauk SSSR Fiz. Atm. Oceana, 11: 1078-1083. (in Russian, with
English abstract).
Belyaev, V.S., Ozmidov, R.V. and Pyjevich, M.L., 1976. O mnogoobrazii
rejimov melkomasshtabnoi turbulentnosti v okeane. Okeanologiya,
16: 229-233 (in Russian, with English abstract).
Korchashkin, N.N. and Ozmidov, R.V., 1975. Prostranstvenno-vremen-
naya izmenchivost tonkoi structury polya temperatury na meridio-
nalnom razreze Tasmaniya-Antarctida. Okeanologiya, 15: 820-825.
(in Russian, with English abstract).
Ozmidov, R.V. (Editor), 1973. Issledovanie okeanskoi turbulentnosti.
"Nauka".
Ozmidov, R.V. (Editor), 1974a. Issledovanie izmenchivosti gidrophi-
zicheskikh polei v okeane. "Nauka".
Ozmidov, R.V., 1974b. 11-yi reis nauchno-issledovatelskogo sudna
"Dmitry Mendeleev". Okeanologiya, 14: 947-951 (in Russian, with
English abstract).
Ozmidov, R.V. and Belyaev, V.S., 1972. Nekotorye osobennosti turbu-
lentnosti v stratifitsirovannom okeane. Mejdunarodnyi Simposium
po stratifitsirovannym techeniyam. Izd. VTs SO AN SSSR, Novosibirsk.
Vasilenko, V.M., Lubimtsev, M.M. and Ozmidov, R.V., 1975. O fluctua-
tsiyakh skorosti dissipatsii turbulentnoi energii i structurnykh
funktsiyakh vysshykh poryadkov polya skorosti v okeane. Izv. Akad.
Nauk SSSR, Fiz. Atmos. Oceana, 11: 926-932 (in Russian, with English
abstract).

Reprinted from *Marine Turbulence*, edited by J.C.J. Nihoul
© 1980 Elsevier Scientific Publishing Company, Amsterdam — Printed in The Netherlands

SOME FUNDAMENTAL ASPECTS OF TURBULENCE WITH IMPLICATIONS IN GEOPHYSICAL FLOWS

J. L. LUMLEY

Sibley School of Mechanical and Aerospace Engineering, Cornell University, Ithaca,
New York 14853 (U.S.A.)

INTRODUCTION

In the last several decades a great deal of attention has been given to turbulence in geophysical flows, some aspects receiving more attention than others.
Two dimensional turbulence, in particular, has been extensively studied, since it
describes the larger scales of turbulent motion in the atmosphere and ocean.
Excellent reviews of this work exist (see, for example, Rhines, 1979) and I will
not touch on it here.   There are many other interesting areas which deserve
attention, some of which have been studied (and may be mentioned in this volume),
and some not, but which I will also not examine; these are, for example, fossil
turbulence, the parametrization of turbulence production by waves, the interaction
of radiation and turbidity, radiation of turbulent energy from the vicinity of the
thermocline by internal waves, absorption of pollution at the air-sea interface
(and its transport into the surface layer of the sea), turbulence in turbidity currents (in which the suspended matter creates a stable stratification), and probably
many others that I have overlooked.

Rather, I will restrict my attention to small scale turbulence, particularly in
the vicinity of a thermo- or halocline, and the simultaneous transport of temperature and salinity fluctuations; the influence of buoyancy on this transport, and
finally, the treatment of such transport near the interface, where intermittency
becomes an important contribution.

An aspect of this small scale turbulence in the ocean that I find most fascinat-
ing are the various double diffusive phenomena; is it necessary to incorporate
these phenomena in a comprehensive treatment of oceanic turbulence, and if so, how?
It seems clear that if the turbulence is fully developed, that is, that the values
of the critical parameter, either Reynolds, Richardson or Rayleigh number (suitably
defined) are sufficiently large, so that the large scales are inertially dominated,
then the influence of the double diffusive phenomena will be only on the dissipation scales, and hence will have no influence on the overall character of the
motion, in common with other anomalies in the dissipation process under such circumstances.   However, as the value of the critical parameter is reduced, the
turbulence gradually approaches a regime in which it is barely turbulent,

and in the case of low Rayleigh number in particular, with the right combination of salinity and temperature gradients, the double diffusive phenomena can begin to play a role.   I am interested here to see whether the second order models which have been developed to treat buoyancy dominated turbulence in the atmosphere, and which are presumably suitable for the description of the small scale turbulence in the ocean when it is fully developed, can be modified so as to include the double diffusive phenomena when the Rayleigh number is reduced to near the stability boundary.   I will treat, in particular the salt fingering regime, as an example.

When the Rayleigh number is sufficiently reduced, the salt fingering regime is not turbulent in any usual sense.   However, if a salt fingering region is bounded by a truly turbulent region, then it will be disturbed by this region, and the variables will be stochastic; thus the regime can be treated by statistical techniques like turbulence.   For slightly higher values of the Rayleigh number the fingering regime will presumably introduce its own stochastic variability.   In philosophy this attempt to make the second order models work in the fingering regime also is much like constructing second order models that remain valid in a boundary layer as the wall is approached, down into the viscous sublayer, which is disturbed, but not turbulent in the usual sense, since it is stable.

TRANSPORT OF TWO SCALARS

There are several situations in which the transport of two scalars is important: in the atmosphere, the surface mixed layer (particularly, but not exclusively, the marine atmospheric surface mixed layer) is often strongly influenced by the simultaneous transport of sensible heat and water vapor.   These, of course, have nearly the same diffusivity.   In combustion, and chemical reactions generally, one or more species are transported simultaneously, and the development of the correlation between them strongly influences the chemical reaction rate.   In gasses, again the diffusivities will probably be nearly the same.   In liquids, there is at least in principle a possibility of different diffusivities if the molecular weights of the reacting species are quite different, but this is usually not the case.   In considering the ecological balance in the surface mixed layer of the ocean, the concentration of phytoplankton and of zooplankton may be considered as two reacting transported scalar species, again having nearly the same diffusivities.   The situation in mixing regions in the ocean, where the simultaneous transport of salt and of heat is dynamically vital, and where the diffusivities of these two differ markedly, is thus nearly unique.

There are excellent reviews of the type of mixing that takes place when two species are present having markedly differing diffusivities (Sherman et al 1978; Turner 1974).   The studies that have been made of this type of mixing are for the most part either experimental or related to stability theory. Such studies have given a clear indication of the mechanisms involved, and have permitted quantitative deliniation of the stability boundaries and of layer thicknesses. These studies have not, however, provided a way of parameterizing the mixing, so that such mixing could be included in a general dynamical calculation scheme which could be used to describe the overall development of such layers.

I believe that it is possible to develop such a parameterization, although at the moment we can only indicate a few necessary pieces. First, it is important to accept the principle that such a parameterization can be developed within the traditional framework of Reynolds averaging. That is, that mixing and transport of two scalars can be described in terms of the variances, fluxes and correlations of the scalars regardless of the mechanisms responsible for the mixing. Although in the laboratory the two double-diffusive types of mixing produce quite regular structures, in nature they are presumably stochastic, and can hence be treated statistically in the same way as other forms of random mixing - for example, traditional turbulence (of the type not associated with double diffusive phenomena) and random surface and internal waves. The difference in the physical mechanism responsible for the mixing should become evident only in the parameterization.

To convince the reader that it is possible to deal with double diffusive phenomena in this way, let us consider a simplified situation describing stochastic salt fingering. We will take the gradients of mean temperature and of mean salinity to be constant, and the disturbance to be statistically homogeneous. We also take the velocity field to be one dimensional, so that the vertical component is the only one present. We assume in addition a steady situation (which is not likely to be true in practice) and we consider that the fluctuations in velocity, temperature and salinity are extremely well-correlated. In the stability problem, of course, the fluctuations are perfectly correlated, since they are all periodic functions of the same period. It is not unreasonable to assume that the correlation is also excellent in the random case, in which the spacing and intensity of the salt fingers are no longer quite uniform. For a one dimensional disturbance, homogeneous, with salinity and temperature fluctuations, the equations for the variances and fluxes become (with the notation and conventions of Turner, 1974; i.e. $-z$ increases downward)

$$\dot{\overline{w^2}} = -2\overline{wp},_3/\rho - 2g(\alpha\overline{tw}-\beta\overline{sw}) - 2\overline{vw,_j w},_j \quad j = 1,2$$

$$\dot{\overline{tw}} = -\overline{tp},_3/\rho - g(\alpha\overline{t^2}-\beta\overline{st}) - (\kappa_T+\nu)\overline{t,_j w},_j - T'\overline{w^2}$$

$$\dot{\overline{sw}} = -\overline{sp},_3/\rho - g(\alpha\overline{st}-\beta\overline{s^2}) - (\kappa_S+\nu)\overline{s,_j w},_j - S'\overline{w^2}$$

$$\dot{\overline{t^2}} = -2T'\overline{tw} - 2\kappa_T\overline{t,_j t,_j}$$

$$\dot{\overline{s^2}} = -2S'\overline{sw} - 2\kappa_S\overline{s,_j s,_j} \tag{2.1}$$

$$\dot{\overline{ts}} = -T'\overline{sw} - S'\overline{tw} - (\kappa_S+\kappa_T)\overline{t,_j s,_j}$$

If we suppose that the correlations between the three quantities are truly perfect, so that the three quantities are proportional, then we can write for the dissipations and correlations (if the gradients of temperature and salinity are negative):

$$\overline{t,_j w,_j} = \overline{w,_j w,_j}\,(\overline{t^2/w^2})^{1/2}; \quad \overline{t,_j t,_j} = \overline{t^2}\,\overline{w,_j w,_j}/\overline{w^2}$$

$$\overline{s,_j w,_j} = \overline{w,_j w,_j}\,(\overline{s^2/w^2})^{1/2}; \quad \overline{s,_j s,_j} = \overline{s^2}\,\overline{w,_j w,_j}/\overline{w^2} \tag{2.2}$$

$$\overline{t,_j s,_j} = \overline{ts}\,\overline{w,_j w,_j}/\overline{w^2}$$

$$\overline{tw} = (\overline{t^2}\,\overline{w^2})^{1/2}; \quad \overline{ts} = (\overline{t^2}\,\overline{s^2})^{1/2}; \quad \overline{sw} = (\overline{s^2}\,\overline{w^2})^{1/2}$$

We must also use that fact that, for a one dimensional turbulence, the correlations with the pressure gradient vanish:

$$\overline{wp,_3} = \overline{tp,_3} = \overline{sp,_3} = 0 \tag{2.3}$$

Finally, if we use the assumption of stationarity, we obtain

$$(\alpha T'/\kappa_T - \beta S'/\kappa_S)(g/\nu)(\overline{w^2/w,_j w,_j})^2 = 1 \tag{2.4}$$

which is essentially the criterion for salt-fingering obtained from linear stability theory.

Thus we can see that the traditional framework of Reynolds averaging contains the potential for describing the mixing produced by stochastic salt fingering; in fact, our description is not restricted to small disturbances, and so is more general than the linear stability theory. It is clear that if we wish to develop a second order model that will describe mixing and transport in general, including salt fingering, it must behave correctly when the correlation between the various quantities becomes essentially perfect, as well as when the turbulence becomes essentially one-dimensional. In what follows we will use the word turbulence to refer to any stochastic mixing process.

Complete treatment of this phenomenon in the general case involves several pieces: the influence of buoyancy on the turbulent transport of itself and of the heat and salt within the well-mixed part; the influence of intermittency on the transport near the edge of a mixed region; the influence of strong anisotropy on all parameterizations (since the turbulence under these circumstances may become nearly one-dimensional); the proper equations for the dissipation (or equivalently, for length scales) for the two scalars; and finally the modeling of the dissipation terms in the equations for the correlations among the three scalars. In this section we will treat the latter subject, deferring discussion of the others to later sections (or other papers). In all these areas, a beginning has been made in other fields, and there is much that can be adapted to our uses, although there is also much that remains to be done.

The general interaction term in the equation for the correlation between temperature and salinity is the simplest to treat as an example. It can be written as

$$(\kappa_T + \kappa_S)\overline{s,_j t,_j} \qquad j = 1, \cdots, 3 \tag{2.5}$$

This term has been discussed by Zeman & Lumley (1976a) who suggested parameterizing it as

$$(\kappa_T + \kappa_S)\overline{s,_j t,_j} = 2(1 + \kappa_S/\kappa_T)\overline{st}\ \overline{\varepsilon_T}/t^2 \tag{2.6}$$

$$[\overline{\varepsilon_T} = \kappa_T \overline{t,_j t,_j}]$$

The latter parameterization was successful in predicting the transport of pollutant in the surface mixed layer of the Los Angeles basin. However, it appears likely that our requirements are more stringent, and that a successful parameterization for our case will have to be constructed with muc greater care. Lumley (1978a) has suggested a way of constructing a rational parameterization.

Briefly, it is necessary to consider the concept of realizability, as introduced by Schumann (1977), and as modified by Lumley (1978b). That is, any quality which is required to remain positive must obey an equation so constructed that, as the quantity approaches zero, the time derivative also approaches zero (see figure 1). This is a necessary, though not sufficient, condition to assure that the quantity will not at the next instant become negative. This condition can be applied to each type of term in the governing equation separately, since the condition must be satisfied under all situations. We can imagine situations in which one or another of the various types of terms in the equation is absent (e.g.- the transport terms are absent in a homogeneous turbulence, and the production terms are absent in a turbulence without mean gradients); we can also imagine situations in which the various types of terms in the equation may be made to take on arbitrary values relative to one another (e.g.- the magnitudes and orientations of the mean gradients can be manipulated arbitrarily relative to one another, at least initially).

180

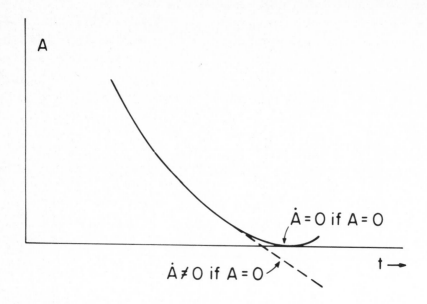

Figure 1. Sketch of a variable which should remain positive approaching zero; the two possibilities indicated correspond respectively to a model that does, and does not, satisfy realizability.

Let us apply this condition to the dissipation term in the equation for the correlation between salinity and temperature; this will, of course, involve also the dissipation terms in the equations for the temperature and salinity variances. The basic condition may be written:

$$(d/dt)(\overline{t^2}\ \overline{s^2} - \overline{ts}^2) = 0 \quad \text{if} \quad \overline{t^2}\ \overline{s^2} - \overline{ts}^2 = 0 \qquad (2.7)$$

Expanding, we have:

$$\overline{\dot{t^2}}\ \overline{s^2} + \overline{t^2}\ \overline{\dot{s^2}} - 2\overline{\dot{ts}}\ \overline{ts} = 0 \quad \text{if} \quad \overline{t^2}\ \overline{s^2} - \overline{ts}^2 = 0 \qquad (2.8)$$

Finally, substituting from the equations of motion, we obtain:

$$(\kappa_T + \kappa_S)\overline{t_{,j} s_{,j}} = (\overline{\varepsilon_T}\ \overline{s^2} + \overline{\varepsilon_S}\ \overline{t^2})/\overline{ts} = \overline{ts}(\overline{\varepsilon_T}/\overline{t^2} + \overline{\varepsilon_S}/\overline{s^2})$$

$$\text{if} \quad \overline{t^2}\ \overline{s^2} = \overline{ts}^2 \qquad (2.9)$$

This condition may be met in several different ways. First, the expression in (2.9) may be multiplied by a function of the correlation coefficient, which should probably also be a function of the diffusivity ratio:

$$(\kappa_T+\kappa_S)\overline{t_{,j}s_{,j}} = \overline{ts}\,(\overline{\epsilon_T}/\overline{t^2}+\overline{\epsilon_S}/\overline{s^2})\,g(1-\overline{ts}^2/\overline{t^2}\,\overline{s^2},\ \kappa_S/\kappa_T)$$ (2.10)

The function is required to obey the condition

$$g(0,\kappa_S/\kappa_T) = 1$$ (2.11)

Nothing is known about the form of this function. By examining the equation for the correlation coefficient, it is possible to show that the function must not be smaller than unity. It is tempting to take a simple linear form:

$$g(x,\kappa_S/\kappa_T) = 1 + A(\kappa_S/\kappa_T)x,\ A > 0$$ (2.12)

and there is some indication that the value of the coefficient may be of order unity (Lumley 1978a). Nothing is known about the dependence on the diffusivity ratios. It would be necessary to do several controlled experiments, in which a homogeneous, isotropic turbulence, with superimposed fluctuations of two different scalars, was allowed to decay, the combinations of scalars being selected to cover a range of diffusivity ratios.

In the expression (2.9), the ratios of dissipation to variance take on particular values when the correlation is perfect:

$$\overline{\epsilon_S}/\overline{s^2} = (\kappa_S/\kappa_T)\overline{\epsilon_T}/\overline{t^2}$$ (2.13)

which can either be obtained directly by assuming that the salinity and temperature are proportional, or from (2.9). In our expression (2.10), we have left these ratios of dissipation to variance (essentially the inverse of eddy time scales) untouched, assuming that the computation scheme will contain equations for the dissipations which will be constructed so that the condition (2.13) will be satisfied when the correlation becomes perfect. Another way of handling the situation is to replace the ratio of the time scales by another unknown function of the correlation:

$$\overline{\epsilon_S}/\overline{s^2} = f(1-\overline{st}^2/\overline{s^2}\,\overline{t^2})(\kappa_S/\kappa_T)\overline{\epsilon_T}/\overline{t^2}$$ (2.14)

which is then required to satisfy

$$f(0) = 1$$ (2.15)

182

Since presumably the time scale ratio will tend toward unity as the correlation tends toward zero, we may expect that this function will tend toward the inverse of the diffusivity ratio as this condition is approached. The approach represented by (2.14) would be required if no equation were being carried explicitly for the dissipation of salinity variance. In the calculation of Zeman & Lumley (1976a), no equation was carried for the dissipation of pollutant concentration variance, which necessitated an assumption of the form (2.6). It can be seen that this does not satisfy realizability, since the coefficient does not go to unity when the correlation is perfect; in addition, the parenthesis dependent on the diffusivity ratio already has the form corresponding to a perfect correlation, which is unlikely to be correct at moderate levels of the correlation.

TURBULENT TRANSPORT

As Woods has underlined in the introduction to the 1979 IAPSO-SCOR workshop on Turbulence in the Ocean, large portions of the ocean are quite stable, and are only intermittently turbulent. Hence, it is particularly important that our understanding of turbulent transport of the variances and fluxes include the strong effect of buoyancy and one- or two-dimensionality. This requirement has already been recognized and studied in the atmosphere, in connection with investigation of the structure of the buoyancy driven surface mixed layer (Zeman & Lumley, 1976b, Lumley et al 1978). There, the thickening of the surface mixed layer, and the erosion of the inversion base, is entirely due to buoyancy; in the vicinity of the inversion base, the turbulence becomes nearly two dimensional, as the vertical component is suppressed by the stable stratification. It has been possible to incorporate the effects of buoyancy in a dynamically correct manner (Zeman & Lumley 1976b), to understand the physics of the mechanism (Lumley et al 1978) and to place the model on a mathematically rigorous basis (Lumley 1978b).

Briefly, the turbulence is taken to be nearly Gaussian, to have the property of relaxing to a Gaussian state in the absence of buoyancy, inhomogeneity, and other disturbing influences. In fact, homogeneous turbulence is observed to be nearly Gaussian in the large scales (for a full discussion see Lumley 1978b). Turbulence is known not to be Gaussian in the small scales (see Batchelor, 1956, p173), and this is a dynamically vital property, since it is associated with spectral transfer of energy. However, we may approximate the large scales as being nearly Gaussian without doing any damage to the dynamics. The extent to which turbulence that approaches two- or one-dimensionality can be regarded as Gaussian is not known; in strongly convective situations in the atmosphere, when the turbulence is dominated by vertical motions, it may still be regarded as nearly Gaussian in the large scales. Near the inversion base, when the vertical component is strongly suppressed, an almost Gaussian model describes entrainment well. The extent to which any turbulent motion may be considered almost Gaussian, even disregarding for a moment the disturbing influences such as inhomogeneity, presumably depends on the amount by which the stability parameter exceeds its critical value (see, for example, Monin & Yaglom, 1971, section 2). That is, when the Reynolds, Richardson, or Rayleigh number just exceeds its critical value, only one mode is present, and the motion is quite regular, and cannot possibly be considered as nearly Gaussian. As the value of the critical parameter increases, however, more and more modes are present, and the character of the motion becomes increasingly complex. We must presume that in the real ocean, stable as it is, parameters are usually sufficiently above their critical values for an assumption of nearly Gaussian behavior to be reasonable. This is certainly true in the nocturnal atmospheric boundary layer, which may be adequately modeled by this assumption (Zeman & Lumley 1979) and which is extremely stable. It is true in addition of the marine atmospheric surface layer, which is often extremely stable (as a result of the moisture transport), with turbulence occuring only in patches, but which may nevertheless be adequately described by use of this type of model (Warhaft 1976). In most flows the increase in complexity that accompanies an increase in the parameter above its critical value happens very quickly, so that the range near the critical value in which an assumption of nearly Gaussian behavior would be unjustified may be regarded as a neglegibly small range, which may safely be ignored in constructing models. In the following section we will discuss one important way in which probability densities near the edge of a mixed region cannot be Gaussian.

As discussed in detail in Lumley (1978b), it is assumed that the length scale of the turbulence is small relative to the length scale of the inhomogeneity. This is, of course, an impossible situation in nature, where the length scale of the turbulence is always comparable with the length scale of the inhomogeneity; it is physically possible, however, in the sense that it is a situation that could be created in the laboratory, by producing the inhomogeneity and the turbulence by different mechanisms. An expansion is carried out in the length scale ratio as a small parameter, and the leading term produces a consistent, realizable expression for the turbulent transport which is correct for weak inhomogeneity. This is essentially a kinetic theory type of approach, and the expressions obtained are gradient transport expressions; for example, neglecting buoyancy, one obtains the the form (for a vertically inhomogeneous, one-dimensional turbulence)

$$\overline{w^2 dt^2}/dz + 2\overline{twd}\,\overline{tw}/dz = -2[2\overline{\varepsilon_T}/\overline{t^2} + \overline{\varepsilon/w^2}]\overline{t^2 w} \tag{3.1}$$

which is very close to an expression used on an ad-hoc basis by Hanjalic & Launder(1972), which is, in its turn, a simple tensorial generalization of classical gradient transport.

When we consider the case of buoyancy, however, we must distinguish two possible cases (Lumley 1878b). In some cases, buoyancy may be considered a relatively small disturbing influence, and the terms arising from the buoyant acceleration may be neglected to first order. This leads to expressions similar to (3.1), which have been successfully used in the atmospheric boundary layer with wind shear. If the turbulence is produced by the buoyancy, however, the buoyant acceleration is of the same order as the other terms in the equation, and must be kept. This then ties the equations for the third moments together, and an expression of the form (again for a one-dimensional, vertically inhomogeneous turbulence without salinity fluctuations)

$$
\left\{
\begin{array}{c}
\overline{w^3} \\[4pt]
\overline{w^2 t} \\[4pt]
\overline{t^2 w} \\[4pt]
\cdot \\
\cdot \\
\cdot
\end{array}
\right\}
= \underline{F}\, \frac{\partial}{\partial z}
\left\{
\begin{array}{c}
\overline{w^2} \\[4pt]
\overline{tw} \\[4pt]
\overline{t^2} \\[4pt]
\cdot \\
\cdot \\
\cdot
\end{array}
\right\}
\; ; \quad
\underline{F} = \frac{\overline{w^2}}{\overline{\varepsilon}}
\left\{
\begin{array}{cccc}
\overline{w^2} + a g \beta \overline{tw}\ \overline{t^2}/\varepsilon_T & \cdots & & \\
& \cdot & & \\
& & \cdot & \\
& & & \cdot \\
\end{array}
\right\}
\qquad (3.2)
$$

results. This is still a gradient transport form, but is now similar to the form of molecular transport one would have in a mixture of several species. This form was very successful in predicting the evolution of the surface mixed layer of the atmosphere, with and without wind shear, Coriolis effects and pollution transport (Zeman & Lumley 1976a, 1976b, 1979).

In the case of stochastic salt fingering in the ocean, we may construct a similar form for the turbulent transport. We suppose that we are dealing with a situation in which the fingers are no longer strictly parallel (see figure 2), as well as being random in strength and spacing. We may apply classical order of magnitude analysis to this situation, since it has the same properties as a boundary layer, wake or other nearly parallel flow. As a result of such an analysis, we find that the pressure gradient term may be neglected in the equation for the vertical velocity:

$$
w w,_z + w,_i u_i = -g(\alpha t - \beta s) + \nu w,_{jj} \quad i,j = 1,2 \qquad (3.3)
$$

The neglected term is (relative to the other terms in the equation) of the order of the square of the ratio of the horizontal and vertical length scales. The substantial derivative terms on the left are assumed small relative to the terms on the right, which are taken to be nearly in balance; the relative magnitude of the substantial derivative is the length scale ratio multiplied by the Reynolds number based on the horizontal length scale; if the fingers have a sufficiently large

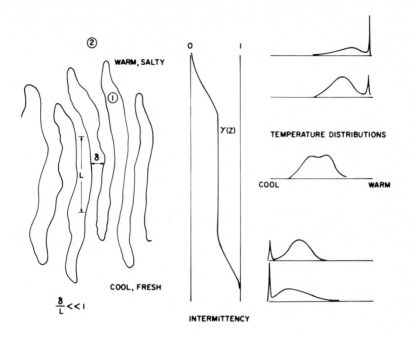

Figure 2.   Sketch of stochastic salt fingers, with a curve of intermittency indicated (defined on the rising fluid), as well as speculative sketches of the probability density for temperature at various levels through the layer.

aspect ratio, the Reynolds number need not be small.   We are neglecting the time derivative for simplicity, considering the case of a slowly varying flow.

It is much simpler to eliminate the pressure gradient term before doing the statistics, than after.   The third order terms to which the pressure gradient gives rise (see Lumley 1978b) are generally the most difficult to model, and are the terms about which the most controversy exists.   In this way, we eliminate them altogether.   We may point out here, that if a model is constructed that is intended to work in general situations as well as in that envisioned here, it will be necessary to construct the terms involving the pressure gradient so that they vanish in the present situation (near one-dimensionality).   While this requirement has been met relative to the models of terms appearing in the second order equations (Lumley 1978b), no attempt has so far been made to meet it in connection with the third order terms.   This is another aspect of realizability.

Beginning from equation (3.3), we may now form equations for all the third order correlations, so as to form a closed system.   We assume a horizontally homogeneous situation, and take the vertical inhomogeneity to be small. There are in all ten such correlations, and the equations are interdependent due to the buoyancy terms. The triple correlations involving only the temperature and the salinity can be eliminated fairly easily, as can the vertical velocity skewness. This leaves a five by five matrix of equations for the mixed moments which can be solved, given a little patience, leading to a form like equation (3.3). The form of all these equations involves only the assumption that the distribution of fluctuations wishes to relax to Gaussian in the absence of disturbing influences.

Although the equations obtained are much simpler than they would be if the pressure terms had been retained, they are nevertheless difficult. We will learn as much about the physics by considering the simplified situation we have already discussed, that in which the vertical velocity, temperature and salinity fluctuations are very well correlated. If we neglect the flux divergence in the equations for the temperature and salinity variance, (i.e.- assume that there is a balance between the production and dissipation of the two variances) we can obtain estimates for the ratio of the variances of the temperature and salinity to that of the vertical velocity. If these are used, and the assumption of large correlation is made, we obtain for the flux of the vertical velocity variance

$$\overline{w^2}d\overline{w^2}/dz = g\overline{w^3}(\alpha T'/\kappa_T - \beta S'/\kappa_S)\overline{vw^2}/\varepsilon - 2\overline{w^3}\,\overline{\varepsilon/w^2} \tag{3.4}$$

while the equation for the vertical velocity variance can be written as

$$d(\overline{w^3})/dz = g\overline{w^2}(\alpha T'/\kappa_T - \beta S'/\kappa_S)\overline{vw^2}/\varepsilon - \overline{\varepsilon} \tag{3.5}$$

Let us designate as

$$\overline{Ra} = (\alpha T'/\kappa_T - \beta S'/\kappa_S)gv\overline{(w^2/\varepsilon)}^2 \tag{3.6}$$

a type of composite Rayleigh number; we can see from equation (3.5) that if

$$\tag{3.7}$$

$$\overline{Ra} > 1$$

then production exceeds dissipation, and there must be export of velocity variance, and vice versa. The equation for the flux of variance, (3.4), may be rearranged as

$$\overline{w^3} = -(d\overline{w^2}/dz)(\overline{w^2}\,/\varepsilon)^2[1-(\overline{Ra}-1)]^{-1} \tag{3.8}$$

If we designate by

$$K_o = \overline{w^2} / \overline{\varepsilon} \qquad (3.9)$$

the transport coefficient when production and dissipation balance, then we may write the transport coefficient in general circumstances as

$$K = K_o / [1-(\overline{Ra}-1)] \qquad (3.10)$$

It can be seen that, if production exceeds dissipation, so that energy is exported (in other words, so that the region is unstable), then the transport coefficient is larger, while if dissipation exceeds production, so that energy must be imported (so that the region must be regarded as stable), then the transport coeffcient becomes smaller. This, of course, is the same mechanism that is at work in the atmosphere, where the vertical transport coefficient is reduced in the stable region at the inversion base. The forms that are obtained in the atmosphere for the transport coefficients in the matrix (equation (3.2)) have much in common with the form (3.10). Physically, the modification of the transport coefficient can be traced to the additional vertical acceleration that is induced by the buoyancy, which either adds to or detracts from the existing vertical acceleration. This has the effect of modifying the vertical Lagrangian integral time scale.

We may draw a qualitative sketch of the structure of a fingering layer between hot, salty water overlying relatively cool, fresh water (see figure 3) (we make no attempt to consider the formation of multiple layers).

INTERMITTENCY

The entrainment which takes place at the top and bottom edges of a stochastic fingering layer involves a type of intermittency(see figure 2). This is not quite like the usual form of intermittency at a turbulent /non-turbulent interface, in that the incursions of each medium penetrate entirely through the layer. Some fingers are wider than others, however, and the widths are not constant, and hence the rate of increase of salinity and of temperature along the fingers has a stochastic component in a given finger, and differs from finger to finger, so that some fingers are more buoyant than others, and penetrate farther into the other fluid. We have also sketched a curve for the average horizontal portion of the layer that is fresh, the intermittency curve. In addition, we have sketched quite speculative forms for the probability densities of the temperature fluctuations at different levels. It is seen that these densities are far from Gaussian in form, as was assumed in section 3. If we assume that there is sufficient randomness in

188

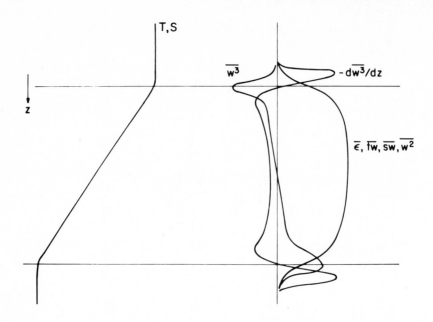

Figure 3. Speculative sketch of the distribution of horizontally averaged values of several variables through the layer, in particular the transport of velocity variance.

the mixing process (i.e.- that the composite Rayleigh number is large enough), then we can imagine near the top of the layer a more or less Gaussian part of the density corresponding to fluid arriving from the bottom of the layer having absorbed different amounts of heat. However, there are also spikes in the densities that correspond to nearly uncontaminated fluid from the layer above. The entire density consists of both parts, and by no stretch of the imagination can we consider the density so constituted to be approximately Gaussian. The technique described in section 3, which is essentially a perturbation about the Gaussian state, can be applied to the continuous part of the density, but cannot be applied to the spike.

A beginning has already been made on dealing with this problem in Lumley (1979). Briefly, the distribution must be split into two parts, and the two handled separately. In figure 4 we show an idealized probability distribution of temperature, labeled appropriately. Essentially, we must define a mean and deviation for each part of the distribution. We show only the distribution appropriate to the upper edge, as an example. We can define averages through the use of an indicator function I, which is unity in the rising fingers (region 1), and zero in the hot salty fluid (region 2). Then, if T is now the instantaneous temperature, we can write

$$T = TI + T(1-I) \qquad (4.1)$$

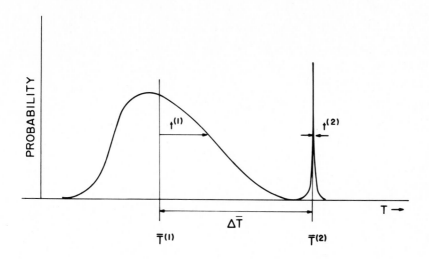

Figure 4. Sketch of a typical probability density for temperature in the upper intermittent region of the layer, with the means and deviations for the two parts of the density labeled.

and the averages in the two regions are defined as

$$\bar{T}^{(1)} = \overline{TI}/\bar{I}, \ \bar{T}^{(2)} = \overline{T(1-I)}/(1-\bar{I}) \tag{4.2}$$

so that the relation among the averages becomes

$$\bar{T} = \gamma\bar{T}^{(1)} + (1-\gamma)\bar{T}^{(2)}, \ \gamma = \bar{I}, \ \Delta\bar{T} = \bar{T}^{(1)} - \bar{T}^{(2)} \tag{4.3}$$

we may also define fluctuations in each region

$$t^{(1)} = T - \bar{T}^{(1)}, \ t^{(2)} = T - \bar{T}^{(2)} \tag{4.4}$$

In this way, we may write the temperature variance and the third moment of the temperature as

$$\overline{t^2} = \overline{t^{(1)}t^{(1)}}^{(1)} + \gamma(1-\gamma)\Delta\bar{T}\Delta\bar{T}$$
$$\overline{t^3} = \overline{\gamma(t^{(1)})^3}^{(1)} + \gamma(1-\gamma)3\overline{t^2}\Delta\bar{T} \tag{4.5}$$

where we have assumed that the jump in properties is small, which may not be the case here. If the jump in mean value is not small, then we have in addition, in the expression for the third moment, terms up to cubic in the value of the jump. In the expression for the third moment of the variance in equation (4.5) we envision obtaining the third moment of the continuous part by a technique like that in section 3.

In order to construct a calculation scheme, we need to have an equation for the intermittency and for the property jump. Let us consider first the jump in velocity between the rising fluid and the fluid as a whole; we expect this velocity to take the fingers upward into the undisturbed fluid, and hence to be in the direction of increasing salinity intermittency, with an unknown coefficient which probably depends on the intermittency (if there is a mean velocity, we can also say that the fluid that comes from a region where the mean velocity is higher, will have a higher velocity - see Lumley 1979)

$$\overline{w}^{(1)} - \overline{w} = -A(\gamma)(d\gamma/dz)(\overline{w^2}/\varepsilon) \qquad (4.6)$$

We can make an entirely similar statement regarding the jump in a scalar property:

$$\overline{T}^{(1)} - \overline{T} = -C(\gamma)(\overline{w}^{(1)}-\overline{w})d\overline{T}/dz(\overline{w^2}/\varepsilon) \qquad (4.7)$$

To close the equation for the intermittency, we need an expression for the intermittency production. This will evidently be proportional to the product of the interfacial area per unit volume, which can be shown to be proportional to the gradient of the intermittency, and to the velocity of propagation of the interface into the undisturbed region. This is just the difference of velocity between the rising fluid and the fluid as a whole, since this determines the rate of advance of the fingers into the undisturbed fluid. Hence, we expect a form like

$$-B(\gamma)(\overline{w}^{(1)}-\overline{w})d\gamma/dz \qquad (4.8)$$

with again a coefficient which is an unknown function of the intermittency. If these expressions are combined, we obtain an equation for the intermittency in the situation we are considering:

$$\dot{\gamma} = (d/dz)[\gamma A(\gamma)(\overline{w^2}/\varepsilon)d\gamma/dz] + B(\gamma)A(\gamma)(d\gamma/dz)^2(\overline{w^2}/\varepsilon) \qquad (4.9)$$

These considerations are highly speculative, and only a beginning has been made on the application of these ideas in other fields; work is currently underway on the incorporation of this type of transport in the intermittent region at the edge of a non buoyant wake or jet. Preliminary indications are that the equation so obtained represents the observed values of intermittency very well. Much experimental and computational work remains to be done before we can say whether the same will be true in the case of stochastic salt fingering.

CONCLUSIONS

We have described several ideas that have been developed in connection with second order modeling of completely different flows, and have shown how they might be applied to the case of stochastic salt fingering in the ocean, as an example. We have worked out a few examples, assuming that the salt fingers are long and thin, and nearly vertical, and that the correlation between the fluctuations of all variables is excellent. Of course, in real situations, this is not likely to be true. At large values of a composite Rayleigh number, the salt fingers will be quite distorted, and the correlation between the variables will drop; as a consequence, the many terms that we have neglected must be retained, and the full set of equations must be used. A calculation of this sort could only be done by machine. Our point is, that this rather special regime of salt fingering can be included in a natural way in a calculation scheme which is already demonstrated to be capable of handling more customary turbulence dominated by buoyancy; presumably a real situation will include temperature gradients in the hot salty region and in the cool fresh region, as well as currents in both, so that in the overlying and underlying layers a more traditional sort of Richardson number turbulence may exist. It is only important that the modeled terms in the equations behave correctly as one moves toward the interface between the regions, so that the transport there, dominated by the salt fingering, can be accurately included.

ACKNOWLEDGEMENTS

Supported in part by the U. S. National Science Foundation, Meteorology Program under Grant Number ATM77-22903, and in part by the U. S. Office of Naval Research, Fluid Dynamics Branch. Prepared for presentation at the Eleventh International Leige Colloquium on Ocean Hydrodynamics "Turbulence in the Ocean", (Leige, May 7-11, 1979).

REFERENCES

Batchelor, G. K., 1956. Homogeneous Turbulence. The University Press, Cambridge UK.

Hanjalic, J. and Launder, B. E., 1972. A Reynolds stress model of turbulence and its application to thin shear flows. J. Fluid Mech., 52:609-638.

Lumley, J. L., 1978a. Turbulent transport of passive contaminants and particles: fundamentals and advanced nethods of numerical modeling. In: Lecture Series 1978-7,

192

Pollutant Dispersal. von Karman Institute for Fluid Dynamics, Rhode-St-Genese, Belgium.

Lumley, J. L., 1978b. Computational modeling of turbulent flows. In: C.-S. Yih (Editor), Advances in Applied Mechanics, vol. 18. Academic Press, New York, pp. 123-176.

Lumley, J. L., 1979. Second order modeling of turbulent flows. In: Lecture Series 1979-1, Prediction Methods for Turbulent Flows. von Karman Institute for Fluid Dynamics, Rhode-St-Genese, Belgium.

Lumley, J. L., Zeman, O and Siess, J., 1978. The influence of buoyancy on turbulent transport. J. Fluid Mech., 84:581-597.

Monin, A. S. and Yaglom, A. M., 1971. Statistical Fluid Mechanics, vol. 1. M.I.T. Press, Cambridge MA.

Rhines, P. B., 1979. Geostrophic Turbulence. In: M. Van Dyke, J. V. Wehausen and J. L. Lumley (Editors), Annual Review of Fluid Mechanics, vol. 11. Annual Reviews Inc., Palo Alto CA, pp. 401-442.

Schumann, U. 1977. Realizability of Reynolds stress turbulence models. Physics of Fluids, 20:721-725.

Sherman, F. S., Imberger, J. and Corcos, G. M., 1978. Turbulence and mixing in stably stratified waters. In: M. Van Dyke, J. V. Wehausen and J. L. Lumley (Editors), Annual Review of Fluid Mechanics, vol. 10. Annual Reviews Inc., Palo Alto CA, pp. 267-288.

Turner, J. S., 1974. Double-Diffusive Phenomena. In: M. Van Dyke, J. V. Wehausen and J. L. Lumley (Editors), Annual Review of Fluid Mechanics, vol. 6. Annual Reviews Inc., Palo Alto CA, pp. 37-56.

Warhaft, Z, 1976. Heat and moisture flux in the stratified boundary layer. Quart. J. Roy. Met. Soc., 102:703-707.

Zeman, O. and Lumley, J. L. 1976a. Turbulence and diffusion modeling in buoyancy driven mixed layers. In: Proceedings of Third Symposium on Atmospheric Turbulence, Diffusion and Air Quality, Raleigh NC. American Meteorological Soc., Boston MA, pp. 38-45.

Zeman, O. and Lumley, J. L. 1976b. Modeling Buoyancy Driven Mixed Layers. J. Atmospheric Sciences, 33:1974-1988.

Zeman, O. and Lumley, J. L. 1979. Buoyancy effects in entraining turbulent boundary layers: a second order closure study. In: F. Durst, B. E. Launder, F. W. Schmidt and J. H. Whitelaw (Editors), Turbulent Shear Flows I. Springer-Verlag, Berlin/Heidelberg/New York, pp. 295-306.

NONLINEAR EVOLUTION AND STRUCTURE OF SALT FINGERS

S. A. PIACSEK[1] and J. TOOMRE[1,2]

[1]Naval Ocean Research and Development Activity

[2]Department of Astro-Geophysics and Joint Institute for Laboratory
 Astrophysics, University of Colorado

ABSTRACT

The finite amplitude growth of salt fingers across the interface in a two-
layer fluid (warm saline over cold fresh) has been investigated by finite difference techniques. The very large mesh required to simulate a collection of
fingers over their full vertical extent has limited the calculations to two
dimensions. Fingers with such planforms are attainable in laboratory experiments
even in the presence of a weak shear. The emphasis is on studying the evolving
shapes of the fingers, their collective growth behavior and the associated vertical fluxes of heat and salt. Most of the calculations involve 8 fingers (4 up
and 4 down), with some involving 16; the fingers grow to a vertical/horizontal
aspect ratio of 30:1. A common feature to these solutions is the termination of
the fingers in bulbous shapes as they penetrate into the quiescent layers. Most
of the vorticity appearing in the fluid is concentrated either in the "bulb
region," where a certain entrainment process operates between the quiescent and
fingering regions, or in the shear zones between the fingers. There appears to
be a cross-modulation of fingers, with the bulbs of later developing fingers
bending and even destroying the stem of tall, fast growing ones. The horizontal
profiles of the temperature and salinity fluctuations show sinusoidal and square
shapes, respectively, somewhat in variance with the assumptions of previous theoretical studies. In the numerical experiments the initial perturbation spectrum
ranges from a purely sinusoidal one to a white spectrum with random phases, and
the stability parameter $\Lambda = \alpha\Delta T/\beta\Delta S$ varies from 2 to 10. The appearance of the
fingers depends somewhat on the initial perturbation. The dependence of the
computed buoyancy flux ratio $\chi = \alpha F/\beta F_S$ on Λ exhibits a linear behavior. The
growth rate of the fingers dh/dt is found to be proportional to $\Lambda^{3/5}$, and so too
the amplitude of the fluctuating temperature field.

I. INTRODUCTION

Some of the microstructure seen in the oceans may result from thermohaline

convection. These structures show up as distinct and sharp steps in certain

vertical profiles of salinity and temperature. The steps seem to reveal well

mixed horizontal layers of nearly uniform salinity and temperature (often tens of

meters in thickness) separated by rather thin interfaces (possibly only a meter

or less in vertical extent); across these horizontal interfaces exist substantial

gradients in both temperature and salinity. The use of increasingly sensitive salinity-temperature-depth (STD) recorders during the past decade led to the discovery of such layering. It has been suggested that thermohaline convection could be at work there; this convection depends crucially upon the differing diffusivities of salt and heat, and such an instability can set in even when the total density increases with depth. Such suggestions are inspired mainly by laboratory experiments that show thermohaline convection organizing itself into a series of layers, with the salinity and temperature profiles having almost a staircase appearance.

a. The oceanic layering

   Various step-like profiles in temperature and salinity, suggestive of layering, have been seen in the Mediterranean outflow into the Atlantic (Tait & Howe 1968, 1971; Hayes 1975), in the Mediterranean itself (Johannessen & Lee 1974), near the bottom of the Red Sea (Degens & Ross 1969), in probing of the San Diego Trough (Gregg & Cox 1972), and in arctic waters probed from a drifting ice island (Neshyba, Neal & Denner 1971). The main thermocline near Bermuda has shown steps (Cooper & Stommel 1968); such structures are also evident in the North Atlantic (Mazeika 1974; Schmitt & Evans 1978), in the central Pacific (Gregg 1976), and in the Weddell Sea (Foster & Carmack 1976). Further, Williams (1974, 1975) and Magnell (1976) report the detection of what appear to be salt fingers by direct optical imaging in the Mediterranean outflow.

   The observed stepped structures in temperature and salinity seem to result from well mixed layers separated by sharp interfaces (e.g. Federov 1978). The horizontal extent of some of these layers is also being established: Stommel & Federov (1967) were able to trace layers in the thermocline over several kilometers horizontally; Pingree (1970) studied horizontal variations in the less regular layered structures seen in the Mediterranean outflow, and Elliot, Howe & Tait (1974) report lateral coherences for deeper layers here of even several tens of kilometers. Lambert & Sturges (1977) have concentrated on the persistence of multi-layered systems in the Caribbean, finding that the thermohaline staircase is nearly invariant over at least four days in time. Although such data on the time evolution and the horizontal extent of the layers are now becoming available, the layering configurations are sufficiently complicated to make it difficult to reliably estimate the transports of heat and salt through such layers.

   These step-like structures in temperature and salinity have various explanations: they may result from the overturning of internal waves or shearing flows, from the horizontal intrusion of water masses, or from thermohaline convective processes. Encouraged by the results of laboratory experiments dealing with thermohaline convection, we have sought to explain some of the above phenomena on the basis of layering mechanisms driven by thermohaline convection.

b. Layering in laboratory experiments

Thermohaline convection depends crucially upon the differing diffusivities of heat and salt, and so is often called "double-diffusive." Provided suitable vertical gradients exist in both salinity and temperature, thermohaline convection can set in even when the total density increases with depth. The overall appearance of the motions however depends upon whether temperature or salinity is the destabilizing agent, and the other the stabilizing one.

With temperature serving to destabilize, the flow in laboratory experiments seems to organize itself into a series of horizontal layers of fairly vigorous convection, each separated from its neighbor by a sharp interface in temperature and salinity (Turner 1965, 1968; Crapper 1975; Marmorino & Caldwell, 1976; Griffiths 1979a). This situation is frequently called "layering," though it is sometimes called the "diffusive" regime since the overall transport of heat and salt is controlled to some extent by diffusion through the sharp interfaces. On the other hand when salinity serves to destabilize, tall and narrow convection cells are established in the experiments, with the flow then consisting of thin fingers of fluid moving alternately up and down. Such "fingering" may terminate above and below in broad horizontal layers of more ordinary convection, and sequences of these fingering and convecting regions have been constructed (Turner 1967; Stern & Turner 1969; Shirtcliffe & Turner 1970; Linden 1971; Lambert & Demenkow 1972; Linden 1973; Turner & Chen 1974; Linden 1978; Schmitt & Lambert 1979). Multiple layers are feasible in both cases, with regions of convection (with nearly uniform temperature and salinity) separated either by narrow diffusive interfaces or by broader zones of fingering, depending on the destabilizing agent [see Turner (1973,1974) for general reviews]. The feature common to both situations is that the mean temperature and salinity possess step-like vertical profiles.

Such an appearance of steps in the laboratory experiments remains the most persuasive argument for thermohaline convection causing some of the oceanic microstructure. But the issue is not settled: although some ocean profiles seem suitable for either fingering or diffusive layering to be occurring, no direct velocity measurements exist in the ocean to either clearly support or refute this idea of double-diffusive convection. It is partly in this spirit that Huppert & Turner (1972) caution that quantitative comparison of ocean data with laboratory experiments is rather premature: they further suggest that the prominent stepped structure in an ice covered antarctic lake (with a minimum of competing mechanisms) may provide a better link between small-scale laboratory thermohaline studies and actual large-scale field results at this stage. Although the use of laboratory results to infer convective transports in observed ocean profiles involves major extrapolations, Lambert & Sturges (1977) and Schmitt & Evans (1978) conclude that salt fingering plays a significant role in the vertical transport of heat and salt in the sites that they have studied.

c. Theoretical background

Previous theoretical studies of thermohaline convection, primarily in support of the laboratory observations, have several facets: First of all, linear stability studies (Stern 1960; Baines & Gill 1969), complemented by finite amplitude studies (Veronis 1965,1968; Straus 1972; Huppert & Moore 1976), clearly established the theoretical basis for either the double-diffusive or fingering instabilities.

Layer formation was taken up by Turner (1968) and Stern & Turner (1969), and they advanced partly dimensional theories for the salt and heat fluxes through such layers. Following on this work, Huppert (1971) studied the stability of an established series of double-diffusive layers, and suggested criteria (based mostly on observed fluxes) for the evolution and melding of the various layers; Linden (1974a) and Linden & Shirtcliffe (1978) have sought to explain the transport across such diffusive interfaces using a series of scaling arguments about boundary-layer structures.

As to fingering, Stern (1969) proposed that a matrix of salt fingers can be collectively unstable to internal gravity waves, and thus predicted a finite vertical extent for the fingering. Huppert & Manins (1973) developed limiting conditions for the appearance of salt-fingering at an interface. More recently, Stern (1976) has predicted the maximum buoyancy flux across a salt finger interface; Huppert & Linden (1976) have presented a theoretical and laboratory picture for the spectral signature of salt fingers; and Griffiths (1979b) has advanced a mechanistic model for the transition between the region of fingering and the large-scale convection that serves to bound it.

Fully nonlinear treatments to predict mean fields, velocities, and transports are still in early stages. Some upper bound estimates were sought by Lindberg (1971) for transports in thermohaline convection. Mean-field modal equations were used by Elder (1969) to show initial stages of layer formation. The most nonlinear solutions are those of Gough and Toomre (1979), who consider the diffusive case using single-mode equations and make some contact with the fluxes observed in the laboratory experiments.

In summary, though the laboratory experiments provide a strong stimulus for arguing that some oceanic layering is double-diffusive in character, our theoretical understanding of such multi-layered convection is presently still rather sketchy, mostly because the motions are complicated and the problem is very nonlinear.

II. THE PHYSICAL EXPERIMENT MODELED

Although our long-range goal in this research is directed toward understanding what role thermohaline convection may play in establishing certain oceanic microstructure, we must at this stage concentrate on trying to explain the results of the much simpler laboratory experiments. These experiments are quite suggestive of how real layers might form, and they provide quantitative data that would still be extremely difficult to obtain in the real ocean.

a. Rationale for simulating laboratory experiments

This study is concerned with the computer simulation of certain aspects of the laboratory experiments. We have chosen this route because we can deal with highly nonlinear behavior of the flow, and also because we can readily vary the experimental boundary conditions and perturbations to explore their sensitivity. But most important, this approach holds out the promise of letting us extrapolate laboratory phenomena into more geophysical settings, thus permitting us to partially bridge the gap between experiments and ocean data. Finally, the relative ease with which computer simulations can scan a broad range of parameters may make it possible to explore thermohaline convection as it might occur under very differing ocean conditions (arctic ice melting, strong evaporative driving, strong advection or shear). Eventually, we would like to be in a position to theoretically predict when such layering will occur, what should be its vertical structure and persistence in time, and what is the transport of heat and salt across such layers.

b. Rationale for two-dimensional numerical simulations

We shall attempt to simulate the fingering laboratory experiments of Turner (1967), where temperature (T) is stabilizing and salinity (S) is destabilizing. In these experiments a region of salt fingers is bounded above and below by layers of large-scale convection. The horizontal width of the salt fingers corresponding to most laboratory values of ΔT and ΔS is ~0.2–0.3 cm and their vertical extent about 10 cm, whereas the dimensions of the tanks are about 50 cm or more. The adequate resolution of the fingers over their full growth cycle by a finite-difference grid imposes a very severe demand on computer storage. This stems from the fact that the location of the entrainment region and sharp gradients near the end of the fingers traverse the whole fluid during the growth cycle, making the use of stationary stretched grids in the vertical ineffectual.

We have sought to make the problem computationally tractable by using two-dimensional simulations in which the fingers appear as sheets of rising and

falling fluid with no variation in one of the horizontal coordinates. Salt fingers usually have a horizontal planform more like that of a square (Shirtcliffe &
Turner 1970), though even a weak imposed horizontal shear flow changes the geometry into one of rolls aligned with the flow. Linden (1974b) finds that the
fluxes and typical horizontal scales in such two-dimensional fingers appear to
be unchanged from the more usual three-dimensional ones. Thus we feel that the
computational convenience of dealing with two instead of three spatial dimensions
does not seriously compromise the physics.

c. <u>Relevant parameters</u>

To make the problem definitive, we have studied motions in a horizontal layer
of overall depth D and width L, bounded by two horizontal planes on which suitable temperature and salinity boundary conditions are imposed. The horizontal
coordinate is x and the vertical z, with the variables in the ranges $0 \leqslant x \leqslant L$
and $0 \leqslant z \leqslant D$. Initially the fluid is divided into two uniform quiescent layers,
differing in temperature by ΔT and in salinity by ΔS, with the upper layer the
hotter and more saline. An incompressible fluid in Boussinesq approximation is
considered, with the density ρ satisfying $\rho = \rho_0[1 - \alpha(T-T_0) + \beta(S-S_0)]$. With
the upper boundary (z = D) maintained at constant temperature $T = T_0 + \Delta T/2$ and
constant salinity $S = S_0 + \Delta S/2$, and the lower boundary (z = 0) likewise at
$T = T_0 - \Delta T/2$ and $S = S_0 - \Delta S/2$ (although flux boundary conditions might be
imposed instead), the problem is entirely described by the four dimensionless
parameters of

$$\text{a Rayleigh number} \qquad R = g\alpha\Delta T d^3/\kappa\nu \qquad , \qquad (2.1a)$$

$$\text{a salinity Rayleigh number} \qquad R_s = g\beta\Delta S d^3/\kappa_s\nu \qquad , \qquad (2.1b)$$

$$\text{a Prandtl number} \qquad \sigma = \nu/\kappa \qquad , \qquad (2.1c)$$

$$\text{and a diffusivity ratio} \qquad \tau = \kappa_s/\kappa \qquad . \qquad (2.1d)$$

Further, a stability parameter can be defined as

$$\Lambda = \alpha\Delta T/\beta\Delta S \qquad , \qquad (2.2)$$

and this is sometimes called the density anomaly ratio. The buoyancy flux ratio
is

$$\chi = \alpha F/\beta F_s \qquad , \qquad (2.3)$$

where F and F_s are the vertical heat and salt fluxes. Here g is the gravitational acceleration, α the coefficient of thermal expansion, β the salinity
density coefficient, ν the kinematic viscosity, while κ and κ_s are the diffusivities of heat and salt; all these coefficients are assumed constant in the

layer. The diffusivity ratio for this heat-salt system is taken to be $\tau = 10^{-2}$, and with water as the fluid medium, the Prandtl number has the value 6.8.

d. Choice of parameter values

Fingering occurs when $\Delta T, \Delta S < 0$, with the values so chosen that $\Delta \rho < 0$ and $\Lambda > 1$. Our particular choice of ΔT and ΔS is based on the following considerations: It has been observed both in the laboratory and in the numerical experiments that the vertical velocity of the fluid in the fingers $w \propto \Delta S$, hence the time for the fingers to reach a given height is $t \sim \Delta S^{-1}$. On the other hand, a numerical stability criterion limits the size of the time-step with which one can march, yielding $\Delta t \sim N^{-1} \sim \Delta S^{-1/2}$, where N is the Brunt-Väisala frequency of internal waves. The number of time steps one has to calculate until the fingers reach a given height is then $n = t/\Delta t \sim \Delta S^{-1/2}$, leading to the desirability of choosing large values of ΔS.

To keep the numerical simulations relevant to the laboratory experiments, we shall use the same range of the stability parameter Λ, with $2 \leq \Lambda \leq 10$, but generally employ higher values of ΔS (and thus ΔT) than reported in the laboratory studies. An upper bound to ΔT (and thus ΔS) is provided by the dependence of the finger width δ on ΔT, with $\delta \sim \Delta T^{-1/4}$. Thus fingers which are very thin would increase the spatial mesh needed to resolve them and hence rapidly increase the computational effort. Typically, the vertical gradient T_z shall range from about 1°C/cm to 5°C/cm for $\Delta S = 0.30\%_o$.

The horizontal width L of the computational domain is so chosen that either 8 or 16 fingers can be simulated; since these are only a small subset of the hundred or so fingers observed in the laboratory experiments, periodic boundary conditions are employed in the horizontal. The vertical depth D of the domain is chosen to be twice the width, with $D = 2L$.

In order to resolve the very narrow salinity features, we have chosen $\Delta x = \Delta z = 0.02$ cm and $L = 1.25$ cm or 2.50 cm, with $D = 2.50$ or 5.00 cm. The corresponding grid sizes are 64×128 and 128×256.

e. Simplified analysis

Some of the gross properties of salt fingers can be deduced from just considering linearized equations, and the results will be found to be reasonably accurate in the middle of a fingering region. To accomplish this, one balances vertical advection of heat and salt by their respective horizontal diffusion rates, and the buoyancy forces by horizontal viscous dissipation. Thus

$$\overline{wT}_z = \kappa \frac{\partial^2 T'}{\partial x^2} \quad ; \quad \overline{wS}_z = \kappa \frac{\partial^2 S'}{\partial x^2} \quad ; \quad -g(\alpha T' - \beta S') = \nu \frac{\partial^2 w}{\partial x^2} \quad . \quad (2.4)$$

Choosing a form for the vertical velocity of $w(x) = w_o \sin (2\pi x/\delta)$ and similar expressions for the fluctuating temperature T' and salinity S', one can deduce that the horizontal width of the fingers is

$$\delta \sim \left(\frac{\alpha g \overline{T}_z}{\kappa \nu}\right)^{-1/4} \left(1 - \frac{1}{\Lambda}\right)^{-1/4} \quad , \quad (2.5)$$

as in the manner of Linden (1973) or in a slightly modified form by Lambert & Demenkow (1972). The overbar here implies horizontal averaging and therefore $T' = T - \overline{T}$. Thus the finger width should vary as $\overline{T}_z^{-1/4}$, which is comparable to the scale of the fastest growing cells from linear stability theory.

f. Special features of the numerical experiments

There are four specific aspects to the numerical simulations that should be mentioned. First, the spectrum of the initial perturbation is imposed as either a pure sine wave or as white noise with random phases but equal Fourier amplitudes; just which choice is used turns out to have some influence on the finger shapes and widths. Second, the initial thickness of the horizontal interface between the two quiescent layers is taken to be 4 mesh points in the vertical, across which the mean temperature and salinity change linearly with z. Third, in the absence of stirring such an initial step in temperature is not maintained for long; the rapid thermal diffusion rate leads to the temperature field becoming nearly linear over the full height of the domain before much nonlinear fingering activity has been established. This implies that the local value of Λ is decreasing during the early stages of the experiment. Fourth, the relatively modest vertical extent of the computational domain will allow us to only concentrate on the development of the central fingering region; its interaction with large-scale convection layers above and below will be the subject of future investigations.

III. THE NUMERICAL MODEL

a. Equations of motion

The relevant equations for the two-dimensional, nonlinear thermohaline convection problem, using the Boussinesq approximation, are

$$\frac{Du_i}{Dt} = -\frac{\partial p}{\partial x_i} + g \, \delta_{i2}(\alpha T - \beta S) + \nu \nabla^2 u_i \qquad i = 1,2 \qquad (3.1a,b)$$

$$\frac{DT}{Dt} = \kappa \nabla^2 T \quad , \quad \frac{DS}{Dt} = \kappa_s \nabla^2 S \qquad (3.2a,b)$$

$$\frac{\partial u_i}{\partial x_i} = 0 \quad , \quad \frac{D}{Dt} \equiv \frac{\partial}{\partial t} + u_i \frac{\partial}{\partial x_i} \qquad (3.3a,b)$$

where the molecular diffusivities ν for vorticity, κ for heat and κ_s for salt are assumed to be constants.

Eliminating p from the two components of (3.1) and using (3.3a) yields the transport equation for the vorticity, $\xi = \partial u/\partial z - \partial w/\partial x$, of

$$\frac{D\xi}{Dt} = g\left(\alpha \frac{\partial T}{\partial x} - \beta \frac{\partial S}{\partial x}\right) + \nu \nabla^2 \xi \quad . \qquad (3.4)$$

Introducing the streamfunction ψ such that

$$\frac{\partial \psi}{\partial z} = u \quad , \quad - \frac{\partial \psi}{\partial x} = w \quad , \qquad (3.5)$$

we get the relation between ξ and ψ of

$$\nabla^2 \psi = \xi \quad . \qquad (3.6)$$

With the help of (3.5) the advection terms in (3.4) may be rewritten as a Jacobian

$$(\vec{u} \cdot \vec{\nabla})\xi \rightarrow \frac{\partial \psi}{\partial z} \frac{\partial \xi}{\partial x} - \frac{\partial \psi}{\partial x} \frac{\partial \xi}{\partial z} \rightarrow J(\psi,\xi) \quad . \qquad (3.7)$$

We now introduce a finite-difference formulation in space and time of the transport equations (3.2a,b) and (3.4); in each of the advective terms the expression (3.7) is substituted. For the purpose of illustration we confine ourselves to a discussion of the vorticity equation (3.4), since the transport equations for heat and salt (3.2a,b) contain similar terms.

b. Time and space differencing

Let the time and spatial coordinates be discretized as

$$t \rightarrow t^n = n\Delta t \quad , \quad x \rightarrow x_i = i\Delta x \quad , \quad z \rightarrow z_j = j\Delta z$$

so that $\xi(t,x,z) \rightarrow \xi(n\Delta t, i\Delta x, j\Delta z) = \xi_{i,j}^n$ will be a short-hand notation for all dependent variables. In our approach the pairs of variables ξ,ψ and T,S are defined on different sets of grid points, called the "staggered" grid approach. This grid is displayed in Figure 1. The actual horizontal boundaries of the fluid occur where $\xi=\psi=0$, with an extra ring of T,S points outside the $\xi=\psi=0$ lines. The vertical boundaries represent periodic surfaces; actually, two columns of points are needed at each end to specify periodicity for both a variable and its gradient (this holds for both the ξ and T grids).

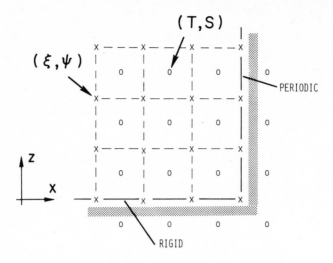

Fig. 1. Staggered finite-difference spatial grid in x and z. Vorticity (ξ) and streamfunction (ψ) are computed on the × points; temperature (T) and salinity (S) are evaluated on the o points.

We begin by considering the time marching scheme. The well known "leap-frog" scheme is applied to the advection and buoyancy terms (Lilly 1965):

$$\frac{\xi_{i,j}^{n+1} - \xi_{i,j}^{n-1}}{2\Delta t} = -J(\psi,\xi)_{i,j}^{n} + g(\alpha T - \beta S)_{i,j}^{n} \quad . \tag{3.8}$$

The friction terms are differenced by the so called "DuFort-Frankel" scheme (Richtmyer & Morton 1967, p. 190). This scheme depends inherently on the spatial differencing as well:

$$\frac{\xi_{i,j}^{n+1} + \xi_{i,j}^{n-1}}{2\Delta t} = \nu \left\{ \frac{\xi_{i+1,j}^{n} + \xi_{i-1,j}^{n} - \xi_{i,j}^{n+1} - \xi_{i,j}^{n-1}}{(\Delta x)^2} \right.$$

$$\left. + \frac{\xi_{i,j+1}^{n} + \xi_{i,j-1}^{n} - \xi_{i,j}^{n+1} - \xi_{i,j}^{n-1}}{(\Delta z)^2} \right\} \quad . \tag{3.9}$$

Because the left-hand sides of (3.8) and (3.9) are identical, we can simply add the time schemes to get the complete time marching scheme.

We must further indicate the spatial differencing used in the Jacobians for the advection of T, S, and ξ. A complete discussion of the various conservative formulations of (3.7) can be found in Lilly (1965), Arakawa (1966) and Williams (1967). For $J(\psi,T)$ and $J(\psi,S)$ a form is chosen that conserves T, T^2 and S, S^2, respectively. For $J(\psi,\xi)$ a form is chosen (actually a linear combination of three forms) that conserves ξ, ξ^2 and total kinetic energy $u^2 + w^2$.

c. Truncation error

The truncation error associated with the leap-frog time differencing is $O[(\Delta t)^2]$, the leap-frog being a centered, second-order scheme. The errors associated with the spatial differencing of the advection Jacobians J_2 and J_3 are shown by Arakawa (1966) to be $O[(\Delta x)^2]$, since they are centered, three-point differences. The error associated with the DuFort-Frankel scheme is $O[(\Delta t)^2/(\Delta x)^2]$.

d. Time step limitations

In the numerical solution of initial value problems that are governed by partial differential equations, stability of the solutions often requires that certain relations concerning the time step Δt be observed. The Von Neumann stability condition (Richtmyer & Morton 1967, p. 263) for the system (3.2) and (3.4) reduces essentially to two time step constraints; one involving the advection terms and one involving internal gravity waves. These conditions are:

$$\Delta t \leqslant \frac{\Delta x}{|u| + \frac{\Delta x}{\Delta z} \cdot |W|} \tag{3.10}$$

$$\Delta t \leqslant \frac{\sqrt{2}\,\pi \cdot \Delta x}{\left(\alpha g\, |\frac{\partial T}{\partial z}| + \beta g\, |\frac{\partial S}{\partial z}|\right)^{1/2}} \cdot \tag{3.11}$$

However, the DuFort-Frankel scheme used for the diffusion terms has no stability condition associated with it.

e. Solution of the Poisson equation

The solution of the Poisson equation (3.6) for ψ is accomplished by a direct inversion scheme based on Fourier series. The method is discussed in detail by Williams (1969). One expands both ψ and ξ in Fourier series

$$\begin{Bmatrix} \psi \\ \xi \end{Bmatrix} = \sum_{m=1}^{\frac{M}{2}-1} \begin{Bmatrix} a^m(z) \\ c^m(z) \end{Bmatrix} \cdot \sin \frac{2\pi m x}{L} + \sum_{m=0}^{\frac{M}{2}} \begin{Bmatrix} b^m(z) \\ d^m(z) \end{Bmatrix} \cdot \cos \frac{2\pi m x}{L} \tag{3.12}$$

and substitutes into (3.6) to obtain a set of M ordinary differential equations

$$\frac{d^2 a^m}{dz^2} - \frac{4\pi^2 m^2}{L^2} a^m = c^m \qquad m = 1,2,\ldots,\frac{M}{2}-1 \tag{3.13a}$$

$$\frac{d^2 b^m}{dz^2} - \frac{4\pi^2 m^2}{L^2} b^m = d^m \qquad m = 0,1,\ldots,\frac{M}{2} \cdot \tag{3.13b}$$

On a finite grid, one can simply employ discrete Fourier series, e.g. $\sin[2\pi m(i\Delta x)/L]$ with $L = M\Delta x$. Thus the total number of modes utilized in the expansion in the x (horizontal) direction is equal to the number of mesh points M on which ξ, T and S are defined. The expansion in (3.12) corresponds to periodic boundary conditions, since we are only sampling a small portion of the finger structure.

Equations (3.13) in turn are solved by finite differences, using a matrix inversion technique. Thus, (3.13a) becomes

$$\frac{a_{j+1}^m + a_{j-1}^m - 2a_j^m}{(\Delta z)^2} - \frac{4\pi^2 m^2}{L^2} a_j^m = c_j^m \tag{3.14}$$

for $j = 2,3,\ldots,N-1$ and $m = 1,2,\ldots,(M/2) - 1$. These can be written in matrix form as $A_{kj}^m a_k^m = c_j^m$, where A has the elements

$$\begin{bmatrix}
\beta_2 & \gamma_2 & & & & \\
\alpha_3 & \beta_3 & \gamma_3 & & & 0 \\
& \alpha_4 & \beta_4 & \gamma_4 & & \\
& & \cdot & \cdot & \cdot & \\
0 & & \alpha_{N-2} & \beta_{N-2} & \gamma_{N-2} \\
& & & \alpha_{N-1} & \beta_{N-1}
\end{bmatrix} \tag{3.15}$$

with

$$\alpha_i = \gamma_i = \frac{1}{(\Delta z)^2} \quad , \quad \beta_i = -\left[\frac{2}{(\Delta z)^2} + \frac{4\pi^2 m^2}{L^2}\right] \quad . \tag{3.16}$$

The matrix A is said to be of "tridiagonal" form, with non-zero elements only on the diagonal and two lines adjacent to it on either side. The system is closed by specifying a_1^m and a_N^m from the boundary conditions. We have chosen the stream-function $\psi = 0$ at the top and bottom, with no mass flow through them, so that $a_1^m = a_N^m \equiv 0$ for all m. An especially efficient numerical method to solve the system of equations (3.14) where A has the form (3.15) is given by Varga (1962, p. 95).

f. Mesh size and resolution requirements

The convective motions in salt fingers are driven by the net buoyancy resulting from the large differences in the conductivities of heat and salt. Thus it is essential that the thin diffusive layers between the adjacent fingers, and between the tips of the fingers and the bounding layers, be adequately resolved by the finite difference grid. A grid spacing of $\Delta x = \Delta z = 0.02$ cm yields such resolution, for this places about 8 mesh points laterally across each salt finger in our simulations. We use comparable grid spacing in the vertical to ensure that the bulbous ends of the fingers are suitably resolved.

g. Boundary conditions

The boundary conditions at the top and bottom provide fluxes of heat and salt but are impermeable to fluid motions; in addition, the surfaces act as frictionless lids on which no stress (and hence no vorticity) can be exerted. Thus

$$T = \pm \Delta T/2$$
$$S = \pm \Delta S/2 \qquad \text{at} \quad z = 0, D \qquad (3.17)$$
$$\xi = \psi = 0$$

where ΔT and ΔS are the initial temperature and salinity jumps across the interface. At $x = 0$, L periodic boundary conditions are imposed on all four dependent variables. For a finite difference scheme, this implies that both the function and its derivative are equal at the two ends for second order accuracy. This is accomplished for instance by setting

$$T^n_{1,j} = T^n_{M-1,j}$$
$$T^n_{M,j} = T^n_{2,j} \qquad \cdot \qquad (3.18)$$

Since the interior points obey the usual forecast scheme, such as (3.8) and (3.9) for vorticity, the use of (3.18) completes the time-step procedure.

h. Initial conditions

The initial state of the system consists of two layers of fluid at rest, each homogeneous at the value of T and S indicated in (3.17). The change in both T and S across the two layers is accomplished by means of linear vertical gradients spread over 4 mesh vertical points. At $t = 0$ a perturbation is applied to the temperature and salinity fields that consist of one or more horizontal wavenumbers with equal amplitude but random phase distribution. The vertical distribution is a half sine wave peaked at the interface. We can represent it as

$$T'(x,z) = \sin \left(\frac{\pi z}{d}\right) \cdot \sum_{m=\ell_1}^{\ell_2} A_T \sin \left(\frac{2\pi m x + \phi_m}{L}\right) \qquad (3.19)$$

where ℓ_1 and ℓ_2 are so chosen that the shortest wavelength $\lambda_{\ell_2} \simeq 2\Delta x$, and the largest $\lambda_{\ell_1} \simeq L/2$, with L being the horizontal extent of the box. Typically, the amplitude of the perturbations in temperature A_T and in salinity A_S is ~0.5% of the interfacial jumps ΔT and ΔS. We will find that physically interesting solutions can be obtained by even employing only one perturbation wavenumber.

IV. RESULTS AND DISCUSSION

The aim of this paper is to investigate the growth cycle of salt fingers in a
two-layer system, thus determining the evolution of the finite-amplitude motions
and the associated fluxes of heat and salt. Critical parameters for the onset of
convection and the preferred wavelengths are not investigated, although indirect
information is obtained on the latter by varying the spectrum of the initial perturbations. Nor are we able to study the formation of large-scale convection
layers above and below the fingering regions; it is hoped that the results of
the current study will enable us to design such a numerical experiment in the
future. What we have done is to carry out about 40 initial value calculations
for a variety of thermal and salinity Rayleigh numbers R and R_S, spanning a range
of stability parameters Λ from about 2 to 10, while imposing different initial
perturbations.

We shall begin by presenting some of the qualitative properties of the salt
fingers, illustrating this with a few typical examples of our numerical simulations. We shall find for instance that the shape of the fingers helps to clarify
how such motions entrain quiescent fluid. The contours of salinity serve as an
excellent indicator of the various stages in the time evolution. We then turn
to quantitative details revealed by the horizontally averaged vertical profiles
of temperature and salinity, the associated convective and total fluxes of heat
and salt, and the dependence of the buoyancy flux ratio χ on the stability parameter Λ.

a. Time evolution of fingers and mean profiles

We shall illustrate the growth of salt fingers by showing a time sequence of
salinity contours and vertical profiles of temperature and salinity. Figure 2
depicts such a sequence for $\Lambda = 3$ in which the initial perturbation is monochro-
matic; here $R = 10^6$ and the domain extends 2.50 cm in the vertical. The contours
and profiles are displayed on adjacent frames to depict the action of the fingers
in distorting the profiles. The solutions are displayed at 10 sec intervals from
t = 60 sec to t = 90 sec after starting the calculations from quiescent conditions. We note that the mean temperature profile, $\bar{T}$, is not as severely distorted as that of salinity, $\bar{S}$. This comes about because of the very significant
effect of thermal diffusion which has served to establish a nearly linear temperature field. There is a considerable penetration of the thermal front into
the quiescent regions above and below the fingers, so that there is no detectable
trace of the end of the fingers on the mean temperature profile $\bar{T}$.

As the fingers grow, they develop irregularities near their extremities. In
time the tips of the fingers deform into bulbous shapes that eventually separate

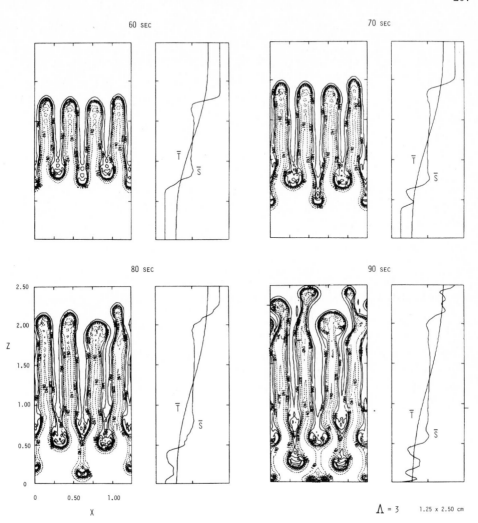

Fig. 2. Evolution in time of salt fingers when the stability parameter $\Lambda = 3$. Shown are salinity contours (on the left) and vertical profiles of mean temperature $\bar{T}$ and mean salinity $\bar{S}$ (on the right) at the four times indicated. Here the imposed temperature difference across the layers is $\Delta T = 4.42°C$ and the salinity difference is $\Delta S = 0.30\%$; the computational domain has dimensions $D = 2.50$ cm and $L = 1.25$ cm. The strong mixing effect of the fingers on the salinity field should be noted, making the mean field $\bar{S}$ nearly isohaline. In contrast, thermal leakage between the fingers contributes to produce a mean $\bar{T}$ which is more linear in the fingering region. Thermal diffusion also causes rapid penetration of the initial temperature jump into the isothermal layers above and below the fingers.

and move away from the leading edge of the fingers; we shall denote this region as the entrainment interface. Because the number of fingers explicitly calculated is small, the presence of each bulb is evident in the $\bar{S}$ profile. We note that the mean salinity field $\bar{S}$ is nearly isohaline in the region occupied by the fingers. However, the contour plots reveal that the fluctuations in salinity between adjacent fingers possess the full contrast ΔS. This suggests that the advection of salt is the primary transport mechanism, with very little diffusion of salt between the fingers. Close inspection of the mean salinity profiles reveals the presence of even a weak negative gradient just inside the entrainment zones.

The next two experiments contrast finger evolution associated with different values of the stability parameter $\Lambda = \alpha\Delta T/\beta\Delta S$. Figure 3 illustrates another $\Lambda = 3$ simulation, but here a multi-mode initial perturbation was used to start the calculations, unlike that imposed in Figure 2. Figure 3 shows successive salinity contour plots at 10 sec intervals from t = 70 sec to t = 100 sec for $\Lambda = 3$. On the other hand, Figure 4 presents a sequence at 20 sec intervals from t = 50 sec to t = 110 sec for the contrasting $\Lambda = 5$ example. A much faster growth rate for the fingers is evident in the $\Lambda = 3$ experiment, but this is not surprising since the stabilizing temperature gradient there has a smaller value. We stopped the calculations once the fingers arrived at the upper and lower boundaries. Not only are the vertical velocities greater in the $\Lambda = 3$ example, but the appearance of the fingers is also considerably more irregular.

b. Bulb and drop formation

After the fingers have grown to about half the height of our computational domain in Figure 3, the leading edges of the fingers begin to develop decidedly bulbous shapes. After t ≃ 80 sec some of the bulbs even detach themselves from the fingers and become distinct drops or elements which travel in advance of the main fingering region. At least a partial explanation for the bulbs lies in the fact that the fluid inside the main body of the fingers has a larger vertical velocity than the movement of the entrainment interface, thus either piling up salt or depleting it from the ends of the fingers. Once bulbs begin to separate from most fingers, they constitute a new turbulent entrainment process whose net transport properties we have so far only determined approximately. Because of the small number of bulbs involved in our experiments, almost every bulb makes a distinct imprint on the horizontally averaged $\bar{T}$ and $\bar{S}$ profiles, and their collective transport has not been calculated in detail. It is however clear that such bulb formation is particularly prominent at the smaller values of Λ.

It appears that both a broad spectrum of initial perturbations as well as small values of Λ contribute to bulb formation and more irregular finger growth: The experiments with $\Lambda = 3$ reveal vigorous bulb formation, whereas the $\Lambda = 5$ case shows much less of such a tendency. However, the modest vertical extent of the computational domain has probably interfered with the development of the bulbs in the $\Lambda = 5$ simulation of Figure 4, for we have found that layers of twice the

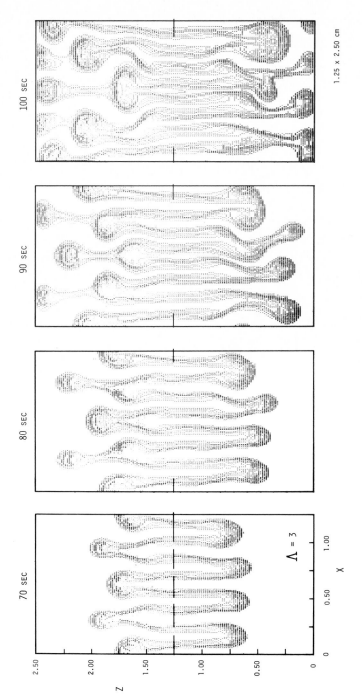

Fig. 3. Time sequence of salinity contours for a numerical simulation with
Λ = 3 and parameters much as in Fig. 2, but the initial perturbation at t = 0
involved several wavelengths. Note the vigorous bulb formation at the tips of
the advancing salt fingers, culminating in the separation of distinct elements
or drops.

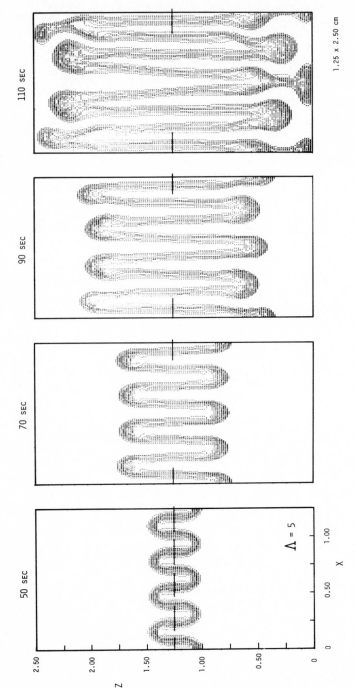

Fig. 4. Time sequence of salinity contours for a numerical simulation with
Λ = 5. Here ΔT = 7.35°C; the other parameters are as stated in Fig. 2. Note
the delayed formation of the bulbous regions, in contrast to the Λ = 3 case.

depth do display distinct bulb formation. It appears that this process at higher
Λ values is delayed until the fingers attain a greater length, though once the
process starts it proceeds with considerable vigor. Such variation with Λ is
readily explained. Larger values of the stabilizing temperature contrast ΔT (and
hence Λ) serve to decrease the vertical velocities in the middle of the fingers,
for the effective buoyancy force is reduced by the presence of larger temperature
fluctuations within the fingers. Consequently there is less fluid piling up at
the ends at any given time.

c. Vorticity distribution

Figure 5 displays vorticity contours that accompany the t = 60, 80 and 90 sec
time frames for the Λ = 3 time evolution presented in Figure 2. The pictures reveal that vorticity is concentrated in the shear regions between the upward and
downward directed fingers and around the bulbs at the ends of the fingers. Based
on the (relatively) high diffusion rate of momentum, vorticity contours might be
expected to be even more diffuse than the isotherms, but in fact they have much
of the appearance of the salinity field. Their concentrated nature can be explained by a close coupling between the buoyancy driving force and the diffusion
of vorticity. For heat the balance is predominately $\kappa \nabla^2 T \sim (\vec{u} \cdot \vec{\nabla}) T$ and for vorticity, $\nu \nabla^2 \zeta \sim g\beta(\partial S/\partial x)$. As a source term for the Laplacian, $(\vec{u} \cdot \vec{\nabla}) T$ gives rise
to no sharp gradient regions but $(\partial S/\partial x)$ certainly does, as can be judged from
the salinity contours.

d. Finger shape and width

The horizontal width of the fully developed fingers, δ, may be expected to be
proportional to $\bar{T}_z^{-1/4}$. The finger widths obtained in the numerical experiments
shown in Figures 2, 3 and 4 are all approximately 0.16 cm. Our changes here in
the value of Λ (and hence ΔT) represent only a factor of about 3, so we are not
able to meaningfully test the sensitivity of δ to Λ, but the values are close to
those reported by Linden (1973).

Figure 6 shows typical profiles of vertical velocity w and of temperature T'
and salinity S' fluctuations in a horizontal cut across the fingers. The profiles are those of the Λ = 5 solution in Figure 4 at t = 70 sec, with the cross
section formed just above midlayer at the level z = 1.50 cm. It is quite evident that whereas w and T' show sinusoidal variation with x, the salinity field
S' has a much squarer profile. This arises primarily from the small diffusion
rate for salt, the full ΔS contrast having been impressed upon the fingers from
their ends. Such a profile for S' is different from the form usually assumed in
deriving solutions to equation (2.4).

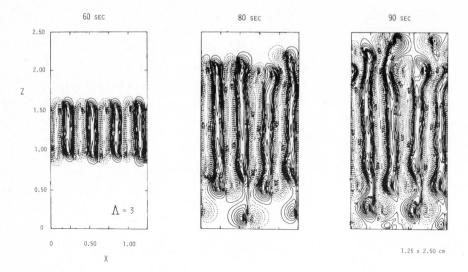

60 SEC   80 SEC   90 SEC

Λ = 3

1.25 x 2.50 cm

Fig. 5. Vorticity contours at three instants in time to accompany the Λ = 3 simulation in Fig. 2. The vorticity has an appearance somewhat similar to the salinity contours, with the vorticity concentrated in the shear regions between the fingers and near the separated drops.

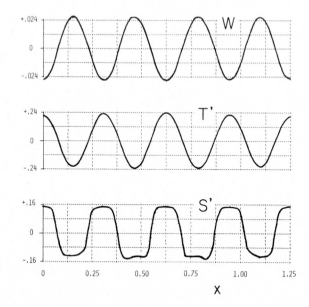

Fig. 6. Horizontal cross section through the fingers just above midlayer (at z = 1.50 cm) in the t = 70 sec frame of the Λ = 5 simulation presented in Fig. 4. Shown are the vertical velocity w and the fluctuating temperature T' and salinity S' as a function of horizontal position x. The profiles for w and T' in the fingering region are largely sinusoidal, but that of S' is distinctly more square in shape.

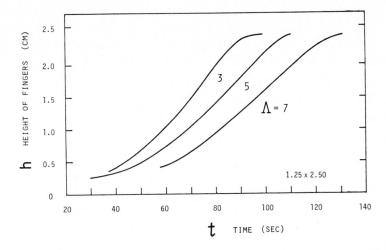

Fig. 7. Growth of the height of the fingers, h, with time in the simulations shown in Figs. 3 and 4, and for an additional one at $\Lambda = 7$. Note the initial exponential growth, tending toward a linear growth rate dh/dt which is maintained until the fingers impact the top and bottom boundaries. Growth rates decrease with increasing Λ as $dh/dt \sim \Lambda^{-3/5}$.

e. Growth rate of fingers

Figure 7 displays the growth rate of the fingers as a function of time for the experiments with $\Lambda = 3$ and 5 shown in Figures 3 and 4, and for a further simulation with $\Lambda = 7$. The finger length h is defined here as the vertical separation of the two entrainment interfaces, which can be determined from the vertical extent of the nearly isohaline region. A relation $dh/dt \sim \Lambda^{-3/5}$ can be deduced from the results shown in Figure 7. Finger growth in each case was terminated by the fingers reaching the top and bottom boundaries. We have not yet determined the height to which such two-dimensional fingers can grow if unimpeded by our bounding surfaces. Even our calculations with vertical domains of D = 5.0 cm did not reveal any collective instabilities that limited finger growth, though bulb formation was very much in evidence. Thus our calculations so far have not yet clarified whether the limiting mechanism to finger growth is the internal gravity wave instability advocated by Stern (1969), or whether the increasing formation of bulbs contributes to the onset of large-scale convection above and below.

f. Buoyancy flux ratio variation with Λ

One of the significant relations reported in the laboratory experiments is the dependence of the heat and salt buoyancy flux ratio $\chi = \alpha F/\beta F_S$ on the stability parameter $\Lambda = \alpha\Delta T/\beta\Delta S$. The total flux ratio χ in the experiments of Turner

(1967) appears to vary linearly with Λ. However, the convective flux ratio χ_c, with the effects of heat conduction subtracted (the diffusion of salt being negligible), appears to be nearly independent of Λ. Turner reports this value to be $\chi_c = 0.56 \pm 0.2$ over the range $4 \leqslant \Lambda \leqslant 9$. Linden (1971) has confirmed this value at $\Lambda \sim 4$; his method differed in that he did not use mechanical stirring to maintain a sharp interface. In contrast, Schmitt (1979) suggests that χ_c may decrease with Λ, for he measures values for χ_c of 0.2 ± 0.1 near $\Lambda = 10$, having started out at 0.65 near $\Lambda = 2$. The differences in the experimental results for χ_c cannot be easily resolved, though it is likely that the nature of the large-scale turbulent layers bounding the fingering regions plays a major role.

Figure 8 presents typical χ and χ_c values attained in the time evolutions discussed earlier for $\Lambda = 3$, 5 and 7. The total and convective flux ratios χ and χ_c both exhibit a linear behavior with Λ. The reader should note that values of χ are plotted on the right side of the figure, and values of χ_c on the left. This dependence of χ on Λ is not inconsistent with the experimental observations, but that of χ_c is discrepant.

Our results for the variation of χ_c with Λ are largely controlled by the temperature field. Due to the significant diffusion of heat, increasing ΔT or Λ in these numerical simulations (while keeping ΔS fixed) serves to proportionally increase $\bar{T}_z$ in the fingering region. The temperature fluctuations T' in the fingers are produced by the vertical advection of $\bar{T}_z$; thus $|T'| = \gamma \Delta T$, with the scaling factor γ fairly insensitive to Λ since the typical vertical velocity, say w_o, varies only slowly with Λ when ΔS is kept fixed. If the fingers were assumed to have a simple sinusoidal horizontal structure, then

$$\chi_c = \frac{\overline{\alpha w T'}}{\overline{\beta w S'}} \simeq \frac{\alpha w_o \, |T'|}{\beta w_o \, |S'|} \simeq \frac{\alpha \gamma \Delta T}{\beta \Delta S} \simeq \gamma \Lambda \quad . \tag{4.1}$$

The low diffusivity of salt results in a complete distortion and advection of the salinity field from one ambient layer into the other; therefore we have taken $|S'| \sim \Delta S$. Thus relation (4.1) serves to explain why χ_c in Figure 8 varies nearly linearly with Λ. Because of the presence of nonlinear processes, particularly near the entraining bulb regions, we do not expect γ to remain constant for all values of the stability parameter Λ. We have in fact found that the actual slope of χ_c with Λ, say γ_c, in a variety of our simulations is about 0.04 near $\Lambda = 3$ and 0.03 near $\Lambda = 10$, and similar values are attained for γ.

How do we account for such differences in the behavior between our χ_c and that of the laboratory experiments? Although the average value of our χ_c is not unpalatable, its increase with Λ has no counterpart in the laboratory results. The difference must rest with the fact that the fluid above and below our growing fingers is quiescent, whereas that in all the experiments is vigorously stirred, whether by convection or by oscillating screens. The coupling between the very

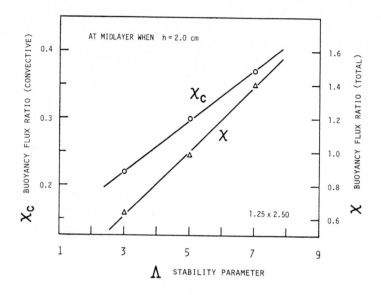

Fig. 8. Variation of the two buoyancy flux ratios with the stability parameter Λ in the three numerical simulations of Fig. 7. These typical values of χ and $χ_c$ were evaluated at midlayer when the fingers obtained a height of 2.0 cm. Both $χ_c$ and χ display a linear variation with Λ.

small scales of motion of the fingers to the large scales of the convection appears to occur through a transition region involving intermediate scales of motion (e.g. Stern 1976, Griffiths 1979b). The transition region may serve to modulate the proportion of ΔS that is available to be converted to S' by the fingers, doing so by effectively mixing fluid between ascending and descending fingers at their ends. The effects on ΔT and thus T' by the transition region would be less due to the importance of diffusion. The near constancy of $χ_c$ with Λ may arise from the enhanced mixing achieved at lower values of the stability parameter, with consequent reduction in the typical S' and in the convective salt flux as Λ decreases. Such a process is entirely absent in our present simulations. Several variants on the role of the transition region have been proposed, but at this stage they are all largely mechanistic models which try to explain the invariance of $χ_c$ with Λ. If instead $χ_c$ decreases with Λ, as Schmitt (1979) has indicated, then the scaling relations in the transition region must be re-examined; also, the variation of $χ_c$ with diffusivity ratio τ is a matter not fully resolved. Detailed dynamical modeling of the fingers and transition region is clearly required, though our numerical simulations so far have not yet been able to have a matrix of salt fingers coexist with large-scale convection.

V. CONCLUSIONS

We have performed a series of exploratory numerical experiments to study the nonlinear nature of salt finger evolution in two-layer systems. The following conclusions can be drawn:

1. The variation of both the total and convective flux ratios χ and χ_c is linear with the stability parameter Λ. The slope γ_c of the line $\chi_c \sim \gamma_c \Lambda$ has approximately the same values as the ratio γ of the amplitude of the perturbation temperature to the temperature contrast between the layers, $\gamma = |T'|/\Delta T$. Values for γ and γ_c range from 0.04 for $\Lambda = 3$ to 0.03 for $\Lambda = 7$.

2. After salt fingers grow to a certain height, they develop bulbs and drops near their ends that can detach and proceed ahead of the fingers. The tendency for bulb formation weakens as Λ increases. This process appears to result from the vertical velocities in the main body of the fingers being faster than that of the extension velocity dh/dt, leading to a piling up of salt at their ends.

3. The early growth rate of finger heights appears to be exponential, followed by a nearly constant growth rate. The latter growth rate $dh/dt \sim \Lambda^{-3/5}$.

4. The amplitude of the temperature perturbations T' and of the convective flux $\overline{wT'}$ vary as $\Lambda^{3/5}$.

5. The horizontal profiles of the fingers are sinusoidal in w and T', but decidedly more square in shape for S'.

Future work will attempt to initiate large-scale convection in layers above and below the fingering region. The differences in the buoyancy flux ratios between our present numerical results and those of laboratory experiments must arise from our evolving salt fingers simply terminating in quiescent regions of fluid. The transition regions between the fingers and the surrounding convection must serve to control the overall fluxes through the stepped structures in $\overline{T}$ and $\overline{S}$. However, an understanding of how the adjustment of scales of motion is accomplished is a matter of considerable difficulty. The numerical simulations provide some promise of helping to clarify this issue.

VI. ACKNOWLEDGMENTS

We are indebted to D.A. Moore for making available to us a fast Poisson solver for the streamfunction. Useful discussions with R.W. Griffiths, H.E. Huppert, P.F. Linden and J.S. Turner are gratefully acknowledged. We also thank L.B. Lin and G.O. Roberts for providing us with some of their plotting routines. This research was supported in part by the Office of Naval Research, first through the Naval Research Laboratory, and presently through the Naval Ocean Research and Development Activity. This work was also supported in part by the National Science Foundation Physical Oceanography Program through Grant DES74-14439 to the University of Colorado.

VII. REFERENCES

Arakawa, A. 1966 Computational design for long-term numerical integration of the equations of fluid motion: Two-dimensional incompressible flow. Part I. J. Comp. Physics, $\underline{1}$, 119-143.

Baines, P.G. & Gill, A.E. 1969 On thermohaline convection with linear gradients. J. Fluid Mech., $\underline{37}$, 289-306.

Cooper, J.W. & Stommel, H. 1968 Regularly spaced steps in the main thermocline near Bermuda. J. Geophys. Res., $\underline{73}$, 5849-5854.

Crapper, P.F. 1975 Measurements across a diffusive interface. Deep-Sea Res., $\underline{22}$, 537-545.

Degens, E.T. & Ross, D.A. 1969 (eds) <u>Hot brines and recent heavy metal deposits in the Red Sea</u>. Berlin: Springer-Verlag.

Elder, J.W. 1969 Numerical experiments with thermohaline convection. Phys. Fluids, $\underline{12}$, II, 194-197.

Elliot, A.J., Howe, M.R. & Tait, R.I. 1974 The lateral coherence of a system of thermohaline layers in the deep ocean. Deep-Sea Res., $\underline{21}$, 95-107.

Federov, K.N. 1978 <u>The Thermohaline Fine Structure of the Ocean</u>. (Technical ed.: J. S. Turner), Pergamon Press.

Foster, T.D. & Carmack, E.C. 1976. Temperature and salinity structure in the Weddell Sea. J. Phys. Oceanogr., $\underline{6}$, 36-44.

Gough, D.O. & Toomre, J. 1979 Single-mode theory of diffusive layers in thermohaline convection. J. Fluid Mech., submitted.

Gregg, M.C. 1976 Temperature and salinity microstructure in the Pacific Equatorial Undercurrent. J. Geophys. Res., $\underline{81}$, 1180-1196.

Gregg, M.C. & Cox, C.S. 1972 The vertical microstructure of temperature and salinity. Deep-Sea Res., $\underline{19}$, 355-376.

Griffiths, R.W. 1979a The transport of multiple components through thermohaline diffusive interfaces. Deep-Sea Res., $\underline{26A}$, 383-397.

Griffiths, R.W. 1979b The physics of salt finger interfaces. Preprint.

Hayes, S.P. 1975 The temperature and salinity fine structure of the Mediterranean water in the western Atlantic. Deep-Sea Res., $\underline{21}$, 1-12.

Huppert, H.E. 1971 On the stability of a series of double-diffusive layers. Deep-Sea Res., $\underline{18}$, 1005-1021.

Huppert, H.E. & Linden, P.F. 1976 The spectral signature of salt fingers. Deep-Sea Res., $\underline{23}$, 909-914.

Huppert, H.E. & Manins, P.C. 1973 Limiting conditions for salt-fingering at an interface. Deep-Sea Res., $\underline{20}$, 315-323.

Huppert, H.E. & Moore, D.A. 1976 Nonlinear double-diffusive convection. J. Fluid Mech., $\underline{78}$, 821-854.

Huppert, H.E. & Turner, J.S. 1972 Double-diffusive convection and its implications for the temperature and salinity structure of the ocean and Lake Vanda. J. Phys. Oceanogr., $\underline{2}$, 456-461.

Johannessen, O.M. & Lee, O.S. 1974 A deep stepped thermohaline structure in the Mediterranean. Deep-Sea Res., $\underline{21}$, 629-639.

Lambert, R.B. & Demenkow, J.W. 1972 On the vertical transport due to fingers in double diffusive convection. J. Fluid Mech., $\underline{54}$, 627-640.

Lambert, R.B. & Sturges, W. 1977 A thermohaline staircase and vertical mixing in the thermocline. Deep-Sea Res., $\underline{24}$, 211-222.

Lilly, R. 1965 On the computational stability of numerical solutions of time-dependent nonlinear geophysical fluid dynamic problems. Monthly Weather Rev., 93, 11-26.

Lindberg, W.R. 1971 An upper bound on transport processes in turbulent thermohaline convection. J. Phys. Oceanogr., 1, 187-195.

Linden, P.F. 1971 Salt fingers in the presence of grid-generated turbulence. J. Fluid Mech., 49, 611-624.

Linden, P.F. 1973 On the structure of salt fingers. Deep-Sea Res., 20, 325-340.

Linden, P.F. 1974a A note on the transport across a diffusive interface. Deep-Sea Res., 21, 283-287.

Linden, P.F. 1974b Salt fingers in a steady shear flow. Geophys. Fluid Dynam., 6, 1-27.

Linden, P.F. 1978 The formation of banded salt finger structures. J. Geophys. Res., 83, 2902-2912.

Linden, P.F. & Shirtcliffe, T.G.L. 1978 The diffusive interface in double-diffusive convection. J. Fluid Mech., 87, 417-432.

Magnell, B. 1976 Salt fingers observed in the Mediterranean outflow region (34°N, 11°W) using a towed sensor. J. Phys. Oceanogr., 6, 511-523.

Marmorino, G.O. & Caldwell, D.R. 1976 Equilibrium heat and salt transport through a diffusive thermohaline interface. Deep-Sea Res., 23, 59-67.

Mazeika, P.A. 1974 Subsurface mixed layers in the northwest tropical Atlantic. J. Phys. Oceanogr., 4, 446-453.

Neshyba, S., Neal, V.T. & Denner, W. 1971 Temperature and conductivity measurements under ice island T-3. J. Geophys. Res., 76, 8107-8120.

Pingree, R. 1970 In situ measurements of salinity, conductivity, and temperature. Deep-Sea Res., 17, 603-610.

Richtmyer, R.D. & Morton, K.W. 1967 Difference Methods for Initial-Value Problems, 2nd ed., Interscience Publishers.

Schmitt, R.W. 1979 Flux measurements of salt fingers at an interface. J. Marine Res., submitted.

Schmitt, R.W. & Evans, D.L. 1978 An estimate of the vertical mixing due to salt fingers based on observations in the North Atlantic Central Water. J. Geophys. Res., 83, 2913-2920.

Schmitt, R.W. & Lambert, R.B. 1979 The effects of rotation on salt fingers. J. Fluid Mech., 90, 449-463.

Shirtcliffe, T.G.L. & Turner, J.S. 1970 Observations of the cell structure of salt fingers. J. Fluid Mech., 41, 707-719.

Stern, M.E. 1960 The 'salt fountain' and thermohaline convection. Tellus, 12, 172-175.

Stern, M.E. 1969 Collective instability of salt fingers. J. Fluid Mech., 35, 209-218.

Stern, M.E. 1976 Maximum buoyancy flux across a salt finger interface. J. Mar. Res., 34, 95-110.

Stern, M.E. & Turner, J.S. 1969 Salt fingers and convecting layers. Deep-Sea Res., 16, 497-511.

Stommel, H. & Fedorov, K.N. 1967 Small-scale structure in temperature and salinity near Timor and Mindanao. Tellus, 19, 306-325.

Straus, J.M. 1972 Finite amplitude doubly diffusive convection. J. Fluid Mech., 56, 353-374.

Tait, R.I. & Howe, M.R. 1968 Some observations of thermohaline stratification in the deep ocean. Deep-Sea Res., 15, 275-280.

Tait, R.I. & Howe, M.R. 1971 Thermohaline staircase. Nature, 231, 178-179.

Turner, J.S. 1965 The coupled turbulent transports of salt and heat across a sharp density interface. Int. J. Heat Mass Transf., 8, 759-767.

Turner, J.S. 1967 Salt fingers across a density interface. Deep-Sea Res., 14, 599-611.

Turner, J.S. 1968 The behaviour of a stable salinity gradient heated from below. J. Fluid Mech., 33, 183-200.

Turner, J.S. 1973 Buoyancy effects in fluids. Cambridge Univ. Press.

Turner, J.S. 1974 Double-diffusive phenomena. Ann. Rev. Fluid Mech., 6, 37-56.

Turner, J.S. & Chen, C.F.   1974   Two-dimensional effects in double-diffusive con-
    vection.   J. Fluid Mech., 63, 577-592.
Varga, R.S.   1962   Matrix Iterative Analysis.   Prentice-Hall.
Veronis, G.   1965   On finite amplitude instability in thermohaline convection.
    J. Mar. Res., 23, 1-17.
Veronis, G.   1968   Effect of a stabilizing gradient of solute on thermal convec-
    tion.   J. Fluid Mech., 34, 315-336.
Williams, A.J.   1974   Salt fingers observed in the Mediterranean outflows.
    Science, 185, 941-943.
Williams, A.J.   1975   Images of ocean microstructure.   Deep-Sea Res., 22, 811-
    829.
Williams, G.P.   1967   Thermal convection in a rotating fluid annulus. Part I.
    The basic axisymmetric flow.   J. Atmos. Sci., 24, 144-161.
Williams, G.P.   1969   Numerical investigation of the three-dimensional Navier-
    Stokes equations for incompressible flow.   J. Fluid Mech., 37, 727-750.

FOSSIL TEMPERATURE, SALINITY, AND VORTICITY TURBULENCE IN THE OCEAN

CARL H. GIBSON[1]

[1]Department of Applied Mechanics and Engineering Sciences and Scripps
 Institution of Oceanography, University of California at San Diego, La Jolla,
 California 92093

ABSTRACT

Small scale fluctuations of temperature, salinity, and vorticity in the
ocean occur in isolated patches apparently caused by bursts of active turbulence. After the turbulence has been dampened by stable stratification the
fluctuations persist as "fossil turbulence". The persistence times and internal
structure of fossil turbulence are investigated by considering the evolution of
a patch of very strong turbulence in a stratified fluid after the source of
turbulent kinetic energy has been removed. A variety of parameters, scales,
and spectral properties are inferred from this model. Comparison with observations reveals that oceanic temperature microstructure below the mixing layer
is fossil temperature turbulence or a combination of active and fossil temperature turbulence; fully turbulent patches are not observed. Even though the
mixing rate is greatest in the most active temperature patches, the degree of
fossilization is also greatest. The persistence of the fossil patches increases
with the parameter γ_o/N, which is the rate of strain of turbulence at the time
of fossilization compared to the Brunt-Väisälä frequency. Information about
the original turbulence, such as γ_o and the size and location of the turbulent
region, is preserved in the fossil turbulence structure.

1. INTRODUCTION

The term "fossil turbulence" refers to remnants of turbulence in fluid which

is no longer turbulent. References to fossil turbulence have been in the liter-

ature for nearly ten years, but remarkably little has been done to document the

existence of the phenomenon or to develop a theoretical description of its

evolution or internal structure. Existing descriptions are obscure because

discussions of turbulence in stably stratified fluids are often vague and

ambiguous about the definition of turbulence and make no clear distinction

between turbulence and internal wave motions at small scales. In the following,

an attempt will be made to clarify the terminology and definitions of various

forms of fossil and active turbulence in stratified fluids by referring to a

simplified model of the actual process, i.e., the evolution of a patch of strong

turbulence in a region of uniform vertical density gradient.

A variety of length, time, and scalar scales emerge from a comparison of buoyancy, inertial, and viscous forces as well as the convection and diffusion velocities of the scalar fields. These scales represent universal similarity laws for fossil turbulence fields analogous to the generally accepted similarity laws for unstratified turbulent mixing. A theory is developed for velocity and scalar gradient spectra at various stages of evolution of the model patch of stratified turbulence. Based on the model spectra, a method for determining the degree of activity of turbulence in a region of fluid containing microstructure is suggested. Methods of discriminating between active and fossil turbulence are important for estimating vertical transport rates from local dissipation rates and local mean vertical gradients, or for characterizing the dominant physical processes in the region where the microstructure is observed.

The first published use of the term "fossil turbulence", and the only published comparison with data, may be found in the December 1969 issue of Radio Science which is entirely devoted to the Proceedings of a Colloquium on Spectra of Meteorological Variables, Stockholm, June 1969. Radar backscattering from the optically clear atmosphere, Ottersten (p. 1247) and Richter (p. 1261), indicated extensive volumes of small scale refractive index fluctuations apparently associated with clear air turbulence (CAT). The braided, wavelike geometry of the scattering volumes suggested that breaking internal waves were the source of turbulence. The strong, stable stratification and wide distribution of the scattering volumes suggested the possibility of fossil turbulence. Recent acoustic backscatter measurements reveal similar features associated with oceanic internal waves (Orr 1978).

A workshop on fossil turbulence was organized at the Colloquium and chaired by Woods. The panel report is included in the Proceedings (Woods, Ed. p. 1365). By noting that the CAT scattering layers had considerable reflectivity over a wide range of scattering angles, the panel inferred that fossil turbulence may be approximately isotropic at scales of order 10 cm. Observation of forward scattering, however, showed considerable anisotropy at scales of several meters. Bussinger and Kaimal reported observations of velocity fluctuations in fossil temperature turbulence in the stable atmospheric boundary layer formed in the evening. Vertical fluctuations were strongly suppressed compared to horizontal fluctuations. Apparently, it was inferred that one property of fossil turbulence is that after the turbulence velocity fluctuations decay, the remaining fluid motions are small and have no effect on the fossil turbulence structure. The report concludes with the statement that "It seems unlikely that fossil turbulence will have a unique spectrum, even immediately after it has been formed". As will be shown in the present paper, the assumption that fossil turbulence is frozen into a stationary fluid and evolves only by molecular diffusion may be incorrect. According to the present analysis, small

scale, internal wave-like motions play a crucial role in determining the small
scale structure of fossil turbulence, and may actually produce a universal
equilibrium spectral form under rather general circumstances. Some evidence
for this interpretation is given by Schedvin (1979), who reviews available
microtemperature spectra in the ocean for comparison with his towed measurements.

Stewart (p. 1269) and Bretherton (p. 1279) presented papers at the
Colloquium, both of which addressed the issue of how to distinguish turbulence
from internal wave motions in a stratified medium. Stewart reviews the considerable range of measurements made in the atmosphere and ocean by his students
and associates to demonstrate various observed characteristics of the phenomena
and the statistical tools which might be brought to bear. However, he concludes
with the statement that "there is probably no clear-cut distinction between turbulence and waves", partly because a strong exchange of energy occurs between waves
and turbulence when they both exist in a stratified medium. Bretherton attempts
to define waves as "essentially a linear phenomena, or a weakly nonlinear
one, . . . ", but later recognizes that nonlinearities in waves may grow,
"leading to transient patches of true turbulence in which the conventional
Kolmogoroff cascade dominates". It would appear that a grey area exists in
Bretherton's definitions between linear wave motions and "true" Kolmogorovian
turbulence where buoyancy forces are negligible.

Stewart also emphasizes the tendency of stratified turbulence to appear in
patches, and provides simultaneous records of temperature and streamwise
velocity gradients, measured from towed bodies and from a submarine in the
ocean, which show very patchy bursts followed by long stretches of quiet
isothermal water. In some cases the patches of temperature gradient fluctuations are unaccompanied by perceptible velocity gradients, suggesting the
possibility of fossil turbulence. Spectra for the active and fossil temperature gradient regions were presented, and showed similar amplitudes and forms,
except that the "fossil" spectrum had a cutoff wavenumber that was smaller
by a factor of 2 to 3, indicating rates of strain about an order of magnitude
less. Since the velocity signals were in the noise, it was not possible to
provide conclusive evidence that weak turbulence was not actually present.
Measurements made with more sensitive velocity gradient sensors on quieter
platforms, by Crawford (1976), seem to indicate that increased temperature
gradient activity is always accompanied by increased velocity gradient
activity, but, as discussed in Section 4, it is not clear from the data
whether the velocity activity is "true turbulence" by Bretherton's definition,
or some other variety of small scale, buoyancy-affected motions from the grey
area between waves and turbulence.

Since the 1969 Colloquium, very little mention of fossil turbulence has
appeared in the literature. Turner's authoritative book, <u>Buoyancy Effects in
Fluids</u> (1973), makes only one reference to fossil turbulence (p. 316): it is
"temperature microstructure remaining after the turbulence has decayed". As
evidence of the phenomenon, Turner presents some unpublished data by Nasmyth,
from a towed body at 200 m depth in the permanent pycnocline, which shows strong
temperature gradient patches unaccompanied by measurable velocity gradient activity, similar to the data (also from Nasmyth) presented by Stewart (1969).
Monin's book, <u>Variability of the Oceans</u> (1977), makes no mention of fossil turbulence except for a reference to Woods' (1969) article. Federov's book,
<u>The Thermohaline Finestructure of the Ocean</u> (1978), mentions the term only in
reference to Woods (1969). However, Federov refers to "finestructure of turbulence origin" (p. 97), "nonstationary turbulent mixing" (p. 126), and
"inversions of density in the sea" (p. 19) which are all distinguishing
characteristics of fossil turbulence. In fact, Federov strongly emphasizes
the observation by Dunbar (1958) that density inversions in the ocean have had
difficulty achieving scientific respectability among oceanographers and suggests
that many oceanographers have resisted continuous high-frequency measurement
techniques in favor of conventional bottle sampling and arbitrarily smoothed
monotonic vertical profiles. Temperature inversions have either been viewed
as measurement errors, or it has been assumed that stable stratification is
maintained by a compensating variation in salinity. More recent high resolution vertical temperature gradient records by Gregg (1976a, b, 1977) show that
patches of high frequency temperature fluctuations exist at all depths to a
few kilometers, and almost certainly contain density inversions in many cases.

Such strong temperature gradient activity is generally taken as prima
facea evidence of turbulence (Gargett 1976). Measured mean-square
gradients have often been used to estimate vertical heat flux by setting the
turbulent production equal to the diffusive dissipation of thermal variance
(Osborn and Cox 1972) and assuming steady, active, horizontally homogeneous
turbulence. It seems likely that most observed microstructure is generated
by turbulence, whether past or present, especially when the Cox number is
large and the microstructure is isotropic. Salt fingering may produce nonturbulent microstructure, at small scales, but will produce turbulence, and
turbulent microstructure, at large scales. Other mechanisms for producing
nonturbulent microstructure are conceivable, but have not been put forth.

The evidence that all microstructure is actively turbulent at all (or any)
scales is very weak. Observations by Crawford (1976) and Osborn (1978), that
temperature microstructure is generally accompanied by velocity microstructure,
are suggestive, but their shear probe data bandwidth is so narrow, and the record

length from a single cut through a microstructure patch is so short, that it is difficult to demonstrate whether the velocity fluctuations are actively turbulent or not (see Section 4). Indeed, by using two thermistors moving at different angles of attack through the patches, Gregg, et al. (1973) have shown that many microstructure patches are quite anisotropic at large (and even small) scales. This observation is substantiated by Schedvin (1979) with towed body data for various angles of attack in statistically homogeneous microstructure layers. It seems likely that such patches of strong temperature gradient activity may once have been turbulent, but it seems unlikely that the fluid is actively turbulent if the temperature gradients are anisotropic. As discussed in Section 4, it has been found (Schedvin 1979) that the rate of strain, γ, in-ferred from the diffusive cutoff wavenumber of the temperature gradient spectra, is generally less than N when the microstructure is anisotropic, but is greater than or equal to N when it is isotropic.

2. PHYSICAL PROCESSES

Fossil turbulence is easily observed by pouring cold milk into hot coffee. The initial turbulence is generally dampened to internal wave motions before mixing is complete, leaving the fossil milk (and temperature) turbulence at the bottom of the cup (double diffusive effects complicate the experiment if cream is used). Skywriting and high altitude jet contrails are also familiar examples of fossil turbulence in stably stratified fluids.

The physical process leading to fossil turbulence is that buoyancy forces remove turbulent kinetic energy at large scales, but no comparable mechanism removes the large scale fluctuations of scalar fluid properties, such as temperature, produced by the original turbulence. Internal restratification will remove some scalar fluctuations if the initial turbulence is weak or has operated for only a short time. As the Reynolds number of the initial patch of turbulence increases, however, a much wider range of fossil scalar fluctua-tions will exist after the turbulence is dampened; the persistence time of the fossil should increase correspondingly.

Suppose a fully developed turbulent velocity field is suddenly imposed upon a region with a constant stable density gradient, $\partial \rho / \partial z = |\text{constant}|$, where z is downward. Consider the buoyancy forces resisting the overturn of a turbulent eddy of diameter L, compared to the inertial forces which keep it going. The inertial forces will be

$$F_I \approx \rho V^2(L) L^2 \approx \rho (\varepsilon L)^{2/3} L^2 = \rho \varepsilon^{2/3} L^{8/3} \tag{1}$$

where the velocity differences, $V(L)$, are estimated by the Kolmogoroff-Obukhov law, and ε is the viscous dissipation rate of the turbulence. The buoyancy forces will be

$$F_B(L) \approx g\left(\frac{\partial \rho}{\partial z}\right)L\ L^3 = g\left(\frac{\partial \rho}{\partial z}\right)L^4 \tag{2}$$

where g is the acceleration of gravity. It is clear from Equations (1) and (2) that the postulated turbulence cannot exist when $F_B > F_I$ and that the largest scales of the turbulence will be most strongly affected since $F_B \sim L^4$ while $F_I \sim L^{8/3}$. The largest scale turbulence should occur at a critical length scale L_R where $F_B = F_I$, which from (1) and (2) occurs at

$$L_R = \left[\frac{\varepsilon^{2/3}}{\frac{g}{\rho}\frac{\partial \rho}{\partial z}}\right]^{3/4} = \left[\frac{\varepsilon}{N^3}\right]^{1/2} \tag{3}$$

where $N = \left[\frac{g}{\rho}\frac{\partial \rho}{\partial z}\right]^{1/2}$ is the Brunt-Väisälä frequency. The smallest scale turbulence will be limited by viscous forces

$$F_V(L) = \mu\left[\frac{\varepsilon}{\nu}\right]^{1/2} L^2 \tag{4}$$

where μ is the coefficient of viscosity, ν is the kinematic viscosity, and $(\varepsilon/\nu)^{1/2}$ is the local rate of strain. Setting $F_I = F_V$ gives the familiar Kolmogoroff length scale

$$L_K = (\nu^3/\varepsilon)^{1/4} = (\nu/\gamma)^{1/2}. \tag{5}$$

Thus, by comparison of buoyancy, inertial, and viscous forces in a stratified medium, we arrive at the conclusion that turbulence cannot exist unless the following three equivalent criteria are satisfied:

Criteria for the existence of active turbulence in a stratified medium

$$\begin{cases} L_R \geq L \geq L_K; & \text{possible length scales} \\ \varepsilon \geq \nu N^2; & \text{dissipation rate} \\ \gamma \geq N; & \text{rate of strain, undetermined} \end{cases} \tag{6}$$

universal proportionality constants are required in cases of "equality" to produce equality conditions.

Conversely, when both of the equality conditions of Equation (6) are satisfied, we might expect a buoyancy dominated regime of flow on the verge of becoming turbulent, which might be called saturated internal wave motion. For such flows the gradient Richardson number is a constant of order 1, and the flow has the following properties:

Properties of saturated internal waves

$$
\begin{cases}
(\nu/N)^{1/2} \equiv L_{KF} \sim L_R \sim L_K; \text{ smallest scale} \\
\varepsilon_F \sim \nu N^2; \text{ viscous dissipation rate} \\
\gamma_F \sim N; \text{ rate of strain.}
\end{cases}
\tag{7}
$$

Again, the proportionality constant has been omitted. To estimate the proportionality constant in (7), we may assume the criterion for transition to turbulence is that the local Richardson number $\overline{N^2}/\overline{(\partial u/\partial z)}^2 = 1/4$, where the averaging is over a scale $(\nu/N)^{1/2}$ in the region of transitional turbulence. Assuming isotropy, $\varepsilon = (15/2)\nu(\partial u/\partial z)^2$, which gives $L_R = (\varepsilon/N^3)^{1/2} = 5.5 \ (\nu/N)^{1/2} = 12.8 \ L_K$. Thus, $L_{KF} \equiv (\nu/N)^{1/2} = L_R/5.5 = L_K \ 2.3$, $\varepsilon_F = 30 \ \nu N^2$, $\gamma_F = (\varepsilon/\nu)^{1/2} = 5.5N$, and $L_R = 12.8L_K$. Experimental determination of the actual constant is needed, since the preceding estimates are only approximate.

In view of the previously described attempts to define turbulence and waves in a stratified medium, the term "active turbulence" will be used to indicate fluid motions which satisfy the criteria of Equation (6). Active turbulence is defined as a regime of three-dimensional, random, eddy-like fluid motions in which a range of velocity length scales exists for which inertial forces are larger than either buoyancy or viscous forces. The statistical laws describing active turbulence are identical to the statistical laws describing nonstratified turbulence, except for the difference in the range of length scales. This includes the approach to local homogeneity and isotropy and the same universal similarity, when normalized with Kolmogoroff length and time scales, as has been found to exist for many laboratory and atmospheric turbulent flows. Active turbulence is equivalent to "true turbulence" as described by Bretherton (1969), and is distinct from two-dimensional turbulence or various buoyancy ranges of turbulence described in the literature.

An important and distinguishing property of active turbulence is its ability to mix scalar fluid properties. The possibility of fossil turbulence arises if the partially mixed properties persist longer than the time required to damp out active turbulence. More generally, we may define fossil turbulence as a remnant in any hydrophysical field of an actively turbulent flow in fluid which is no longer actively turbulent. Hydrophysical fields include any measureable property of the fluid, such as temperature, salinity, or species concentration, and may also include vector properties such as velocity or vorticity, since active turbulence may leave detectable remnants in these fields. We indicate which field preserves the active turbulence information by the terminology "fossil temperature turbulence", "fossil salinity turbulence", or, as discussed below, "fossil vorticity turbulence".

3. SPECTRAL DESCRIPTION

In order to understand the properties of fossil turbulence in general, it is useful to consider a particularly simple case. Suppose a patch or layer of very strong turbulence is created in a uniformly stratified stationary fluid. How would the patch evolve, and what would be the form of the velocity and temperature gradient spectra within the patch at various stages of its evolution?

Suppose that the initial vertical dimension of the patch is L_p, where L_p is less than $L_R = (\varepsilon/N^3)^{1/2}$. From Equations (1) and (2) we know that the inertial forces of the turbulence in the patch are larger than the buoyancy forces of the stratified fluid. Therefore, the size of the patch will increase by entraining additional nonturbulent fluid, and the dissipation rate will decrease as the kinetic energy is distributed to a larger volume of fluid or lost to viscous friction.

3.1 Velocity Gradient Spectrum

The initial velocity gradient spectrum will have the universal form, shown in Fig. 1, from wave number $2\pi L_p^{-1}$ to a peak at about $0.1\ L_K^{-1}$ with slope $+1/3$ in the inertial subrange.

Neglecting losses to heat and internal waves, the kinetic energy per unit mass of the turbulence, $\underline{u}^2 \sim (k\phi)_p$, times the mass, $\sim L_p^3$, of the patch, should be constant, where ϕ is the velocity spectrum and $k_p = 2\pi L_p^{-1}$; therefore,

$$k_p^{-2}\phi(k_p) = \text{constant}. \tag{8}$$

Since the velocity gradient spectrum of Fig. 1 is given by $k^2\phi$, the peak of the turbulent kinetic energy spectrum, $\phi(k_p)$, forms a locus of points, $(k^2\phi)_p/k_p^4 = \text{constant}$, with slope $+4$ on the log-log plot shown in Fig. 1. This represents the envelope of maximum spectral values, assuming no losses of turbulent kinetic energy, only redistribution by entrainment. To illustrate the approach of the buoyancy scale, L_R, to the scale of the turbulence patch, L_p, as the turbulence weakens (decreasing L_R) and the patch grows (by entrainment), a point is plotted on Fig. 1 at the same spectral level $(k^2\phi)_p$ as the energy scale of the turbulence, but at the buoyancy wavenumber $2\pi L_R^{-1}$. Since $(k^2\phi)_p \sim k_p^4$ from (8), and is also $= \alpha\varepsilon^{2/3}k_p^{1/3}$, we find $(k^2\phi)_p \sim \varepsilon^{8/11}$. Therefore, the locus of such buoyancy scale points has slope $-16/11$, and intersects the locus of energy scales of the patch turbulence at an intermediate scale $L_{R_o} = (\varepsilon_o/N^3)^{1/2}$ which is the size of the turbulent patch when fossilization begins.

As the patch grows, ε will decrease and the peak of the spectrum, $\alpha\varepsilon^{3/4}\nu^{-1/4}10^{-1/3}$ at wavenumber $\varepsilon^{1/4}\nu^{-3/4}10^{-1}$, follows a locus of points with slope +3, where α is the universal Kolmogoroff constant ($\approx 1/2$), as shown in Fig. 1.

Eventually the buoyancy length, $L_R = (\varepsilon/N^3)^{1/2}$, will equal the patch size, L_p, and the largest scale eddies in the patch will begin to be affected by buoyancy. The dissipation rate of the turbulence at this point, where fossilization begins, has decreased to a value of ε_o, the rate of strain $\gamma = \gamma_o = (\varepsilon_o/\nu)^{1/2}$, and the size of the patch is $L_{R_o} = (\varepsilon_o/N^3)^{1/2}$. Information about the initial source of turbulence in the patch has mostly been destroyed by the turbulent mixing, but ε_o, γ_o, and L_{R_o} may be preserved by the fossil turbulence.

The turbulence spectrum, at the time when fossilization begins, is shown in Fig. 1. It should have a universal inertial subrange, $\alpha\varepsilon_o^{2/3}k^{1/3}$, between about $2\pi/L_{R_o} < k < 0.1\,L_K^{-1}$, ranging from $\alpha(\varepsilon_o N)^{1/2}(2\pi)^{1/3}$ to $\alpha\varepsilon_o^{3/4}/\nu^{1/4}10^{1/3}$. Taking

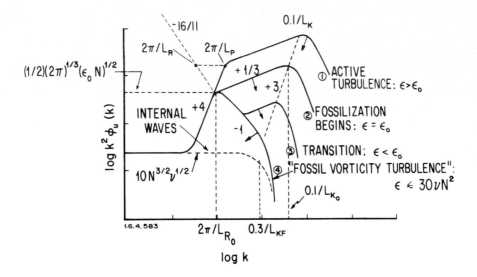

Figure 1.

Velocity gradient spectra for a patch of active turbulence in stably stratified medium after source of turbulence is removed

(1) Active turbulence: $L_R \gg L_K$; $L_R \to L_p$ as $\varepsilon \to \varepsilon_o < \varepsilon$.

(2) Fossilization begins: $L_R = L_{R_o} \simeq L_p$; $\varepsilon = \varepsilon_o$
      (active turbulence).

(3) Transition: $L_{R_o} > L_T > L_K$; $30\nu N^2 < \varepsilon < \varepsilon_o$
      (active + fossil turbulence).

(4) Fossil vorticity turbulence: $\varepsilon \leq 30\nu N^2$
      (no active turbulence).

the proportionality constant to be $\simeq 1.0$, as found by Williams and Gibson (1974), it has been assumed that the wavelength corresponding to the beginning of fossilization is equal to L_{R_O}. Clearly, this constant should be determined when appropriate velocity measurements can be made.

Since the dissipation rate in the patch must continue to decrease, it is clear that the patch fluid can not remain turbulent at large scales for subsequent times. The actively turbulent eddy size is limited to $\leq L_{R_T} = (\epsilon_T/N^3)^{1/2}$ which is less than $L_{R_O} = (\epsilon_O/N_O^3)^{1/2}$ because the dissipation rate during transition, ϵ_T, is less than the rate at the beginning of fossilization, ϵ_O. Active turbulence will be confined to scales of size L_{R_T} and less contained within the fossil turbulence fluid. Such velocity fluctuations should still obey universal similarity; that is, $k^2\phi = \alpha\epsilon_T^{2/3}k^{1/3}$ for $2\pi/L_{R_T} < k < 0.1\ L_{K_T}^{-1}$, as shown in Fig. 1. The low wavenumber beginning of the universal inertial subrange has a value $\alpha(\epsilon_T N)^{1/2}(2\pi)^{1/3}$ at wavenumber $2\pi(N^3/\epsilon_T)^{1/2}$ which forms a locus of points with slope -1 in Fig. 1. This line intersects at $\sim(0.1 - 0.3)\ L_{KF}^{-1}$, with the locus of points, with slope $+3$, corresponding to the peak of the velocity gradient spectrum of the small scale active turbulence within the fossil turbulence patch during dampening, at $0.1\ L_{K_T}^{-1}$.

As the average ϵ of the patch decreases, the largest eddies of the active turbulence will be so weakened that they can no longer "turn over", producing a random internal wave-like motion. Since the wave motion comes directly from weakened energy-containing eddies of the turbulence, it will be on the same scale as the turbulence and nearly isotropic. In addition, the wave motions will have roughly the same kinetic energy per unit mass as the turbulence, as well as an equal amount of gravitational potential energy. However, the dissipation processes for this turbulence-produced internal wave motion are much less efficient than for turbulence, so we can expect that the energy spectral level of the velocity fluctuations at each wavenumber will remain the same during the transition period. This gives the double peaked spectrum shown in Fig. 1, where the low frequency peak is at $2\pi L_{R_O}^{-1}$ and the high frequency peak is at $0.1 L_{K_T}^{-1}$, corresponding to the waves and turbulence, respectively. Since the vorticity of the internal wave motion was produced by active turbulence and is preserved as a remnant of the active turbulence in nonturbulent fluid, it satisfies our definition of fossil turbulence. Therefore, we may refer to internal wave motions caused by the dampening of active turbulence as "fossil vorticity turbulence."

During the transition phase, buoyancy and viscous forces will continue to damp out the small scale turbulence as its average dissipation rate decreases, finally leaving only the patch of internal wave motion with largest scale L_{R_O} and with spectral form labelled "fossil vorticity turbulence", as shown in Fig. 1. The wave spectrum has a viscous cutoff at the buoyancy analog of the Kolmogoroff scale, $(\nu/N)^{1/2} \equiv L_{KF}$. We can determine the frequency, and also the shear, of the waves by dividing the velocity of the turbulence eddies, at their time of dampening, by their size. That is, $\omega \sim (k\phi)_p^{1/2}/L_R \sim N$, which shows that the frequency and the shear for each wave motion in the **fossil** vorticity spectrum of Fig. 1 has a value approximately equal to the Brunt-Väisälä frequency, N. The propagation velocity of linear internal waves approaches zero as the frequency approaches N. This is consistent with our assumption that the fossil vorticity turbulence patch will not substantially increase in size during the transition period which should be a time of order N^{-1}.

The evolution of the fossil vorticity turbulence patch, at times greater than N^{-1}, will depend very much on the **state** of motion and stratification of the surrounding fluid. Presumably the frequency of the motions will decrease and the flow will become more horizontally polarized. If the vorticity concentrates on nearly horizontal layers with thickness $L_{KF} = (\nu/N)^{1/2}$ and mean square vorticity $\sim N^2$, the spectrum will approach a constant value $(k^2\phi)_B = $ constant, such that $(k^2\phi)_B k_{KF} \sim N^2$. Then, $(k^2\phi)_B \sim \nu^{1/2}N^{3/2}$. This background internal wave spectrum is shown in Fig. 1. It differs somewhat from the Garrett-Munk (1975) **empirical** spectrum since its spectral level is $\sim N^{3/2}$ rather than $\sim N^2$ and is flat rather than $\sim k^{-1/2}$. If the cutoff wavenumber is $0.3(N/\nu)^{1/2}$, and the Richardson number $N^2/(\partial u/\partial z)^2$ is about 1/4, then the proportionality constant c is about 10, since $\overline{(\partial u/\partial z)^2} = cN^{3/2}\nu^{1/2}0.3(N/\nu)^{1/2} = N^2/Ri = 4N^2$, as shown in Fig. 1.

It should be pointed out that measurements of small scale internal waves and turbulence in the ocean (or any other stratified medium) are in a very early stage of development. Little data is available for comparison with the predictions of Fig. 1. Some recent measurements of $\partial u/\partial z$ spectra made by Oakey in the JASIN 1978 experiment (personal communication) using aerofoil shear probes on a dropsonde, seem to show a high frequency dampening, similar to the fossil vorticity turbulence spectrum of Fig. 1, for regions of very weak velocity microstructure, and show a high frequency peak, similar to the active turbulence spectra, where the velocity shear levels were much higher. Similar results from Crawford (1976) are described in Section 4.

It is interesting to compare the persistence times for the active and fossil turbulence spectra of Fig. 1. Both have the same kinetic energy per unit mass $\overline{u^2} \sim (k\phi)_p \approx \alpha(2\pi)^{-2/3}(\varepsilon_0/N)$. However, since $\phi_F \sim k^{-3}$ and $\phi_0 \sim k^{-5/3}$ for $k > k_p$, the dissipation rates are quite different; $\varepsilon_F/\varepsilon_0 = \nu N^2/\varepsilon_0 = N^2/\gamma_0^2 \ll 1$, using the subscript F to indicate fossil turbulence and o to indicate the state of active turbulence when fossilization begins. In general, the persistence time, τ, for an actively turbulent patch, in the absence of buoyancy effects, is of order L/u' ($\tau \sim \overline{u^2}/\varepsilon \sim u'^2/(u'^3/L)$) corresponding to a turnover time for the energy-containing eddies with scale L. For the spectrum in Fig. 1 at the point of fossilization, the active turbulence persistence time $\tau \sim N^{-1}$. After fossilization, the dissipation rate has been reduced from ε_o to only νN^2, so the persistence time of the fossil vorticity turbulence is of order $\alpha(2\pi)^{-2/3}(\varepsilon_o/N)/(\nu N^2) \sim (\gamma_o/N)^2 N^{-1}$: increased by the factor $(\gamma_o/N)^2$.

We have shown that the ratio of the turbulence rate of strain at the time of fossilization, γ_o, to the Väisälä frequency, N, is a crucially important parameter of fossil turbulence. For the coffee cup fossil milk turbulence experiment, and for most laboratory experiments, γ_o will not differ greatly from N, and N^{-1} will be small. However, in the ocean and atmosphere γ_o/N may be >> 1, and N^{-1} very large, so that the kinetic energy of fossil vorticity turbulence may be quite persistent to viscous dissipation, and will probably be propagated to the surrounding fluid with minor losses. This suggests the possibility that turbulence patches may prove to be a significant source of small scale internal wave energy in some regions of the ocean and atmosphere.

The fossil vorticity turbulence subrange between $2\pi L_{R_o}^{-1}$ and $0.1L_{KF}^{-1}$ has the form

$$k^2\phi = (2\pi)^{4/3}\alpha N^2 k^{-1}, \tag{9}$$

which is the same as the "buoyancy-inertial subrange of turbulence" derived by Lumley (1964) and Turner (1973, p. 144, Eq. 5.2.10). Velocity spectra with slopes close to -3 have been observed by aircraft in stable layers of the atmosphere, as shown by Stewart (1969), but have not been observed in the laboratory, according to Turner (1973, p. 144), so the constant, $\alpha(2\pi)^{4/3} \doteq 5.8$, of Eq. (9) cannot presently be tested. An excellent review of laboratory experiments in stratified shear flows, with photographs showing the evolution of active patches of turbulence damped out by stratification, is given by Thorpe (1973). He also gives some measurements of the energy balance. Further discussion of the physical processes of such "forced, strongly inter-acting internal waves" are given by Turner (1973, pp. 46-47, 144-145). Perhaps the first published discussion of the buoyancy length scale, L_R, of Eq. (9) is by Dougherty (1961), although it was known to the Russian school of turbulence many years earlier (Obukhov, personal communication). It was used by Ozmidov (1965) to estimate vertical diffusivity in the ocean, and is often called the Ozmidov scale.

3.2 Scalar Spectra

Fig. 2 shows the evolution of the temperature gradient spectral form for the patch of active turbulence discussed in the previous section. It is assumed that the fluid is stratified because the temperature decreases with

depth. Initially, buoyancy forces are irrelevant to the turbulence and turbulent mixing, so the spectrum should approach the universal form expected
for a scalar fluid property, with Prandtl number $\nu/D \approx 10$, when normalized
with turbulence length, time, and scalar scales $L_B = (D/\gamma)^{1/2}$, $T_B = \gamma^{-1}$, and
$\Sigma_B = (\chi/\gamma)^{1/2}$, respectively, where χ is the dissipation rate of temperature
fluctuation variance, and D is the molecular diffusivity (see Gibson, 1968 a.,
b.). As shown, the peak of the temperature gradient spectrum $k^2\phi_T$ should occur
at wavenumber $k = 0.3\ (\gamma/D)^{1/2}$ with level $0.3\chi/(D\gamma)^{1/2}$. If the Reynolds number of the active turbulence patch is sufficient, the spectrum should exhibit
a viscous-convective subrange $k^2\phi_T = \beta_B\chi\gamma^{-1}k^{+1}$ for $(\gamma/\nu)^{1/2}/30 < k < 0.3\ (\gamma/D)^{1/2}$,
and a scalar inertial subrange $k^2\phi_T = \beta_K\chi\varepsilon^{-1/3}k^{+1/3}$ for $2\pi L_p^{-1} < k < (\gamma/\nu)^{1/2}/30$,
where β_B is a universal constant which should be in the range $\sqrt{3}$ to $2\sqrt{3}$ (Gibson,
1968 a, b), and β_K is another universal constant with value about 0.5. The subranges have slopes +1 and +1/3, respectively, on the log-log plot of Fig. 2,
and should have the same small wavenumber limit, $2\pi/L_p$, as the velocity gradient
spectrum.

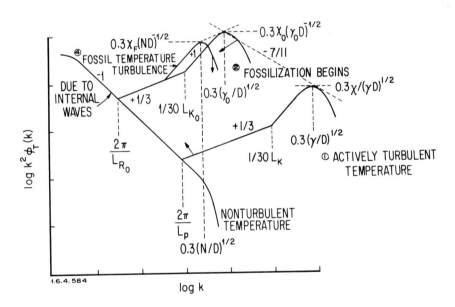

Figure 2. Temperature gradient spectra corresponding to patch of active turbulence in region stratified by temperature. Same stages as Figure 1.

The amplitude of the temperature fluctuations in the turbulent patch should increase as the size of the patch increases, since the source of the fluctuations is the assumed vertical temperature gradient. The variance, δT^2, should be $\stackrel{\sim}{\sim} (L_p \partial T/\partial z)^2 \stackrel{\sim}{\sim} (k\phi_T)_p$, so that $(k^2\phi_T)_p = c(\partial T/\partial z)^2 k_p^{-1}$, where c is a constant which must be determined empirically. The point follows a locus with slope -1, shown in Fig. 2. Note that the amplitude of the scalar inertial subrange increases as the size of the turbulence patch increases, whereas the amplitude of the velocity inertial subrange decreases.

When the patch reaches its maximum size by entrainment, and fossilization begins, the low wavenumber limit of the scalar inertial subrange will be at $k = 2\pi L_{R_o}^{-1}$, corresponding to the low wavenumber limit of the velocity inertial subrange. Thus,

$$c \left[\frac{\partial T}{\partial z}\right]^2 \frac{L_{R_o}}{2\pi} = \beta_K \chi_o \varepsilon_o^{-1/3} \left[\frac{2\pi}{L_{R_o}}\right]^{1/3}$$

which can be solved for ε_o by substituting $L_{R_o} = (\varepsilon_o/N^3)^{1/2}$ to give

$$\varepsilon_o = \frac{\beta_K}{c} (2\pi)^{4/3} \frac{\chi_o N^2}{(\partial \bar{T}/\partial z)^2} \tag{10}$$

where χ_o is the scalar dissipation rate at the point of fossilization. If the temperature field is isotropic, $\chi = 6D \overline{(\partial T/\partial x)^2}$, where x is any direction of measurement, so (10) can be written

$$\varepsilon_o = \frac{6\beta_K (2\pi)^{4/3}}{c} D \frac{\overline{(\partial T/\partial x)_o^2}}{(\partial \bar{T}/\partial z)^2} N^2 = 13 \, DC_o N^2; \tag{11a}$$

$$\gamma_o = 3.6 \, (C_o/Pr)^{1/2} N \tag{11b}$$

where β_K is assumed to be 0.5 and c is taken as 0.9. In (11a) the ratio of mean square gradient at fossilization to square mean vertical gradient is the Cox number, C_o. Gregg (1977) normalizes vertical temperature gradient spectra, from a variety of depth ranges, with $(\partial \bar{T}/\partial z)^2$ where the average is taken over his 100 m record lengths, and found "fine structure" subranges where $(k^2\phi) = c(\partial \bar{T}/\partial z)^2 k^{-1}$ for most cases, but with wide variation in the value of c. Values of c were from about 0.1 to 0.9. Small values were usually associated with a small microstructure range, and large values with strong microstructure activity. We choose the upper bound of the observed values, c = 0.9, because Eq. (11) represents the transition from active to fossil turbulence. Many of the weaker patches of microstructure observed by Gregg (1977) were probably in an advanced state of decay, with internal wave activity substantially below the saturation level and with correspondingly small c values. Eq. (11) shows that a large measured Cox number implies that the turbulence activity was very strong at the time of fossilization. We shall use (11) in a later section as a means of testing whether microstructure is in an active or fossil turbulence state, or some combination of the two.

During the entrainment process the patch size grows such that $(k^2\phi)_p \sim k_p^4$. Hence, $\chi \sim (\partial T/\partial z)^2 \, L_p^2 u'/L_p \sim (\partial T/\partial z)^2 \, L_p (k_p\phi_p)^{1/2} \sim (\partial T/\partial z)^2 L_p^{-1/2}$ decreases slightly, even though the spectral levels at low wavenumbers increase. The peak of the scalar gradient spectrum is at $0.3\chi/(D\gamma)^{1/2}$. Since $L_p \sim \varepsilon^{-2/11}$ and $\gamma \sim \varepsilon^{1/2}$, the peak level is $\sim \varepsilon^{-7/44}$, and the wavenumber of the peak, $0.3(\gamma/D)^{1/2}$, is $\sim \varepsilon^{1/4}$, so that the locus of the peak lies on a line with slope -7/11, as shown on Fig. 2. Therefore,

$$\chi_o/\chi_p \simeq (L_p/L_{R_o})^{1/2} \simeq (\varepsilon_o/\varepsilon_p)^{1/11} \tag{12}$$

which should be very slightly less than a value of 1.0.

The scalar spectrum at the point of fossilization should have peak $0.3\chi_o/(D\gamma_o)^{1/2}$ at wavenumber $0.3(\gamma_o/D)^{1/2}$, as shown in Fig. 2, with the universal active turbulent scalar spectral form for smaller wavenumbers greater than $2\pi L_{R_o}^{-1}$. During the period of time that the active turbulence is dampened, the low frequency portion of the scalar spectrum should be relatively unaffected since, in the absence of restratification or straining due to external shears, the large scale fluctuations of temperature will simply bob back and forth with the small scale internal wave motion set up by the dampened turbulence and will not be much affected in amplitude or spatial distribution during several wave periods N^{-1}. However, the small scale temperature structure in the turbulence patch will be affected as the active turbulence disappears, since the local rate of strain will decrease from γ_o to N during the transition period.

As shown by Gibson (1968a), the smallest structure of scalar fields mixed by turbulence is determined by an equilibrium between convection and diffusion at points of maximum and minimum scalar value. Such zero gradient points are characteristic of turbulent mixing and cannot be formed by laminar flows. Analysis of the dynamics of the hot and cold spots show that they tend to move with the fluid velocity and align with the local rate of strain tensor. The plane of the minimum radius of curvature of the scalar distribution around a hot spot orients in the direction of fluid convergence, and an equilibrium is rapidly established between the molecular diffusion, which tends to increase the radius, and the rate of strain, which tends to decrease it. Therefore, the local radii are $\sim(D/\gamma)^{1/2}$ (for any Pr) where γ is proportional to the rms rate of strain of the turbulence $\sim (\varepsilon/\nu)^{1/2}$. The same scalar microscale is predicted by Batchelor (1959) using a different physical model valid only for Pr >> 1.

Using the zero gradient point model for the dynamics of fossil temperature turbulence is a straightforward extension of the turbulent mixing case. Since (in the absence of double-diffusive instabilities, distributed sources, or other complexities) zero gradient points in the scalar field are only produced by active turbulence, they indicate fossil temperature turbulence when they persist after the turbulence is dampened. Consequently, the existence of scalar zero gradient points in a nonturbulent fluid might be considered a necessary and sufficient condition for the existence of fossil scalar turbulence. The eventual disappearance of the remnant hot and cold spots by molecular diffusion may be identified with the disappearance of the fossil temperature turbulence.

In turbulent mixing of high Prandtl number scalars, the large scale structure is determined by the turbulence, and the small scale structure by the local strain. The transition between the scalar inertial subrange and viscous convective subrange occurs at about $k = (\gamma/\nu)^{1/2}/30$. For fossil turbulence the large scale structure is essentially constant during the transition period when $N \leq \gamma \leq \gamma_0$, but the small scale mixing process is nearly identical to that of active turbulence, except that the strain rate is now $\gamma \leq N << \gamma_0$. Therefore, we might expect the form of the high wavenumber spectrum of fossil scalar turbulence, during both the transitional and completely fossil phases, to be close to the universal similarity form for active turbulence, except that the rate of strain will be less. For this to be true, and yet have no effect on the level of the low frequency fossil scalar spectrum, the points of the viscous-convective subrange and diffusive cutoff subrange must translate along loci with slope +1/3, as shown in Fig. 2. For the fossil versus active temperature turbulence forms, this gives a peak value of $0.3\chi_F/(DN)^{1/2}$ at wavenumber $0.3(N/D)^{1/2}$, compared to $0.3\chi_0/(D\gamma_0)^{1/2}$ at $0.3(\gamma_0/D)^{1/2}$, with transition from viscous-convective to inertial subranges at wavenumber $(N/\nu)^{1/2}/30$ rather than $(\gamma_0/\nu)^{1/2}/30$.

The dissipation rate of the fossil, χ_F, may be estimated in terms of the dissipation rate at the time of fossilization, χ_0, from this spectral translation. Since $\chi = 6D(\partial T/\partial x)^2 \sim (k^3\phi)_B$, then $\chi_F/\chi_0 = (k^3\phi)_{B_F}/(k^3\phi)_{B_0} = (k_F/k_0)_B (k^2\phi)_{B_F}/(k^2\phi)_{B_0} = (k_F/k_0)_B (k_F/k_0)_B^{1/3} = (k_F/k_0)_B^{4/3} = (N/\gamma_0)^{2/3}$, since $k_{B_F} = 0.3(N/D)^{1/2}$ and $k_{B_0} = 0.3(\gamma_0/D)^{1/2}$. The persistence time of the active turbulent temperature, at the time of fossilization, should simply be N^{-1}, the eddy turnover time. Therefore, the persistence time, τ_T, of the fossil temperature turbulence should be

$$\tau_T = (\chi_o/\chi_F)N^{-1} = (\gamma_o/N)^{2/3}N^{-1} \ .$$

The subscript B refers to the peak of the temperature gradient spectrum.

Thus, we see that the fossil scalar turbulence persistence time also depends on the parameter γ_o/N, and is less than the persistence time of fossil vorticity turbulence, which is $(\gamma_o/N)^2N^{-1}$. Note that the persistence of fossil turbulence is independent of the molecular diffusivities of either the scalar property, D, or the vorticity, ν. Mixing occurs in fossil scalar turbulence, due to non-turbulent straining by the internal wave motions created when the active turbulence is dampened, or by ambient non-turbulent straining; not by simple molecular decay. The decay rate of fossil vorticity turbulence (the internal waves produced when the active turbulence is dampened) is independent of molecular viscosity, just as the decay rate of active turbulence, from which it derives its energy, is independent of ν.

Another important parameter of turbulent mixing is the scalar scale $\Sigma_B = (\chi/\gamma)^{1/2}$ (Gibson 1968). Since the diffusive cutoff of turbulent scalar spectra occurs at length scale $L_B = (D/\gamma)^{1/2}$, and since the dissipation rate $\chi \sim D(\partial T/\partial x)^2$, it is clear that Σ_B represents the magnitude of the temperature fluctuations on scale L_B, since $\delta T \sim L_B[(\partial T/\partial x)^2]^{1/2} \sim (D/\gamma)^{1/2}(\chi/D)^{1/2} = \Sigma_B$. The initial value of Σ_B will be of order $L_p(\partial T/\partial z)Re^{-1/4}$, where L_p is the initial patch size, $Re = L_p u'/\nu$ is the Reynolds number, and u' is the characteristic velocity fluctuation in the initial patch. It is interesting to estimate whether the smallest scale temperature fluctuations in the patch will increase or decrease during the process of fossilization. The ratio $\Sigma_{B_o}/\Sigma_B = (\chi_o/\chi)^{1/2}(\gamma/\gamma_o)^{1/2}$. The scalar dissipation rate should be $\sim L_p^2[(\partial T/\partial z)]^2 u'/L_p$. If the kinetic energy is conserved during entrainment, $u'^2 L_p^3 = $ constant, so that $\varepsilon \sim u'^3 L_p^{-1} \sim L_p^{-11/2}$. Therefore, $\Sigma_{B_o}/\Sigma_B = [(u'_o L_R)/(u'L_p)]^{1/2}[\varepsilon/\varepsilon_o]^{1/4} = [L_R/L_p]^{-1/4}[\varepsilon/\varepsilon_o]^{1/4} = (\varepsilon/\varepsilon_o)^{9/44}$, which shows that both the magnitude and length scale of the smallest temperature fluctuations tend to increase as the turbulence patch grows to a scale where buoyancy forces become important. The Reynolds number at fossilization $Re_o \sim L_{R_o}(\varepsilon_o/N)^{1/2}/\nu \sim (\gamma_o/N)^2 \sim (C_o/Pr)$.

The fossilization process may change the smallest temperature scale. $\Sigma_{BF}/\Sigma_{B_o} = (\chi_F/\chi_o)^{1/2}(\gamma_o/\gamma_F)^{1/2} = (N/\gamma_o)^{1/3}(\gamma_o/N)^{1/2} = (\gamma_o/N)^{1/6}$, using the result that $\chi_F/\chi_o = (N/\gamma_o)^{2/3}$. Thus, fossilization also increases both the magnitude and length scale of the smallest temperature fluctuations, since γ_o

is greater than N. It would seem that fossil temperature turbulence should be easier to detect than active temperature turbulence because both the spectral amplitude and the minimum length scale increase, whereas fossil vorticity turbulence may be more difficult to detect than active turbulence, since the amplitude of velocity gradient fluctuations is decreased at all length scales.

3.3 Summary of Parameters

Table 1 summarizes the most important parameters and scales of stratified and nonstratified turbulence and mixing, which have been discussed in the preceding section. The list is not exhaustive but may serve as a useful reference and reminder of definitions.

Table 1

Important Parameters and Scales of Stratified Turbulence

Parameters:

ν = kinematic viscosity, cm^2/sec

D = molecular diffusivity, cm^2/sec

ε = viscous dissipation rate, $=2\nu e_{ij}^2$; $i=1,2,3$, cm^2/sec^3

e_{ij} = rate of strain tensor = $(\partial u_i/\partial x_j + \partial u_j/\partial x_i)/2$

γ = turbulence rate of strain parameter = $(\varepsilon/\nu)^{1/2} = \gamma_e \beta_B^{+1} = \gamma_B$, sec^{-1}

χ = diffusive dissipation rate = $2D(\partial\theta/\partial x_i)^2$, θ^2/sec

N = Brunt-Väisälä frequency = $[g(\partial\rho/\partial z)/\rho]^{1/2}$, rad/sec

$\chi_0, \varepsilon_0, \gamma_0, L_{R_0}, L_{K_0}, L_{B_0}, \Sigma_{Bo}, T_{K_0}$ = parameters of active turbulence at the time fossilization begins

χ_F, ε_F = dissipation rates of fossil turbulence

$$
\begin{cases}
\chi_F = \chi_0 (N/\gamma_0)^{2/3} \quad ; \quad \varepsilon_F = \varepsilon_0 (N/\gamma_0)^2 \\
\\
\chi_0 = (\partial T/\partial z)^2 (\varepsilon_0/N^2); \quad \varepsilon_0 \cong 13 D C_0 N^2; \quad C_0 = \dfrac{(\nabla T)_0^2}{(\partial\bar{T}/\partial z)^2} \\
\\
Re_0 = L_{R_0} u'/\nu \sim (\gamma_0/N)^2 \sim (C_0/Pr)
\end{cases}
$$

some formulae from fossil turbulence model

Table 1. Important Parameters and Scales of Stratified Turbulence - cont'd.

Length Scales:

$L_K = (\nu/\gamma)^{1/2}$ = Kolmogoroff scale; viscous forces equal inertial forces of active turbulence

$L_{KF} = (\nu/N)^{1/2}$ = viscous scale of internal waves; viscous forces equal inertial forces of waves

$L_R = (\varepsilon/N^3)^{1/2}$ = Buoyancy (Osmidov) scale; inertial forces of active turbulence equal buoyancy forces of local stratification

Note: $L_R = L_{KF}(\gamma/N) = L_K(\gamma/N)^{3/2}$ criterion for existence of active turbulence; $L_R \gg L \gg L_K$; i.e., $\gamma \gg N$.

$L_B = (D/\gamma)^{1/2}$ Batchelor scale; smallest feature in turbulent scalar θ field, molecular diffusion balances fluid convection of θ in direction of compressive strain near points of maximum and minimum θ giving radius of curvature L_B.

$L_{BF} = (D/N)^{1/2}$ = diffusive scale of fossil θ turbulence; (max, min) θ points persist as remnants of active turbulence after $L_R \rightarrow L_{KF}$; diffusion balances strain rate, N, convection at points , to give θ distribution with radius of curvature $L_{BF} = \theta'/(\partial^2\theta/\partial x^2)'^{1/2} \approx (D/N)^{1/2}$. Primes indicate values at the zero gradient point.

Note: $L_R = L_{BF}Pr^{1/2}(\gamma/N) = L_B Pr^{1/2}(\gamma/N)^{3/2}$, $Pr = \nu/D$

Table 1.   Important Parameters and Scales of Stratified Turbulence - continued

Time Scales:

$$T_K \;=\; \gamma^{-1} \;=\; \text{turbulence time}$$

$$T_{KF} \;=\; N^{-1} \;=\; \text{buoyancy time}$$

$$\left.\begin{array}{l} \tau_\omega \;=\; (\gamma_o/N)^2 N^{-1} \\[12pt] \tau_\theta \;=\; (\gamma_o/N)^{2/3} N^{-1} \end{array}\right\} \quad \begin{array}{l}\text{viscous and diffusive}\\ \text{persistence times for}\\ \text{fossil } (\omega,\theta) \text{ turbulence}\end{array}$$

Scalar Scales:

$$\Sigma_B \;=\; (\chi/\gamma)^{1/2} \;=\; \text{active turbulence scalar scale; amplitude}$$

of θ fluctuations at L_B scale.

$$\Sigma_{BF} \;=\; (\chi_F/N)^{1/2} \;=\; \text{fossil turbulence scalar scale; amplitude}$$

of θ fluctuations at L_{BF} scale.

from the
fossil
turbulence
model
$$\left\{\begin{array}{l} \dfrac{\Sigma_{B_o}}{\Sigma_B} \;=\; (\varepsilon/\varepsilon_o)^{9/44} \;\geq\; 1 \\[20pt] \dfrac{\Sigma_{BF}}{\Sigma_{B_o}} \;=\; \left(\dfrac{\gamma_o}{N}\right)^{1/6} \;\geq\; 1 \end{array}\right.$$

4. EVIDENCE FOR FOSSIL TURBULENCE IN THE OCEAN

Small scale fluctuations of temperature and velocity in the ocean are typically isolated in patches which occupy a small fraction of the water sampled.
Often the patches of microtemperature fluctuations are accompanied by an
apparent increased activity in the velocity fluctuations, as illustrated in
Figure 3a. The upper trace is a bandpassed (1-12 Hz) microbead thermistor
temperature signal from a body towed at 32 meters depth in the seasonal
thermocline, taken during the 1977 Mixed Layer Experiment (MILE). The lower
trace shows the velocity fluctuation signal from a microbead thermistor heated
by constant current to about 30°C above ambient. The signal shows increased
activity in the same frequency band for this 80 meter patch, with indicated
fluctuations about 2 cm/sec.

Figure 3b shows a patch of microconductivity fluctuations in the lower trace
(probably also due to temperature) that has no perceptible fluctuations in the
velocity signal above the instrument noise level. The traces of 3a and 3b are
similar to the qualitative evidence for fossil turbulence in Nasmyth's data
presented by Stewart (1969). Analysis of the MILE data is not complete, so
the present interpretation of such evidence of turbulent mixing activity is
tentative. Preliminary analysis indicates that even though the active patches
occupy only a small percentage ($\approx$ 5% for the layers shown in Fig. 3) of the
total volume, they must be included in any sample record or the average of the
sample will not be representative. For example, spectral analysis of the apparent velocity signal from the active patch shown in Figure 3a indicates that
ε is 10^{-1} cm^2/sec, compared to background values of less than 10^{-4} cm^2/sec^3.
This makes the average value for the layer 50 times larger than the average
value observed in the water outside the patches. Similarly, the scalar
dissipation rates, χ, in the patches are as large as 10^{-4} °C^2/sec which is 3
or 4 orders of magnitude larger than χ **in the surrounding fluid. The**
average, including the patches, is 5 x (10^1 to 10^2) times the background values.
Such active temperature fluctuations may contaminate the velocity signals,
even with very high overheat temperature (Gibson and Deaton, 1979), so the
value of ε is uncertain until careful evaluation of various effects such as
temperature sensitivity can be made.

Clearly, in such an intermittent process, it is crucial that a sufficiently
large number of vertical samples through a particular layer, or a sufficiently
long horizontal record from the layer, be accumulated for the estimates of
average values to converge. For the example shown in Fig. 3, about 16 kilometers of horizontal record are required to see 10 active patches. If the
dominant patches for a given depth (or isopycnal) are thin, a correspondingly
narrow range of depth (or density) control must be maintained to accumulate an

242

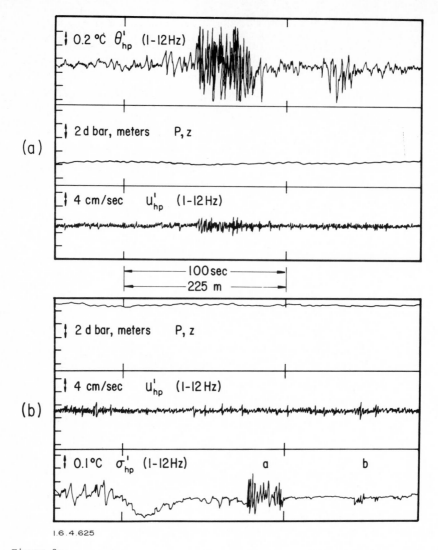

I.6.4.625

Figure 3a.

Patch of temperature microstructure accompanied by a patch of apparent velocity
microstructure measured by UCSD towed-body during MILE experiment. θ is microbead thermistor temperature, u is a heated microbead anemometer velocity, σ is
a microconductivity sensor output; all bandpassed from 1 - 12 Hz; P is pressure
in decibars; depth, z, is about 30 meters in seasonal thermocline, $\bar{u} \sim 4$ knots
(temperature sensitivity effects have not been removed from velocity signal,
and may be important).

Figure 3b.

Patch of conductivity microstructure (treated as if it is due to temperature)
at point a, not accompanied by perceptible velocity fluctuations. Note active
patch at b with corresponding apparent velocity activity.

adequate towed sample. A larger number of vertical dropsonde samples are required to give the same total record length and statistical confidence in the
estimates of average quantities. Assuming the thickness of the patch shown in
Fig. 3 is 3 meters (corresponding to $(\varepsilon/N^3)^{1/2}$ = 3m with N $\simeq 10^{-2}$ rad/sec),
then over 5000 dropsondes would be needed to give 16 km of data in the depth
range of interest, and 58 would be needed to be 95% sure of hitting at least
one patch.

Measurements of simultaneous velocity and temperature microstructure made by
Crawford (1976) in the equatorial undercurrent using a thermistor and an aerofoil shear probe, show that temperature microstructure is always accompanied
by velocity microstructure, as shown by Fig. 4a (from Fig. 11, Crawford 1976).
However, velocity spectra from the most active regions of Profile 23 show somewhat
better agreement with the form given in Fig. 1 for fossil vorticity turbulence
where $k^3\phi$ = constant, than with the active turbulence form, $k^3\phi \sim k^{+4/3}$, as
shown in Figure 4c (from Figure 24, Crawford 1976). The less active spectra
are even more strongly dampened. The spectra in Figure 4c have been corrected
for the high pass filter with 3db point at 1 Hz.

Since the bandwidth is narrow, the scatter is high, and the level is close
to the noise, it is not possible to use these spectra to demonstrate the existence of fossil vorticity turbulence or confirm the fossil spectral form predicted in Figure 1. On the other hand, the departure from the active turbulence
form in Fig. 4c is rather substantial, so neither is it possible to state that
the velocity field associated with the active temperature patches is actively
turbulent. Crawford describes the activity in his measured velocity microstructure as "turbulence", but it may not be active turbulence as defined in
this paper, even at small scales.

Dissipation rates computed from the velocity gradient spectra shown in
Figure 4b seem poorly correlated with the large scale mean density profile on
the left. However, a better correlation may exist between local ε and the local
mean density gradients (averaged over 1-2 meters). Most of the patches show
skewed temperature gradients, suggesting they contain local, strong, stable
density gradients. Positive correlation between local stratification and ε
is a characteristic of fossil vorticity turbulence since $\varepsilon_F \sim \nu N^2$, from
Equation (7). Thus, for fossil temperature turbulence in a region stratified
by temperature, one would expect to see $(\partial u_1/\partial x_3)^2 \sim \partial T/\partial x_3$, much like in the
observed records in Figure 4a and in Crawford's thesis (see Fig. 17 p. 65 for
some high resolution records of Profile 23). Fossil temperature turbulence
spectral levels should also be positively correlated with the local mean temperature gradient, so fossil temperature and velocity activity should be positively
correlated with each other.

244

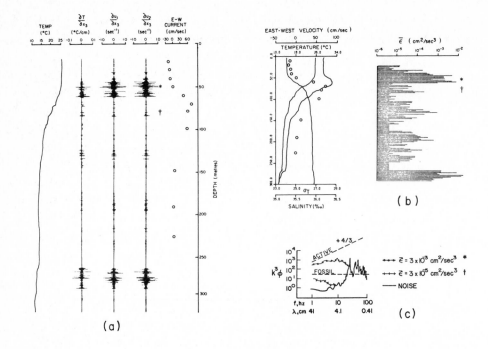

(a)

(b)

(c)

Figure 4.

Velocity shear and temperature gradient microstructure in Atlantic equatorial undercurrent by Crawford (1976).

(a) Note coincidence and positive correlation of $\partial T/\partial z$ and $\partial u/\partial z$ fluctuation amplitudes: a characteristic property of fossil turbulence. Amplitudes of $\partial T/\partial z$ and $\partial u/\partial z$ tend to be negatively correlated in active turbulence.

(b) Distribution of ε versus depth from shear probe.

(c) Spectral forms from active (*) and inactive (†) regions, compared to active and fossil turbulence forms for velocity gradient spectra (see Figure 1). Velocity fluctuations may be fossil.

In contrast, active turbulence unaffected by buoyancy will reduce the local temperature fluctuations as its strength increases (recall that $\Sigma_B \sim Re^{-1/4}$), so the correlation of velocity and temperature activity will be negative. Figure 4a shows a remarkably positive correlation between the local gradients of velocity and temperature, suggesting that the turbulence implied by the observed microstructure is in various stages of fossilization and usually affected by buoyancy.

ε values measured from the dropsonde are much less than average values from towed bodies at most depths. Even Crawford's largest value of 10^{-2} cm^2/sec^3, observed at about 60 meters depth, was 6 times less than the minimum average value reported by Belyaev et al. (1974) in the same region of the Atlantic equatorial undercurrent; most values were at least 3 orders of magnitude less. Belyaev et al. (1974) show that the dissipation rate from their horizontal measurements is quite patchy, with extreme values of ε as large as 2 cm^2/sec^3, but with average values for most reported depths ranging from 3 x 10^{-2} to 4 x 10^{-1} cm^2/sec^3. Kurtosis values for the local ε estimates were as large as 38 for the (most patchy) layers with minimum $\bar{\varepsilon}$ values. Little information is given regarding the possible effects of vibrational noise or contamination by temperature sensitivity. For such large signal levels these effects should be small. It seems likely that the discrepancy between the dropsonde and towed body estimates of $\bar{\varepsilon}$ arises because the dropsondes undersample at each layer. Undersampling a very intermittent quantity like ε will, with high probability, produce an underestimate of $\bar{\varepsilon}$.

A similar discrepancy exists between tow-body and dropsonde estimates of $\bar{\chi}$. Gregg (1976) reports estimates of χ, from various depths in the Pacific equatorial undercurrent, which range from 2 x 10^{-10} to 1 x 10^{-7} $°C^2/sec$, with a value of 2 x 10^{-8} for the depth range 90-140 m compared to 10^{-5} $°C^2/sec$ measured by Williams and Gibson (1974) from a towed body at 110 m. Belyaev et al. (1974) report microconductivity spectra from which one can estimate χ values. It is assumed that the conductivity is dominated by temperature, and that $\phi = \beta_K \varepsilon^{-1/3} \chi k^{-5/3}$. Values of 2.4 x 10^{-6} to 1.5 x 10^{-5} $°C^2/sec$ are indicated for the depth range 80 - 176 m, which are also 2 to 3 orders of magnitude greater than the Gregg (1976) dropsonde estimates. From the diffusive cutoff of (very noisy) temperature spectra, Williams and Gibson (1974) found $\bar{\varepsilon}$ to be about 0.1 cm^2/sec^3, which is consistent with the Russian tow-body estimates, but much larger than the Crawford (1976) dropsonde values (Fig. 4).

Although the available data seems to indicate that the apparent discrepancy between tow-body and dropsonde estimates of mean dissipation rates may be explained in terms of the difference in sampling methods, the data is still too meager for a definite conclusion to be drawn. Certainly the evidence suggests that extreme caution should be exercised when interpreting a single dropsonde record, or even a short towed body record, as giving a representative estimate of the average dissipation rates. It also appears that the presence of velocity microstructure in a temperature microstructure patch is not definite evidence that the patch is actively turbulent.

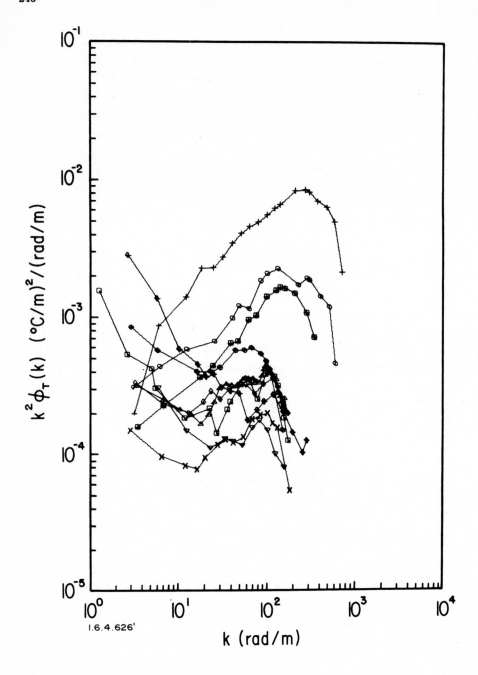

Figure 5.

Temperature gradient spectra below base of mixed layer by Schedvin (1979)
showing variety of forms and levels.

Figure 5 shows a variety of temperature gradient spectra measured with a towed body by Schedvin (1979) using a small platinum film resistance thermo-meter. Instrumentation, frequency response, and noise corrections are described by Schedvin (1979). The tow-body was rapidly profiled through the complex layered structure below the base of a wind mixed layer, and the spectra shown in Figure 5 were conditionally sampled to be representative of various identifiable layers or regions. Although only 4 km of data were analyzed, a wide variety of spectral levels were observed, with χ values in the range of about 10^{-8} to 10^{-6} $^\circ C^2$/sec. It should be noted that the high amplitude spectra also tend to extend to the highest wavenumbers and have the widest "microstructure" subrange.

Figure 6 shows the same data, normalized by the buoyancy scales L_{BF}, Σ_{BF}, and T_{BF}. The buoyancy scales have already been discussed and are defined in Table 1. The peak of the turbulent temperature gradient spectrum normalized with Batchelor scales occurs at $[k^2\phi(\gamma D)^{1/2}/\chi]_B \doteq 0.3$ and normalized wavenumber $[k(D/\gamma)^{1/2}]_B \doteq 0.3$. On the buoyancy normalized plot in Figure 6, these tur-bulent spectral peaks are distributed along a locus of points with slope -1, with γ/N as a mapping parameter. If the spectra are actively turbulent at the point of fossilization, the wavenumber marking the transition to the "microstructure" subrange, k_{T_O}, should be about $2\pi L_{R_O}^{-1}$, and the peak wavenumber, k_{B_O}, should be about $0.3\,L_{B_O}^{-1}$. Taking the ratio gives $(k_{B_O}/k_{T_O}) = (0.3/2\pi)Pr^{1/2}$ $(\gamma_O/N)^{3/2}$. If we assume that the minimum value for $(k_{B_O}/k_{T_O}) = \sqrt{Pr}$, then the value of γ_O/N corresponding to marginally active turbulence is $(2\pi/0.3)^{2/3} = 7.5$. This value depends on the proportionality constant in $k_{T_O} \sim L_{R_O}^{-1}$, which we took to be 2π.

As a means of testing a given spectrum for fossilization, we can compare the extent of its microstructure range with that for active turbulence at the point of fossilization. For this purpose a locus of points is formed on Figure 6 for the buoyancy normalized spectral peak, $(k^2\phi)(ND)^{1/2}/\chi$, plotted at the location of the transition wavenumber, k_T, at fossilization, where $\gamma = \gamma_O$ and $\chi = \chi_O$. Since $(k^2\phi(\gamma_O D)^{1/2}/\chi_O)_B = 0.3$, we find $(k^2\phi(ND)^{1/2}/\gamma_O)_B = 0.3(N/\gamma_O)^{1/2}$, and since $k_{T_O} = 2\pi(N/\gamma_O)(N/\nu)^{1/2}$, we find $k_{T_O}(D/N)^{1/2} = 2\pi(N/\gamma_O)Pr^{1/2}$. The resulting line has slope +1/2.

The test is carried out as follows: 1. plot the buoyancy normalized spectrum on Figure 6, 2. locate the transition wavenumber at fossilization, k_{T_O}, by projecting a horizontal line from the spectral peak to the transition wavenumber locus, 3. compare the actual transition wavenumber of the spectrum, k_T, (if any) with k_{T_O}. If k_T and k_{T_O} are equal, the fluid may be turbulent at the point of fossilization. If k_T is larger than k_{T_O}, the inertial forces of the turbulence dominate buoyancy at all scales. If k_T is less than k_{T_O}, then the wavenumbers between them are fossil temperature turbulence. By this method,

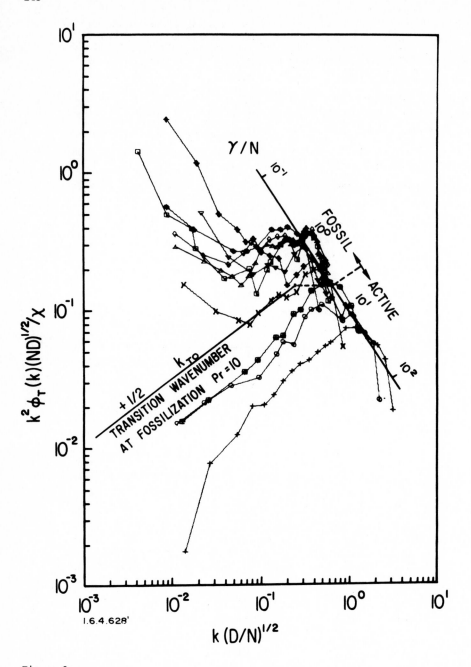

Figure 6.

Spectra of Figure 5 normalized by L_{BF} and Σ_{BF}, (Table 1). "Buoyancy scaling."

all of the spectra of Figure 6 appear to be at least partially fossilized, even though γ/N values up to about 15 are observed. Many of the spectra have γ/N values less than 1, which suggests they are fossil but strained by sub-saturated shear or subsaturated internal wave motions. No fully turbulent regions are indicated.

Published oceanic temperature gradient spectra similarly normalized by Gibson and Deaton (1979) show the same tendency to peak in the range $k(D/N)^{1/2} = 0.2$ to 0.5, as is shown in Figure 6. Most exhibit extensive micro-structure subranges which imply fossil temperature turbulence by the transition wavenumber criteria discussed above. For example, Gregg (1977, p.450) gives a temperature gradient spectrum from an "active patch" with peak wavenumber at 20 cycles/meter in a region where $\partial\overline{T}/\partial z = 2.98 \times 10^{-2}$ °C/m. Assuming thermal

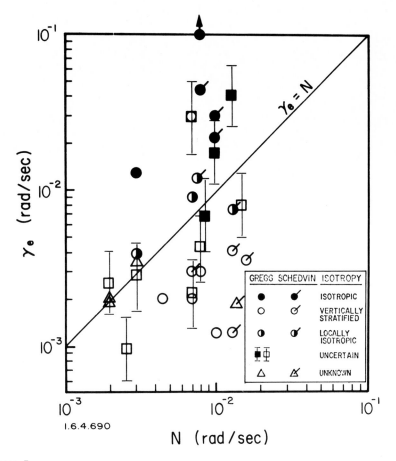

Figure 7.

Strain rate γ_e indicated by diffusive cutoff versus N. Darkened points are isotropic, open are vertically stratified, partially darkened are locally isotropic (See Table 2).

Table 2 (from Schedvin 1979)

Summary of spectra for γ_e, N Comparison

Identifier	Depth (meters)	Cox number	k_p (m^{-1})	γ_e (sec^{-1} x 100)	N (sec^{-1} x 100)	λ_i (m)	$\frac{\gamma_e}{N}$	sym-bol	Ref.	Comments
MC04	38-64	>3300	>600	>10	0.8	≤1.0	>12	●	G76,1	Wing, nose records suggest isotropy scales, <1 meter inversions common
MC07	89-1339	8	144	0.5-1.4	1.5	NA	0.3-1.0	□	G76,1	thermocline region with weak intermittent microstructure.
MC09	305-342	128	73	0.4	0.3	<0.5	1.3	◑	G76,1	many inversions 0.1-0.15 m possible isotropy at 0.5 m
MC09	356-380	2	65	0.12	1.0	<<0.1	0.12	○	G76,1	completely anisotropic
MC10	380-500	21	200	1.1-3.	1.0	NA	1-3	■	G76,1	several intense m.s. patches, largest in high gradient N = 1.3×10^{-3}
TAS11,MSR01	241-393	22	330	3.0	0.7(2)	<0.02	4	○	G77,1	N from dT/dz, max. slope +0.4
TAS11,MSR03	474-589	12	120	0.9	0.7	<0.05	1.4	◑	G77,1	apparent approach to isotropy
TAS1,MSR17	226-374	5	96	0.3-0.7	0.8	NA	0.4-0.9	□	G77,1	irregular steppy profile with relatively constant m.s. level
C2,MSR15	255-416	19	127	0.4-1.2	0.85	NA	0.5-1.5	■	G77,1	many 0.1 m inversions, patch with 0.2-0.3 m overturn
C2,MSR20	653-656	>107	>250	1.7-5.0	0.7(1)	NA	2.5-7.	■	G77,1	active patch, 1.6 m overturn
C2,MSR20	650-652	NA	47	0.06-0.15	0.26(1)	NA	0.2-0.6	□	G77,1	very quiet, little micro-structure
C2,MSR23	857-984	36	82	0.2-0.5	0.3(3)	NA	0.7-1.7	△	G77,1+2	
C2,MSR33	1385-1500	19	69	0.13-0.35	0.2(3)	NA	0.6-1.6	△	G77,1+2	N estimated from STD profile
C2,MSR39	3506-3642	30	68	0.13-0.35	0.2(3)	NA	0.6-1.6	△	G77,1+2	
A9,MSR4	270-373	2.2	87	0.2	0.6(1)	<<0.1	0.3	○	GCH	Complete anisotropy observed
A9,MSR7	550-672	2	92	0.2	0.45(3)	<<0.1	0.4	○	GCH	strong anisotropy to small scales
SDT,MSR1	212-340	>286	131	1.3	0.3	0.1-0.3	4	●	GCH	many inversions <1 m, one >1. m
TAS11,MC11	120-187	33	105	0.3	0.8	<<0.1	0.4	○	G76,2	
TAS11,MC15	35-53	330	≥290	2.6-6.4	1.3(1)	NA	2-5	■	G76,2	transition at base of the mixed layer with entrainment indicated
TAS11,MC15	60-95	NA	80	0.2-0.5	0.3	NA	0.7-1.7	□	G76,2	very quiet, smooth gradient
TAS11,MC29	50-59	24	75	0.16-0.4	0.2(1)	NA	0.8-2.0	□	G76,2	transition base of the mixed layer weak winds, profile suggests no entrainment
TAS11,MC29	115-180	5	70	0.1-0.35	0.7	NA	0.2-0.5	□	G76,2	irregular steppy, quiet profile
1-T2	43.0-45.5	13,200	280	4.5	0.8	≥1.0	5.6	◕	Schedvin (1979)	
2-T2	45.0-50.0	980	215	3.0	1.0	0.2	3.0	◕		
3-T2GH	49.5-57.0	29	100	0.76(5)	1.3	≤0.1	0.6	◑		apparent approach to isotropy at scales less than 0.1 meters
5-T2GH	52.5-58.5	4.2	60	0.19(5)	1.4	NA	0.1	△		
6-T2GH	49.0-54.0	4.2	38	0.12	1.3	<0.1	0.1	◔		anisotropic to smallest scales
7-T2	48.5-51.5	2,790	130	1.2	0.8	0.15	1.5	◕		
8-T2GH	50.5-55.5	30	43	0.4(4)	1.3	<0.05	0.3	◔		anisotropic to smallest scales
9-T2	47.5-53.0	560	180	2.2	1.0	0.3	2.2	●		
10-T2GH	51.0-58.5	4.0	71	0.35	1.6	<0.05	0.2	◔		anisotropic to smallest scales
11-T2GH	53.0-62.5	21	28	0.4(4)	0.7	<0.1	0.4	◔		anisotropic to smallest scales

notes:
(1) N calculated from MSR potential density gradient
(2) N calculated from dT/dz
(3) N estimated from STD profile
(4) γ_e from fit to AA spectrum corrected to value appropriate for $\sin|\theta_s|$ scaling and corrected for complete anisotropy.
(5) γ_e computed from $\gamma_e = D_T (k_p/0.43)^2 = \gamma/2.05$

stratification gives $N \approx 7.6 \times 10^{-3}$ rad/sec, then $k_B (D/N)^{1/2}$ is about 1.45, but with a microstructure subrange of over a decade and a half. The indicated γ/N is only $(0.45/0.3)^{1/2} \approx 1.2$. By the spectral comparisons using Fig. 6, the patch appears to be fossil at all wavenumbers in the "microstructure" subrange.

An important characteristic of active turbulence is the tendency for both velocity and scalar fields mixed by the velocity to approach local isotropy. Figure 7 shows a plot of the 'effective' rate of strain, $\gamma_e = \gamma_B \beta_B^{-1}$, indicated by the diffusive cutoff wavenumber of various oceanic temperature spectra versus N, from Schedvin (1979), where β_B is a universal constant in the range, $\sqrt{3}$ to $2\sqrt{3}$. In Figure 7, tow-body data points have flags; dropsonde points do not. Darkened points indicate that spectra from samples taken at different angles from the horizontal, in the same region, gave the same spectral level. Open points indicate that substantial anisotropy was observed, with stronger gradients in the vertical. Triangles are for spectra with no indication of isotropy, and partially darkened circles indicate only small scale isotropy, with anisotropic large scales. Descriptions of the individual points are given in Table 2 (from Schedvin 1979).

As shown in Figure 7, there is a clear tendency for the points with $\gamma_e \gg N$ to be isotropic, for $\gamma_e \ll N$ to be anisotropic, and for $\gamma_e \sim N$ to be only locally isotropic. There seems to be no strong tendency for points to cluster near the $\gamma_e = N$ line.

As shown in the last section, we have two independent methods of estimating the rate of strain in the fluid, depending on our assumption of whether it is active or fossil turbulence. If we assume only active turbulence, or active turbulence at the point of fossilization, we find from Eq. (11) and (12) that

$$\beta_B \gamma_e = \gamma_B \geq \gamma_o = 3.6 \; (C_o/Pr)^{1/2} N$$

$$= 3.6 \; (C/Pr)^{1/2} \; N \; (C_o/C)^{1/2} \doteq 3.6 \; (C/Pr)^{1/2} N \equiv \gamma'_o \qquad (13)$$

because, for active turbulence, $(C_o/C) = \chi_o/\chi = (\epsilon_p/\epsilon)^{1/11} \doteq 1$. The strain rate can therefore be estimated from the Cox number, C_o, and the Väisälä frequency, N, and should agree with the value, γ_B, estimated from the universal peak wavenumber $k_B = 0.3(\gamma_B/D)^{1/2}$,

$$\gamma_B = D(k_B/0.3)^2 \qquad (14)$$

which pertains to all actively turbulent scalar gradient spectra.

However, if the fluid is in a transitional or fossil state, then the actual rate of strain, $\gamma \approx \gamma_B$, may be much less than $\gamma_o \approx \gamma'_o$ from (13). During transition

$$N \leq \gamma_B \leq \gamma_o , \qquad (15)$$

and for fossil turbulence

$$\gamma_B \leq N \ll \gamma_o. \tag{16}$$

We may use (13), (15), and (16) to define a turbulence activity parameter A_T where

$$A_T \equiv \gamma_B/\gamma'_o = D(k_B/0.3)^2/3.6(C/Pr)^{1/2}N \gtrless \gamma_B/\gamma_o \tag{17a}$$

which can indicate the state of the fluid from measurements of k_B, C, Pr, and N.

The equality and inequality conditions of (17) correspond to various hydro-physical states of the fluid as follows:

$$\gamma_B/\gamma_o > A_T > 1 \qquad\qquad \text{active turbulence} \tag{17b}$$

$$A_T = 1 \qquad\qquad \text{active turbulence at fossilization} \tag{17c}$$

$$1 \geq \gamma_B/\gamma_o \leq A_T \geq (Pr/C)^{1/2}/3.6 \qquad\qquad \text{transition, active and fossil} \atop \text{turbulence combined} \tag{17d}$$

$$A_T = (Pr/C_F)^{1/2}/3.6 \qquad\qquad \text{fossil turbulence, saturated} \atop \text{internal waves} \tag{17e}$$

$$A_T < (Pr/C_F)^{1/2}/3.6 \qquad\qquad \text{fossil turbulence, weak} \atop \text{internal waves.} \tag{17f}$$

Measurements of A_T versus C, by Caldwell and Dillon (personal communication) taken during the MILE expedition, using a high frequency response thermistor on a dropsonde, are shown in Figure 8. The dropsonde was small and could be cycled frequently by means of an **attached line** and fishing pole arrangement. Values of N were computed from the mean vertical temperature gradients of the individual records. Cox numbers ranged from 1 to nearly 10^8. When the turbulence is strong, some of the very large Cox numbers may be underestimated due to inadequate frequency response, or overestimated because the denominator for the record, $\partial \overline{T}/\partial z$, may approach zero and not represent the local stratification. A point from the UCSD tow-body MILE data (also unpublished) is shown, as well as representative values computed from Table 2 for Gregg (MC09/MSR03) and Schedvin (1-T2/7-T2). The various regions of fluid motion defined by Equation (17 a b c d e f) are indicated on Figure 8.

The measured A_T values in Figure 8 are near 1 only for C values near 1, but show a strong tendency to decrease as C increases. Most points lie in the transition range, suggesting a mixture of fossil and active turbulence as indicated by the spectral comparisons described in Figure 6 for the Schedvin (1979) data. The decrease in A_T at large C may reflect the fact that fossils of turbulent patches with large C_o and γ_o/N values are more persistent ($\tau_T \sim (\gamma_o/N)^{2/3} N^{-1}$), and therefore will be encountered more frequently than the original active patches. Patches with small C_o are more numerous and less persistent as fossils than those with large C_o. Therefore, they are more likely to be encountered in an active turbulence state (A_T near 1.0) than in a fossil state (A_T near $C^{-1/2}$).

It is interesting to note that even though the patches with large Cox number have larger γ values, with $\gamma \gg N$, and larger mixing rates, χ, than those with small C values, they actually represent turbulence patches in a more advanced stage of fossilization, as indicated by the turbulence activity parameter $A_T \ll 1$.

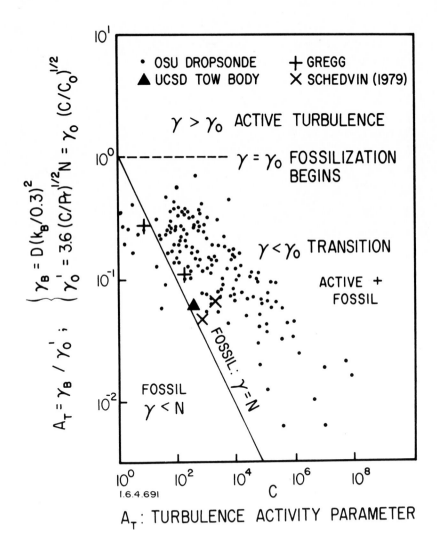

Figure 8.

Turbulence activity parameter versus Cox number. Measured values from Caldwell and Dillon (unpublished) MILE data using a dropsonde thermistor •, UCSD tow-body (unpublished) MILE data Δ, Gregg (Table 2) +, and Schedvin (Table 2) ×.

The Cox number of the turbulence at fossilization $C_o \simeq C^{3/2}/(\gamma/N)$. Also, values of χ_o and ε_o may be estimated, and are made larger than χ and ε in fossilized microstructure regions when C is large and γ/N is small. For example, for Schedvin's 1-T2 spectra from Table 2, C_o, ε_o, and χ_o are an order of magnitude larger than the measured values of C, ε, and χ, respectively.

5. SUMMARY AND CONCLUSIONS

The evolution and decay of the velocity and temperature fields in an isolated region (patch or layer) of very strong turbulence in a thermally stratified fluid after the source of turbulent energy is removed has been analyzed theoretcally. Several important parameters, scales, and spectral properties were identified for this model flow (see Table 1), which should be representative of the small scale structure of other stratified, turbulent velocity and scalar fields. It was found that remnants of the active turbulence should persist in the temperature and vorticity fields as fossil temperature turbulence and fossil vorticity turbulence after the active turbulence has been damped out by buoyancy and viscosity. The persistence time of fossil temperature turbulence should be $\sim(\gamma_o/N)^{2/3}N^{-1}$; the persistence time of fossil vorticity turbulence should be $\sim(\gamma_o/N)^2 N^{-1}$, where γ_o is the rate of strain of active turbulence affected by buoyancy only at the largest eddy size, as fossilization begins, and N is the Brunt-Väisälä frequency. The ratio γ_o/N is an important parameter of stratified turbulence ($\gamma_o/N \sim \sqrt{Re_o} \sim \sqrt{C_o/Pr}$, where Re is the Reynolds number at fossilization, C_o is the Cox number, and Pr is the Prandtl number) since many other parameters depend on this quantity.

Ratios γ_e/N from oceanic temperature microstructure measurements were found to be less than $\sqrt{C/Pr}$, indicating that the microstructure is not fully turbulent, where γ_e is the effective rate of strain inferred from the diffusive cutoff wavenumber. Often $\gamma_e/N < 1.0$, indicating that the smallest microstructure scales are maintained at $(D/\gamma_e)^{1/2}$ by nonturbulent internal wave or shearing motions. Although sometimes $\gamma_e/N > 1$, necessary for active turbulence, γ_e/N was never $> \sqrt{C/Pr}$, necessary for the turbulence to be fully active. Spectra with $\gamma_e/N > 1$ were found to be isotropic; $\gamma_e/N = 1$, locally isotropic; $\gamma_e/N < 1$, anisotropic and vertically stratified (Schedvin 1979): a pattern of isotropy-anisotropy consistent with the active-fossil turbulence interpretation.

The hydrodynamic state of microstructure may be determined from the ratio $\gamma_B/3.6\ N\ \sqrt{C/Pr} \equiv A_T \doteq \gamma_e/N\ \sqrt{C/Pr}$, the turbulence activity parameter. When $A_T \gtrsim 1$ the fluid is actively turbulent; when $1 > A_T > (Pr/C)^{1/2}/3.6$ the fluid is partially active turbulence and partially fossil turbulence, and when $A_T \lesssim 1$ the fluid is nonturbulent and the microstructure is fossil temperature turbulence at all scales.

Values of A_T for all available oceanic microtemperature measurements were plotted versus C in Figure 8. By this criterion, most microstructure is a mixture of active and fossil turbulence, some is nonturbulent, and none is fully turbulent. Since regions of fully active turbulence required to produce the observed fossil turbulence are not included in the microstructure measurement records, the records undersample the turbulent mixing and diffusion phenomena. Estimates of average dissipation rates and vertical turbulent diffusion rates from measurements in regions which are partially or completely fossil, but which do not take the possibility of fossil turbulence into account, may be very inaccurate. The strong tendency of stratified turbulence to be patchy and intermittent requires that very long data records (much longer than have been collected so far) be collected so that the mixing and diffusion parameters of the record will be representative of the layer.

Because of the extreme patchiness and intermittency of active turbulent events in the ocean, measurements which include a representative number of active turbulent events may be practically impossible. Fossil turbulence can preserve information about the active turbulent events such as ε_o and χ_o, and the location and volume fraction of the turbulence patches. If the persistence times and evolution of fossil turbulence structure can be determined for a wide range of γ_o/N values, it may be possible to refine the present model and improve methods of estimating the mixing and diffusion by active turbulence in a stratified medium from microstructure measurements by taking the effects of patchiness and fossil turbulence into account. If measurements covering the active turbulence regions are impossible, a scheme must be devised based entirely on evidence from fossil turbulence measurements and models. Such a system of "hydropaleontology" may be vital to any complete study of fluxes of conserved properties in the ocean and atmosphere, especially in interior regions away from fully turbulent boundary interfaces.

Acknowledgements

Funding for this work was provided by the Office of Naval Research, ONR Contract Grant Number N00014-75-C-0152, and the National Science Foundation, NSF ENG 27398 (Cal Tech P. O. # 28-464865). The author is grateful to Doug Caldwell and Tom Dillon for useful discussions and access to their unpublished MILE results, and to his students John Schedvin and Libe Washburn for analyzing some of the data and for many useful discussions of the ideas presented.

REFERENCES

Batchelor, G.K., 1959. Small-scale variation of convected quantities like temperature in a turbulent fluid. Journal of Fluid Mech., 5:113-133.

Belyaev, V.S., A.N. Gezentsvey, A.S. Monin, R.V. Ozmidov and V.T. Paka, 1974. Spectral characteristics of small-scale fluctuations of hydrophysical fields in the upper layer of the ocean. Journal of Physical Oceanography, 5:492-498.

Belyaev, V.S., M.M. Lubimtzev and R.V. Ozmidov, 1974. The rate of dissipation of turbulent energy in the upper layer of the ocean. Journal of Physical Oceanography, 5:499-505.

Bretherton, F.P., 1969. Waves and turbulence in stably stratified fluids. Radio Science, 4 No.12:1279-1287.

Crawford, W.B., 1976. Turbulent energy dissipation in the Atlantic equatorial undercurrent. Thesis, The University of British Columbia, Canada.

Dougherty, J.P., 1961. The anisotropy of turbulence at the meteor level. Journal of Atmos. and Terrestrial Physics, 21:210-212.

Dunbar, M.J., 1958. Physical Oceanographic results of the "Calanus" expedition in Ungava Bay, Frobisher Bay, Cumberland South, Judson Strait and Northern Hudson Bay, 1945-1955. Journal of Fish. Res. Board of Canada, 15:115-201.

Federov, K.N., 1978. The thermohaline finestructure of the ocean. Pergamon Press, 170.

Gargett, A.E., 1976. An investigation of the occurrence of oceanic turbulence with respect to finestructure. Journal of Physical Oceanography, 6:139-156.

Garrett, C., and W. Munk, 1975. Oceanic mixing by breaking internal waves. Deep Sea Res., 19:823-832.

Gibson, C.H., 1968. Fine structure of scalar fields mixed by turbulence:Part I. Zero-gradient points and minimal gradient surface. Physics of Fluids, 11:2305-2315. Spectral Theory. Physics of Fluids, 11:2316-2327.

Gibson, C.H. and T. Deaton, 1979. Hot/cold sensors of oceanic microstructure. Instruments and Methods of Air-Sea Interaction, Russ Davis, Ed., NATO Textbook, to be published.

Gregg, M.C., 1976. Finestructure and microstructure observations during the passage of a mild storm. Journal of Physical Oceanography, 6:528-555.

Gregg, M.C., 1976. Temperature and salinity microstructure in the Pacific equatorial undercurrent. Journal of Geophysical Res., 81:1180-1196.

Gregg, M.C., 1977. A comparison of finestructure spectra from the main thermocline. Journal of Physical Oceanography, 7:33-40.

Gregg, M.C., C.S. Cox, and P.W. Hacker, 1973. Vertical microstructure measurements in the central north pacific. Journal of Physical Oceanography, 3:458-469.

Lumley, J.L., 1964. The spectrum of nearly inertial turbulence in a stably stratified fluid. Journal of Atmospheric Science, 21:99-102.

Orr, Marshall, 1978. Paper presented at the American Physical Society, Fluid Dynamics Division Meeting, University of Southern California, Bull. APS, 11, 1978.

Osborn, T.R., 1978. Measurements of energy dissipation adjacent to an island. Journal of Geophysical Res., 83:C6:2939-2957.

Osborn, T.R. and C.S. Cox, 1972. Oceanic finestructure. Geophysical Fluid Dynamics, 3:321-345.

Ottersten, Hans, 1969. Mean vertical gradient of potential refractive index in turbulent mixing and radar detection of CAT. Radio Science, 4:No.7:1247-1249.

Ozmidov, R.V., 1965. On the turbulent exchange in a stably stratified ocean. Izv. Atm. Ocean Phys., 1:853-860.

Richter, J.H., 1969. High resolution tropospheric radar sounding. Radio Science, 4:No.12:1261-1268.

Schedvin, J.C., 1979. Microscale temperature measurements in the upper ocean from a towed body. Thesis, University of California at San Diego, 422.

Stewart, R.W., 1969. Turbulence and waves in a stratified atmosphere. Radio Science, 4:No.12:1269-1278.

Thorpe, S.A., 1973. Turbulence in stably stratified fluids: A review of laboratory experiments. Boundary-Layer Meteorology, 5:95-119.

Turner, J.S., 1973. Buoyancy effects in fluids, Cambridge Press, 367.
Williams, R.B. and C.H. Gibson, 1974. Direct measurements of turbulence in the
    Pacific equatorial undercurrent. Journal of Physical Oceanoαraphy, 4:104-108.
Woods, J.D. Ed., Report of working group (V. Hogstrom, P. Misme, H. Ottersten
    and O.M. Phillips): fossil turbulence, Radio Science 4:1365-1367.

VARIATIONS WITH HEIGHT OF THE TURBULENCE IN A TIDALLY-INDUCED BOTTOM
BOUNDARY LAYER.

K. F. BOWDEN and S. R. FERGUSON

Oceanography Department, University of Liverpool (England).

ABSTRACT

Turbulent fluctuations of velocity were measured in the bottom boundary layer
at several sites in the eastern Irish Sea.   The measurements were made using
two-component electromagnetic flowmeters with a frequency response extending
to 2Hz.   Signals from three sensors at heights of 50 cm, 100 cm and 200-210 cm
from the sea bed have been analysed to observe possible variations with height
of the turbulence structure.   The signals analysed covered a range of mean
flow velocities, U, extending up to approximately 70 cm.s^{-1}, and included data
obtained at different stages of the semi-diurnal tidal cycle.
The r.m.s. levels of the longitudinal (u) and vertical (w) components
were well correlated with U at each height.   The variations with height in the
level of the u component were generally insignificant but the w component showed
a slightly higher level at the uppermost sensor.   The mean product $\overline{uw}$
correlated well with U^2 at each height, but the variability between estimates
of $\overline{uw}$ was larger than any systematic variation with height.
Although systematic variations with height of the mean values $(\overline{u^2})^{\frac{1}{2}}$, $(\overline{w^2})^{\frac{1}{2}}$ and
$\overline{uw}$ were generally less than the variability at each height, some significant
variations with height were observed in the spectral levels at different
wavenumbers.   At low wavenumbers ($\sim 2 \times 10^{-3}$ cm^{-1}) the w spectral level
increased with height, whereas the u spectral levels were similar.   At higher
wavenumbers ($\sim 2 \times 10^{-1}$ cm^{-1}) the spectral levels of both components decreased
with increasing height.

INTRODUCTION

One of the main points of interest of turbulence in the bottom boundary layer

of the sea is the generation of Reynolds stresses, which are the means by which

the frictional stress at the bed is communicated to the water.   The stress

originating at the boundary has important effects both on the bed, where it

influences the erosion and deposition of sediment and the formation of ripples,

and on the water, in which it determines the profile of the mean velocity

within the boundary layer.   The direct measurement of bottom stress presents

technical difficulties and has only recently become practicable.   More usually

the bottom stress is derived from measurements made in the water of either

(a) the profile of the mean velocity, within the first one or two metres above

the bed or (b) the Reynolds stresses derived from the components of turbulent

velocity measured at a particular height.

Other features of interest in the turbulence arise from the effects of the turbulent velocity components in maintaining material in suspension and their effects on diffusion through the boundary layer.   The diffusive flux plays an essential part in the exchange of dissolved substances between the overlying water and the pore waters of the sediment.   A survey of problems involving bottom boundary layer flow in both deep and shallow waters was given by Bowden (1978).

The present paper arises from a programme of investigation of turbulence in tidal currents in relatively shallow water.   It involves the direct measurement of the three components of turbulent velocity and the component of Reynolds stress parallel to the boundary in the direction of the mean flow. An additional problem in the tidal situation is whether the time variation in the currents, involving periodic reversals in direction, has an important effect on the turbulent intensities and stresses.   One may expect, as a first approximation, that when the currents are near a maximum the turbulent effects will be in quasi-equilibrium with the mean flow and will resemble those in a steady flow of similar magnitude, but significant differences might occur during the accelerating and decelerating phases.   Recent studies of turbulent flow in tidal currents include those by Heathershaw (1976), who made measurements at two heights near the bottom, mostly near times of maximum current, and by Soulsby (1977), who obtained measurements at two heights for a period of over 4 hours, but in a situation where the current speed increased rapidly to a value which then remained almost steady for most of the period.

Particular objectives of the present investigation were to determine

(1) whether there is any significant variation of stress within the first 2 m above the bed and, if so, over what range of height the Reynolds stress measurements are indicative of the stress at the boundary,

(2) the variation of the turbulent velocity components and stress with time over as much of a complete tidal cycle as possible.   This would make it possible to investigate the dependence of the turbulence characteristics on the mean velocity and to compare these characteristics during the accelerating and decelerating phases of the flow.

INSTRUMENTATION

Measurements of fluid velocities near the sea bed were made using two types of sensor.   Turbulence components were measured using four two-component electromagnetic flowmeters.   These were 5 cm diameter discus type units, based on a design developed at the British National Institute of Oceanography and described by Tucker, Smith, Pierce and Collins (1970), Tucker (1972) and Heathershaw (1975).   The sensors were manufactured under licence by Colnbrook

Instrument Developments Ltd.   Mean flow speed and direction were measured at a
height of 100 cm from the sea bed using an impellor type directional current meter
(Braystoke BFM 008).   The four flowmeter sensors, their associated electronics
and the mean flow sensor were mounted on an aluminium frame for lowering to the
sea bed (see Figure 1).   At the rear of the frame a large vane orientated the
flowmeter sensors into the mean flow during lowering.

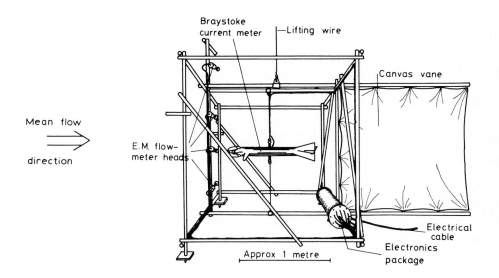

Figure 1.   Sketch of the turbulence rig, showing e.m. flowmeters on the upstream
side, and directional reading Braystoke current meter mounted at a height of 100 cm
above the sea bed.

Signals from the velocity sensors were carried to recording equipment on board
an anchored support vessel by a multicore electrical cable (see Figure 2).   Analogue
signals from the eight flowmeter channels were continuously recorded on an analogue
magnetic tape recorder, using a frequency modulation/multiplexing system.   The
useful frequency range of the flowmeter and associated electronics was DC-5 Hz.
Each of the sensors and its associated electronics were calibrated in the
recirculating flume of the Mechanical Engineering Department, University of
Liverpool, with relative accuracies of $\pm$ 1%.   The absolute value of the flume
speed was known with the same accuracy.

EXPERIMENTAL PROCEDURE.

At a chosen station the ship was anchored and, as soon after slack water as possible, the rig was lowered to the sea bed, where it remained with fixed orientation until the following slack water. (If the mean current direction veered more than about 20° the frame was redeployed). Readings from the Braystoke current meter were taken manually at 5 or 10 minute intervals (2 or 5 minute rotor count with spot direction reading before and after each count). The signals from the flowmeters were monitored on chart recorders during recording.

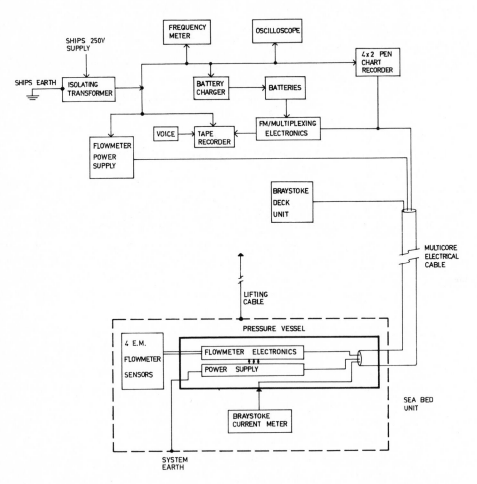

Figure 2. Block diagram of the turbulence recording system as operated at sea. The turbulence signals were monitored on chart recorders, and recorded on magnetic tape using an FM/MX system.

Following each cruise the recorded signals were replayed through a demultiplexing/demodulating unit to reproduce the original flowmeter signals. Nine channels (eight signal plus one reference channel) were sampled cyclically at 180 ms intervals, with 20 ms delay between adjacent channels. The analogue signals were digitised and punched on paper tape, the equivalent r.m.s. noise level of the sampled data being generally less than 1 mm.s^{-1}.   Prior to sampling all signal channels were low pass filtered to reduce aliasing.

DATA ANALYSIS

As some of the digitised data contained large amplitude high frequency spikes, the data analysis began with an examination by a pre-processing routine to identify and replace these spikes.   Most of the spikes are believed to have been caused by oxide dropout in the recording tape, although some noise spikes were also present during recording.   If more than about 20 spikes were identified in a roll of paper tape (containing over 40,000 samples) the roll was rejected for further analysis.

The data was analysed in 'records' 12 min 17 sec long, each record containing 4096 samples in each flowmeter channel.   The corrected raw data series were converted to units of velocity using the known calibration of each flowmeter channel, and the calibrated data became the 'input time series' of Figure 3, where the processing sequence is shown in diagrammatic form for two velocity components (denoted u and w) from a single sensor.   Each data series was detrended by subtracting a least square fitted straight line to leave the 'turbulent' fraction of each series, with zero mean.   All further processing was concerned only with these turbulence series.

The turbulence series were Fourier transformed to obtain the amplitude spectra, to which phase shifts were applied to correct for the non-simultaneity of the sampling process.   The phase shifted spectra were then transformed back to obtain time series with effectively simultaneous sampling of all channels. To obtain an alias-free product series between the two components at each flowmeter head, the phase shifted spectra were augmented with zeroes to form double length spectra, and transformed back to form double length time series with effectively twice the sampling rate of the original time series.   The term-by-term product of these series was formed, low pass filtered to the original Nyquist frequency, and decimated by a factor of two to form a product series with samples coincident with those of the phase shifted component series.   The output component and product series were subjected to standard statistical analysis.

Energy spectral density estimates for positive frequencies were obtained from the phase shifted amplitude spectra in the usual way (see for example Bendat and Piersol 1971).   The raw spectra were smoothed by block averaging to provide 36 smoothed spectral estimates at fixed frequencies ranging from 10^{-2} Hz to 2.5 Hz. The 80% confidence interval for the smoothed estimates G_i decreases from $0.68 \ G_i - 1.72 \ G_i$ at 2.5×10^{-2} Hz to $0.9 \ G_i - 1.1 \ G_i$ at 2.5 Hz.

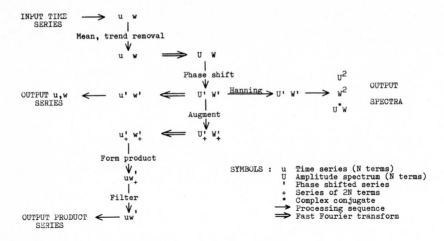

Figure 3. Time series analysis : processing sequence. The diagram indicates the procedures used to generate component and product series with effectively simultaneous samples for all series, and raw frequency spectra.

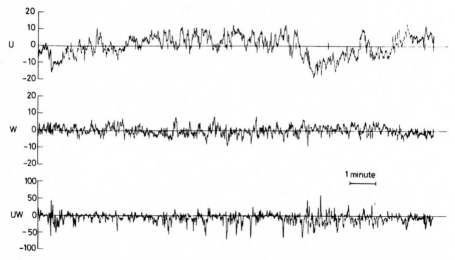

Figure 4. Example of u, w component series and uw product series from a sensor 100 cm above the sea bed. Tick marks are shown along the horizontal axis at 1 minute intervals - the total record length is 12min 17sec. Units of velocity components are $cm.s^{-1}$, of the product series $cm^2.s^{-2}$. Data shown are from Record 17, for which U_{100} = 24.6 $cm.s^{-1}$.

Wavenumber spectral density estimates $G(k_i)$ were derived from frequency spectral estimates $G(f_i)$ by the transformations

$$k_i = 2\pi f_i / U_{100},$$

$$G(k_i) = G(f_i) \, U_{100}/2\pi,$$

where U_{100} is the mean flow speed 100 cm above the sea bed.   The spectral estimates were corrected for the response of the flowmeter electronics and alias filters.

An example of processed time series from one flowmeter head is shown in Figure 4, showing u and w components at a height of 100 cm above the sea bed, and the generated uw product series.

NOTATION

For the purposes of discussion a right-handed cartesian co-ordinate system is used with z measured vertically upwards from the sea bed, and the x direction defined by the mean flow direction at the time of the frame deployment.   The turbulent velocity components in the x, y, z directions are u, v, w respectively. The mean flow speed determined by the Braystoke current meter is denoted U_{100}. A tidal phase ϕ at time T is defined as

$$\phi = \omega(T - T_o)$$

where $\omega = 1.4 \times 10^{-4}$ rad.s^{-1} is the angular frequency of the semi-diurnal tide, and T_o is the time midway between the times of slack water preceeding and following the tidal flow being considered.

RESULTS

Details of records.

The results to be considered in the following discussion were derived from data obtained from three successive ebb tides at station 7 and consecutive ebb and flood tides at station 10 in the Eastern Irish Sea (see Figure 5).   Only a few records have been analysed from the first two ebb tides at station 7 (4 and 3 respectively), but from the third ebb flow a total of 12 records have been analysed, covering a wide range of tidal phase ϕ.   On this tide however the turbulence rig had to be hauled in during the acceleration phase (as the ship was swinging on its anchor), but was redeployed almost immediately.   A plot of the mean flow velocity U_{100} and water depth during these three ebb tides is shown in Figure 6, with the break between trials 35 and 36 indicating the brief period during which the rig was being redeployed.   The times of the analysed records are indicated beneath the curves of U_{100}.

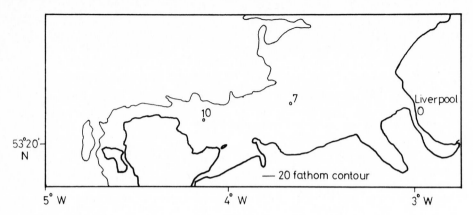

Figure 5. Chart of the Eastern Irish Sea showing stations at which turbulence data have been recorded. The data described here were obtained at stations 7 and 10 off the North Wales Coast.

For the two tides at station 10 the rig could not be deployed until near the end of the acceleration phase, although in both cases recording continued until almost slack water. Plots of U_{100} and water depth for these tides are also shown in Figure 6, together with the times of the analysed records (9 from trial 50, 7 from trial 51). Comparison of U_{100} values in Figure 6 shows that the peak tidal velocities recorded at station 10 were roughly twice those at station 7 – the recordings at station 10 were made very close to the time of spring tides, while those at station 7 were within a few days of neaps.

Most of the records analysed describe signals from three flowmeter heads (each measuring two components), although for some records the signals from one head were rejected as unsuitable for analysis due to the presence of zero checks within the record or occasional bursts of (presumably instrumental) noise. A total of 96 component pairs were considered acceptable and these are listed in Table 1 by trial and orientation. For trial 35/36 two heads were mounted 100 cm above the sea bed, one measuring u and w (mounted directly beneath the head measuring u and w at 200 cm) and one measuring u and v. The two heads at 100cm were separated by 30 cm laterally to avoid interference either by the magnetic fields of the two heads or by fluid dynamic effects. In the other trials all three heads were mounted along the same vertical line.

Turbulent intensities.

An examination of the r.m.s. levels of the turbulence components showed that there was significant correlation between the r.m.s. value of each component at each height, and U_{100} (see Figure 7 – the sloping straight lines in the figure will be discussed later). There is, however, some scatter of the

267

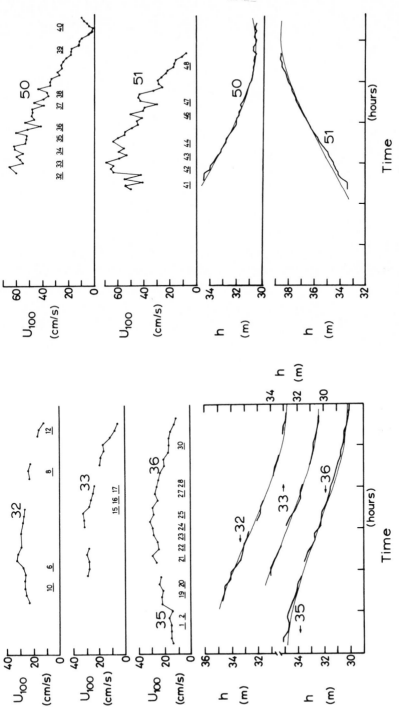

Figure 6. Mean velocity, U_{100}, and water depth, h, during three consecutive ebb tides at station 7 (left panels) and consecutive ebb and flood tides at station 10 (right panels). The data are plotted on common time axes, with approximately coincident times of slack water. Large figures indicate trial number, small underlined figures indicate record number. The underlining indicates the duration of each record. Smooth curves on depth plots represent expected tide (with arbitrary absolute value).

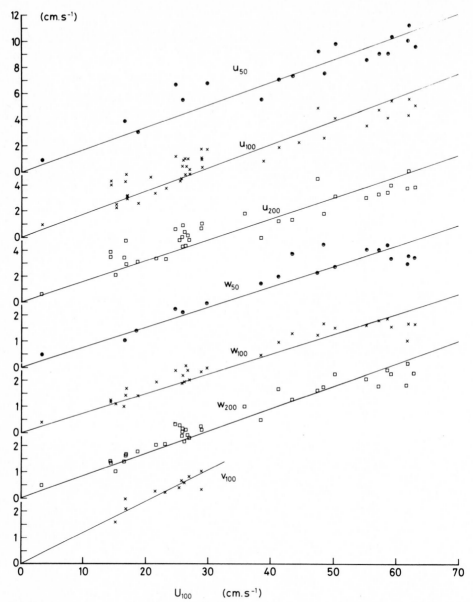

Figure 7. Plots of r.m.s. turbulence levels against U_{100} for all data. Different symbols are used to separate data for the u, v and w components at sensor heights (indicated as subscripts to the component identifier) of 50,100 or 200-210 cm. Scale marks on the vertical axis are continuous, but note change of scale for u plots. Straight lines represent equations of form y = Ax, least squares fitted to the appropriate data groups.

TABLE 1

Number of records analysed, by component and sensor height.

Trial	Components	Sensor height			
		50 cm	100 cm	200 cm	210 cm
32	u, w		5	5	
33	u, w	3	3	2	
35/36	u, w		8		
35/36	u, v		11	11	
50	u, w	9	9		9
51	u, w	7	7		7
all	u	19	43	34	
all	v		11		
all	w	19	32	34	

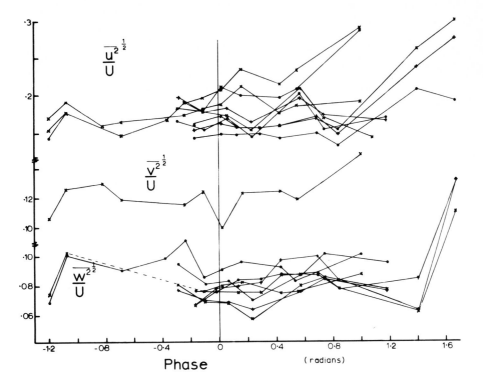

Figure 8. R.M.S. turbulence levels, normalised by U_{100}, plotted against tidal phase for the data from trials 35/36, 50 and 51. Data from a particular sensor are joined by straight lines. Symbols indicate sensor height : +, 50 cm; x, 100 cm; solid circle, 200-210 cm.

270

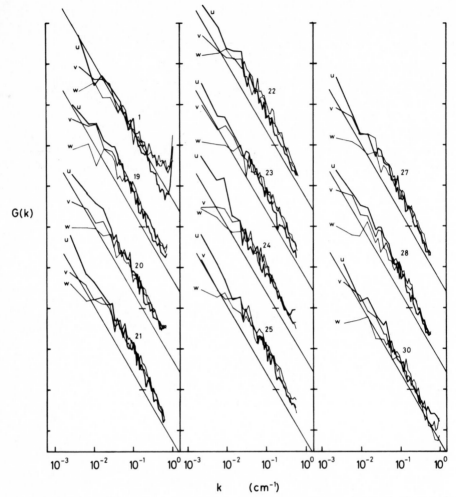

Figure 9. Wavenumber spectra from sensors at 100 cm above the sea bed, for the records from trial 35/36. Spectra are shown for u, v and w components. Sloping straight lines represent a $k^{-5/3}$ power law and have the same relative position for each record -- the horizontal tick mark through each line represents an energy density of 1 $cm^3.s^{-2}$. Vertical scale is logarithmic, with tick marks each decade.

points on the plots of r.m.s. level against U_{100}. As the data were obtained at different states of the tide some of this scatter may have been caused by a systematic variation in r.m.s. level with tidal phase. In order to check this, the dimensionless quantities of form $(\overline{u^2})^{1/2}/U_{100}$ were plotted against tidal phase ϕ (Figure 8). (The normalising velocity was chosen as U_{100} rather than the friction velocity, u_*, usually used in the examination of boundary layers, as the former was directly observed whereas the latter could only be inferred from estimates of the Reynolds stress $-\rho\overline{uw}$ and would be subject to much larger random, and possibly systematic, errors). Although the normalised r.m.s. values show a lot of scatter, there is no apparent systematic variation in level with tidal phase. (The last one, and sometimes two, records in several trials show a marked increase in normalised r.m.s. level. This increase can be attributed to surface wave effects and will be discussed later).

A second approach used to detect possible differences in turbulence structure with tidal phase was to examine the wavenumber spectra of records from different states of the tide. In Figure 9 u, v and w energy spectra are plotted for the records from trial 35/36. Apart from record 1, which appears to contain some high frequency noise, the spectra were very similar in form throughout the trial. In most records the spectral slope is approximately $k^{-5/3}$ for wavenumbers greater than $\sim 3 \times 10^{-2}$ cm^{-1} and for these wavenumbers the u-spectrum level is, on average, slightly lower than that of the v and w spectra, suggesting that the turbulence is approximately isotropic at these wavenumbers. There are of course shifts in absolute level of the spectra as U_{100} varies - the straight lines in Figure 9 representing a $k^{-5/3}$ dependence are in the same relative position for each record, and the horizontal dash crossing these lines near $k = 2 \times 10^{-1}$ cm^{-1} represents an energy density of 1 $cm^3.s^{-2}$.

As there was no apparent dependence of the normalised r.m.s. level on tidal phase, the r.m.s. level of each component, at each height, was parameterised in terms of U_{100} by fitting the data in Figure 7 to equations of the form $y = Ax$, using a least squares fitting criterion. The straight lines in Figure 7 represent the fitted values of the slope parameter A for each data group. By using this method to determine the slope A ($= \Sigma x_i y_i / \Sigma x_i^2$) values with small U_{100} (x in the equation) are given less weighting than for a simple average of terms like $(\overline{u^2})^{\frac{1}{2}}/U_{100}$. Because of this the anomalously high values for large ϕ in Figure 8 do not significantly affect the estimates of A. The resulting values of the slope, (r.m.s. level)/U_{100}; 90% confidence interval; regression coefficient r and number of data points used for each estimate are listed in Table 2. In calculating the slope for the uppermost sensor, values at 200 cm and 210 cm have been combined in a single estimate. The data were also fitted to an equation of the form $y = Ax + B$, but except for some data groups involving u at station 10 the intercept B was not significantly different from zero, so the use of this extra parameter did not seem to be justified.

From Table 2 it can be seen that the r.m.s. levels of u, v, and w were typically 17%, 12% and 8% of U_{100} respectively, with the 90% confidence interval width being about $\pm$ 10% of the central value at each height and somewhat less for the all-height values for u and w. In examining the slope values at each height, there appears to be no significant variation with height for the u estimates (in the sense that each central estimate lies within the 90% confidence interval of the other estimates). For the w component however, the

TABLE 2

Parameterisation of r.m.s. turbulence levels in terms of U_{100}.

Component	Height (cm)	Slope	Correlation coefficient	No. of observations
u	50	0.172 ± 0.025	0.94	19
u	100	0.179 ± 0.016	0.94	43
u	200	0.162 ± 0.018	0.93	34
v	100	0.120 ± 0.033	0.89	11
w	50	0.077 ± 0.011	0.94	19
w	100	0.076 ± 0.007	0.96	31
w	200	0.086 ± 0.007	0.97	34
u	(all)	0.171 ± 0.011	0.94	96
w	(all)	0.080 ± 0.005	0.95	84

estimate at 200 cm lies outside the confidence interval for 100 cm (although within the wider interval for 50 cm) and outside the interval for the all-height estimate, implying that there was possibly a slight increase in r.m.s. w between 100 and 200 cm from the sea bed.

While calculating the regression coefficients for u and w in Table 2, the data were also grouped by trial and by station. Within each trial and each station there were no significant differences in level with height for either u or w (the slight increase in w at 200 cm only becoming significant when data from both stations was combined). In general, however, the r.m.s. levels were somewhat higher at station 7 (20.5 ± 3.7% and 9.1 ± 1.3% for u and w respectively) than at station 10 (16.3 ± 1.2% and 7.78 ± 0.66%). This fact can be seen in Figure 7, where the clusters of points in the range 15 cm.s^{-1} < U_{100} < 30 cm.s^{-1} are mostly due to data from station 7, whereas all points with U_{100} > 40 cm.s^{-1} were obtained at station 10.

Reynolds stresses.

For those sensors measuring the components u and w the Reynolds stress component $-\rho \overline{uw}$ was calculated, where the averaging interval was the same as the record length (12 min 17 sec). The resulting values are shown plotted against U_{100} in Figure 10. To parameterise the Reynolds stress, τ, in terms of U_{100}, the data were fitted to equations $\tau = \rho A U_{100}^2$ in various groupings by trial, station and height. There was no significant difference with height of the coefficient A within each trial, or within each station. However the all-record value of A was slightly higher at 100 cm than at 50 cm or 200 cm (there being no significant difference between the latter two). The stress estimates at station 7 showed

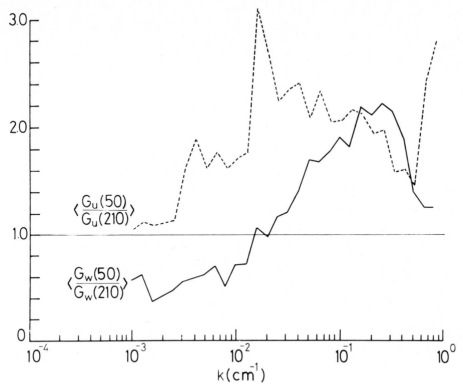

Figure 13. Average value of the ratio of spectral levels at different sensor heights for the u and w components. Averages are shown for sensor heights of 50 and 210 cm, for data from 14 records from trials 50 and 51.

In Figure 14 energy density spectra are shown as functions of frequency for records 32 (U_{100} = 63 cm. s^{-1}) and 40 (U_{100} = 3.3 cm.s^{-1}) and the presence of a spectral peak due to surface waves can be clearly seen in the frequency range 0.1 - 0.2 Hz for record 40, but is not apparent in the spectra for record 32. In the case of record 39 wave energy can be detected in the r.m.s. levels of u, but is not apparent in the r.m.s. levels of w. This can be explained by examining the expected mean velocity at which a surface wave of amplitude a would become apparent in the r.m.s. turbulence signal.

Wave energy will be detectable in the r.m.s. turbulence record for a given component when (r.m.s. wave amplitude) > α (r.m.s. turbulence level), where the turbulence level is taken to be that due to shear alone. The r.m.s. turbulence levels for the u and w components can be written in terms of U_{100} using the data of Table 2. For waves travelling in the x direction with small amplitude a, angular frequency ω and wavenumber k in water of depth h, the r.m.s. value of the u and w velocity components can be calculated using linear wave theory (see e.g. Neumann and Pierson 1966). It can then be shown that for a detectability

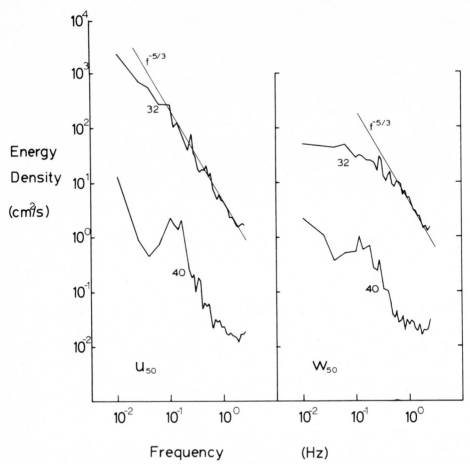

Figure 14. Energy density spectra plotted against frequency for u and w components
50 cm above the sea bed for record 32 (U_{100} = 63 cm.s^{-1}) and record 40 (U_{100} =
3.3 cm.s^{-1}). The spectral peak due to surface wave energy can be seen in the
spectra for record 40.

level α the wave energy will be detectable in the u r.m.s. signal when

$$U_{100} < \frac{a\omega}{0.17\sqrt{2}\,\alpha}\,\frac{\cosh(kz)}{\sinh(kh)}$$

and in the w r.m.s. signal when

$$U_{100} < \frac{a\omega}{0.08\sqrt{2}\,\alpha}\,\frac{\sinh(kz)}{\sinh(kh)} \qquad \text{where } \omega^2 = gk\,\tanh(kh).$$

For α = 10% (the typical uncertainty in the turbulence level parameterisation)
and h = 30 m (the water depth in which our data were recorded) the mean velocities
at which surface waves would just become apparent in the u and w r.m.s. levels
are shown in Figure 15. From the figure, it can be determined that a wave of
period 8 s, with a = 5 cm, would be detectable in the u record at a mean velocity

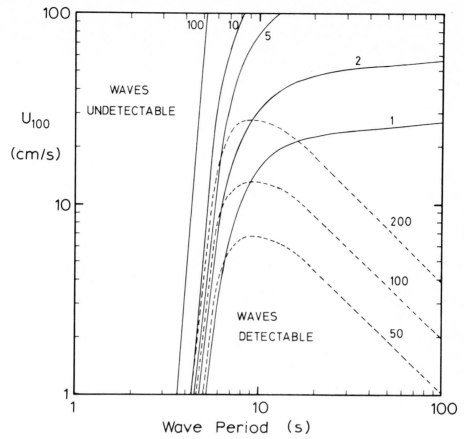

Figure 15. Detectability limits for surface waves. The mean velocity at which surface waves become detectable in the r.m.s. turbulence signal are shown for the u component (solid lines) and w component (dashed lines) records, for a detectable level of 10% and water depth of 30 m. Numbers beside curves for the u component represent wave amplitude in cm, those beside curves for the w component represent sensor height in cm. Curves for the w component are shown for a wave amplitude of 10 cm.

of 50 cm.s^{-1}, but would not become apparent in the w level, even at z = 200 cm, until U_{100} had fallen to 13 cm.s^{-1}. (Of course wave energy might become apparent at much lower mean velocities in spectra than in the r.m.s. signals discussed here).

The detectability of surface waves may become an important factor in the study of the dynamics of tidal boundary layers, particularly when the acceleration effects near slack water are being studied. Maxey (1978) has postulated that the parameter $\tau/\rho q^2$, where q^2 is the turbulent kinetic energy, should be practically constant except during a brief period near slack water when τ decreases to zero and changes sign, but q^2 remains finite due to a relatively slow rate of dissipation. In attempting to examine the behaviour of this parameter near slack water, particular care will be needed to exclude any effects due to surface waves. From Figure 15 it can be seen that even very low amplitude waves, particularly those of long period, are likely to bias estimates of q^2 so that $\tau/\rho q^2$ will exhibit the postulated behaviour for reasons unconnected with the behaviour of the shear induced turbulence.

Tidal hysteresis in turbulence parameters.

In examining the variation of r.m.s. turbulence levels with tidal phase, our data show no apparent dependence of r.m.s. level (normalised by U_{100}) on phase (see Figure 8). The normalised Reynolds stress component $-\overline{uw}/U_{100}^2$ showed an apparent decrease during the deceleration phase (see Figure 11). In contrast, Gordon (1975) found a marked hysteresis in both the Reynolds stress and the turbulent kinetic energy, in both cases the values observed during deceleration of a tidal current being typically twice those at the same mean flow speed during acceleration. However, Gordon's measurements were made at a level $z/\delta \sim$ 0.5, where δ is the boundary layer thickness, whereas the results described here were obtained from values of z/δ much smaller than 0.5. The difference in behaviour between the two sets of results can be explained, at least in part, on the basis of this difference in z/δ.

As no observations were made of the complete velocity profile $U(z)$ during our measurements, the boundary layer thickness δ, defined by $U(\delta) = 0.99\ U_e$, where U_e is the velocity outside the boundary layer and is taken to be equivalent in our case to the near-surface velocity U_s, cannot be directly determined. However, the velocity profile throughout the water column has been observed at half-hour intervals throughout the tidal cycle at a site quite close to station 10 (Bowden, Fairbairn and Hughes, 1959). Although the profile was complicated by acceleration effects, particularly near slack water, over most of the tidal cycle the profiles resembled those of a classical boundary layer with a value of δ close to the total water depth (16 m at their station). Theoretical calculations, treating the flow as a steady turbulent boundary layer over a flat plate (see e.g. Schlichting 1968) or as a neutral boundary layer formed under currents oscillating at the semi-diurnal tidal frequency (see e.g. Bowden 1978), suggest that the boundary layer thickness will develop to fill the entire water depth from perhaps two hours after slack water. On the basis of these estimates of δ our measurements were made at z/δ <0.06.

In considering the effects of measuring height on the behaviour of turbulence parameters at different stages of the tide, we examine the results obtained from some well instrumented engineering flows with external pressure gradients. These flows were steady in time, and in all cases the flow depth was much greater than δ. In the tidal flows the (temporal) acceleration term in the equation of motion is comparable in magnitude to the pressure gradient term, and from the above discussion

$\delta \simeq h$. Both of these factors may cause a modification in the behaviour of the tidally induced turbulence from that of deep steady flows with external pressure gradient, but consideration of the effects of pressure gradient alone satisfactorily explains much of the behaviour of the turbulence in the tidal flows.

The engineering flows considered are those described by Bradshaw (1967) with adverse pressure gradients, and by Coleman, Moffat and Kays (1977) with favourable pressure gradients. Both papers relate the results with applied pressure gradient to those obtained with zero pressure gradient described by Klebanoff (1955). Bradshaw (1967) characterises the pressure gradient by a term $\beta = (\delta_1/\tau_0)\frac{\partial p}{\partial x}$, where δ_1 is the displacement thickness, τ_0 is the wall stress and $\frac{\partial p}{\partial x}$ is the horizontal pressure gradient. For comparison of the pressure gradients in the tidal flow with those of Bradshaw we use a modified term $\beta' = (\delta_1/\tau_0)(\rho\frac{\partial u}{\partial t} + \frac{\partial p}{\partial x})$ to include the effects of acceleration in the effective pressure gradient. Assuming δ_1/δ is similar for the engineering and tidal flows, taking $\tau_0 = 4.3 \times 10^{-3}\rho u_{100}^2$ and estimating $\frac{\partial u}{\partial t}$ and $\frac{\partial p}{\partial x}$ from a simple two dimensional standing wave model of the tide along the North Wales coast, the value of β' has been calculated at different times after maximum current speed (see Table 3). Bradshaw describes flows with $\beta = 0$, ~ 0.9 and ~ 5.4. For the tidal flow it can be seen from Table 3 that the term β' has comparable values, except very near slack water when β' increases rapidly. In the flows described by Coleman et al. (1977) the pressure gradient was characterised by a different parameter, but it is possible to deduce that the equivalent value of β is ~ -0.4, comparable to the value of β' expected perhaps two hours before maximum tidal current.

TABLE 3

Estimated value of the parameter β' at different times after maximum current velocity.

Time (hrs. mins.)	0	0.30	1.00	2.00	2.20	2.40
β'	0	0.10	0.25	0.53	1.29	12.2

Representative profiles of $\overline{u^2}/U_e^2$, $\overline{v^2}/U_e^2$, $\overline{w^2}/U_e^2$ and $-\overline{uw}/U_e^2$ derived from the results from the engineering flows are shown in Figure 16. From these profiles it can be seen that for $z/\delta \sim 0.5$ the effect of a favourable pressure gradient is to decrease the level of all three components of the turbulent kinetic energy and the Reynolds stress. The effect of an adverse pressure gradient is to increase all these quantities. However for $z/\delta < 0.1$ the effect of the pressure gradients on the turbulent kinetic energy components is relatively small, while the effect on the wall stress τ_0 is opposite to that on the Reynolds stress at $z/\delta \sim 0.5$ - τ_0 is increased by a favourable pressure gradient and decreased by an adverse one.

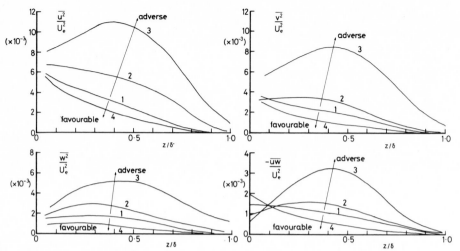

Figure 16. The effects of an external pressure gradient on profiles of the components of the turbulent kinetic energy and the Reynolds stress. Curve 1 : zero pressure gradient. Curves 2 and 3 : increasingly adverse pressure gradient. Curve 4 : favourable pressure gradient. Arrows indicate directions of increasingly strong pressure gradient. (Based on the results of Bradshaw 1967; Coleman, Moffat and Kays 1977; and Klebanoff 1955).

From the results shown in Figure 16, we expect to find that, in a tidal boundary layer, at heights $z/\delta > 0.2$ there will be a hysteresis of both the turbulent kinetic energy q^2 and the Reynolds stress $-\rho\overline{uw}$. Both parameters would be expected to be lower during the accelerated phase (favourable pressure gradient) and higher during the decelerated phase (adverse pressure gradient) for comparable values of the mean flow velocity. This is in agreement with the observations of Gordon and Dohne (1973) and Gordon (1975). At heights $z/\delta < 0.1$ however, hysteresis effects on the turbulent kinetic energy are very much reduced, and would probably be unobservable in a tidal boundary layer. The wall stress τ_o, however, would be expected to be higher during the acceleration phase and lower during the deceleration phase of a tidal flow. The shear stress profiles $F_{zx}(z)$ of Bowden, Fairbairn and Hughes (1959), when plotted as the dimensionless parameters $F_{zx}/\rho U_s^2$ and z/h where U_s is the mean velocity 4 m below the surface and h is the water depth, display many aspects of this expected behaviour. Profiles are shown in Figure 17 for a number of accelerating and decelerating tidal phases. In Figure 17(b) and (d) the increase in wall stress at the beginning of the acceleration phase is apparent, with the stress decreasing away from the wall (and changing sign at times). During the deceleration phase the increase in stress away from the wall is apparent in Figures 17 (a) and (e), although the expected decrease in τ_o with increasingly adverse pressure gradient is less clear. In Figure 17 (c) the expected increase in stress away from the wall with increasing tidal phase does not occur - probably due to acceleration

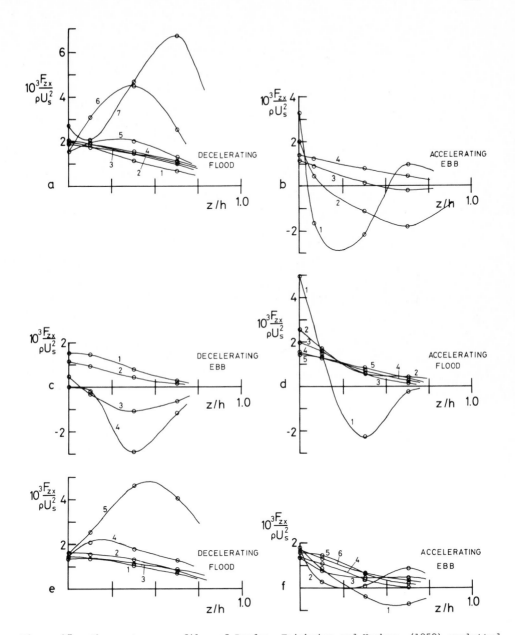

Figure 17. Shear stress profiles of Bowden, Fairbairn and Hughes, (1959) replotted
on dimensionless axes. The profiles on each set of axes are numbered in
chronological order, with 30 minutes between successive profiles (a),(b) Station 1
16 July 1957, 0939 to 1509 G.M.T.; (c), (d), (e), (f) Station 2 19 July, 1957,
0730 to 1800 G.M.T. (a) 3½ h before HW to ½ h before HW; (b) ½ h after HW to 2 h
after HW; (c) 7½ h before HW to 6 h before HW; (d) 5 h before HW to 3 h before HW;
(e) 2½ h before HW to ½ h before HW; (f) ½ h after HW to 3 h after HW.

effects, as velocity profiles shown in the original paper indicate that the mean flow changed direction near the sea bed appreciably in advance of that at the surface near the time that these profiles were measured.

Because our turbulence sensors covered only about 4% of the boundary layer thickness, with an effective resolution in the turbulence parameters of at best 10%, the failure to detect any systematic variation with height is not surprising - this is particularly so for the Reynolds stress estimates in view of the high sampling variability associated with each estimate (Heathershaw and Simpson 1978). The apparent decrease in $-\overline{uw}/U_{100}^2$ observed during the deceleration phase is in agreement with the behaviour expected from the consideration of pressure gradient effects alone, but again other factors such as sampling errors and acceleration effects may have contributed to this.

Variations with height of spectral levels.

The observed variation with height of the u and w component spectral levels, summarised by Figure 13, are in general agreement with the results of Soulsby (1977).The lower wavenumber limit of our spectra is $k \sim 10^{-3}$ cm^{-1}, which for sensor heights of 50, 100 and 200 cm corresponds to Soulsby's parameter k* values of .05, 0.1 and 0.2 respectively. At the lowest wavenumbers, our values of k* are close to Soulsby's u spectral peak and we would not expect to find much variation with height in the u spectral levels. At higher wavenumbers a decrease in spectral level is observed, as expected. For the w spectra our observations extend to wavenumbers well below Soulsby's w spectral peak, so at these wavenumbers the spectral level should increase with increasing height as observed. At higher wavenumbers we were observing above the spectral peak, so the levels are expected to decrease with height, as observed.

CONCLUSION

The main results of this investigation may be summarised as follows :
1. No significant variation was found in the turbulent intensity $\overline{u^2}$ with height over the range 50 to 200 cm but there was an indication that $\overline{w^2}$ was slightly larger at 200 cm than at 50 and 100 cm. The ratios $(\overline{u^2})^{1/2}/U_{100}$, $(\overline{v^2})^{1/2}/U_{100}$ and $(\overline{w^2})^{1/2}/U_{100}$ did not vary significantly with phase of tide. From all records the mean value of these ratios were 0.17, 0.12 and 0.08 respectively.

2. No significant variation of Reynolds stress $-\rho\overline{uw}$ with height was found but the ratio $-\overline{uw}/U_{100}^2$ decreased with decreasing U_{100} on the decelerating phase of the tide. Equating the Reynolds stress at 100 cm with the stress at the boundary, the drag coefficient C_{100}, in the equation $\tau = -\rho\overline{uw} = \rho C_{100}U_{100}^2$, averaged 5.29

$\pm$ 1.7 x 10^{-3} at station 7 and 4.30 $\pm$ 0.53 x 10^{-3} at station 10, with an overall average of 4.35 $\pm$ 0.33 x 10^{-3}. This is somewhat greater than the values 3.6 x 10^{-3} found by Bowden and Fairbairn (1952) and 3.4 x 10^{-3} found by Charnock (1959), in the same area using different methods, but considerably greater than the value 1.73 $\pm$ 0.18 x 10^{-3} found by Heathershaw (1976) using the eddy correlation method in deeper water between 40 km and 70 km to the north of our stations.

3. The energy density spectra $G_u(k)$, $G_v(k)$ and $G_w(k)$ of the u, v and w fluctuations had similar forms throughout the tidal cycle, differing only in their general level.   For wavenumbers k > 6 x 10^{-2} cm^{-1} approximately, the turbulence appeared to be essentially isotropic.   This critical value of k corresponds to the dimensionless parameter kl $\simeq$ 2π where l is the height of measurement.   At lower wavenumbers the spectra of the three components diverged with $G_u > G_v > G_w$. Comparing the spectra at 50, 100 and 200 cm, the spectral density of the u fluctuations decreased with height at all wavenumbers, whereas that of the w fluctuations increased with height for k <10^{-2} cm^{-1} but decreased with height at higher k values.

4. An examination of the probable effects of acceleration and deceleration of the flow indicates that hysteresis effects in the turbulent intensities and Reynolds stress are likely to be small within 0.06 h of the boundary (h = depth of water), where these measurements were made, although they may be appreciable at greater heights, as found by Gordon and Dohne (1973) and Gordon (1975).

5. The effects of particle velocities due to surface waves were evident in some turbulence spectra taken at low mean velocity.   Estimates are given of the mean current speed at which waves of a given period and amplitude would affect significantly the r.m.s. values of the u and w fluctuations.

ACKNOWLEDGEMENTS

    The work described here was supported by a grant from the Natural Environment Research Council, which also provided sea time.

REFERENCES

Bendat, J.S., and Piersol, A.G., 1971. Random Data: Analysis and Measurement Procedures.   Wiley-Interscience, New York, 407 pp.
Bowden, K. F., 1978. Physical problems of the benthic boundary layer. Geophysical Surveys, 3: 255-296.
Bowden, K.F., Fairbairn, L.A., and Hughes, P., 1959. The distribution of shearing stresses in a tidal current. Geophys. J. R. astr. Soc., 2: 288-305.
Bradshaw, P., 1967. The turbulence structure of equilibrium boundary layers. J. Fluid Mech., 29:625-645.

Coleman, H.W., Moffat, R.J. and Kays, W.M., 1977. The accelerated fully rough turbulent boundary layer. J. Fluid Mech., 82:507-528.

Gordon, C.M., 1975. Sediment entrainment and suspension in a turbulent tidal flow. Marine Geology, 18:1757-1767.

Gordon, C.M.and Dohne, C.F., 1973. Some observations of turbulent flow in a Tidal Estuary. J.Geophys. Res., 78:1971-1978.

Heathershaw, A.D., 1975. Measurements of turbulence near the sea-bed using electromagnetic current meters. Proc. Instrumentation in Oceanography Conf. Bangor, Wales. Sept. 1975, 47-58.

Heathershaw, A.D., 1976. Measurements of turbulence in the Irish Sea benthic boundary layer. In: I. N. McCave (Editor), The Benthic Boundary Layer. Plenum Press, New York, pp 11-31.

Heathershaw, A.D. and Simpson, J.H., 1978. The sampling variability of the Reynolds stress and its relation to boundary shear stress and drag coefficient measurements. Estuar. Coastal Mar. Sci.,6:263-274.

Hinze, J.O., 1959. Turbulence: An Introduction to its Mechanism and Theory. McGraw Hill, New York, 586 pp.

Klebanoff, P.S., 1955. Characteristics of turbulence in a boundary layer with zero pressure gradient. N.A.C.A. Report 1247, Washington.

Maxey, M.R., 1978. Aspects of unsteady turbulent shear flow, turbulent diffusion and tidal dispersion. Unpublished Ph.D. thesis, University of Cambridge, England.

Neumann, G., and Pierson, W. J., 1966. Principles of Physical Oceanography. Prentice-Hall, Englewood Cliffs, N.J., 545 pp.

Schlichting, H., 1968. Boundary Layer Theory, 6th Edition. McGraw-Hill, New York, 748 pp.

Soulsby, R.L., 1977. Similarity scaling of turbulence spectra in marine and atmospheric boundary layers. J. Phys. Oceanogr., 7:934-937.

Tucker, M.J., Smith, N.D., Pierce, F.E., and Collins, E.P., 1970. A two-component electromagnetic ship's log. J. Inst. Nav., 23:302-316

Tucker, M.J., 1972. Electromagnetic current meters: an assessment of their problems and potentialities. Proc. Soc. for Underwater Technology, 2:53-58.

SPATIALLY VARYING TURBULENCE PRODUCTION IN TIDAL CHANNELS

M. ROBINSON SWIFT

Mechanical Engineering Department, University of New Hampshire, Durham, NH
03824 (U.S.A.)

ABSTRACT

Effects of spatially varying turbulence production in tidal channels are calculated using an analytical model. The model is developed from the boundary layer form of the equation of motion and an eddy viscosity representation for Reynolds shear stress. The eddy viscosity is calculated using the turbulence kinetic energy equation. A perturbation method is applied to solve for turbulence kinetic energy and the vertical distribution of current. It is found that when turbulence production is spatially varying, advection of the inhomogeneous turbulence field introduces an asymmetric time dependence into the effective viscosity. One result of the hysteresis in the stress/current relationship is the formation of residual currents.

INTRODUCTION

Understanding the relationship between turbulent shear stress and current is essential to modeling the tidal dynamics of estuaries. Experimental studies by Brown and Trask (1979) and theoretical work by Reichard and Celikkol (1978) and LeBlond (1978) have examined the relative importance of the inertial and bottom friction response to forcing by the tidal elevation induced pressure gradient. They found that for shallow, high velocity tidal channels, the principal dynamic balance is between shear stress and pressure gradient with flow inertia playing a relatively minor role. Thus, it is important to model the turbulent shear stress accurately and to be aware of secondary processes which can influence stress. In this paper, the effects of non-uniform turbulent kinetic energy production will be investigated. This problem will be addressed using an eddy viscosity representation for Reynolds stress in the boundary layer form of the equation of motion and continuity equation. These equations will be used to develop an analytical model for the vertical distribution of current and stress.

This approach has been used in a number of previous investigations of tidal flow. Johns and Odd (1966) and McGregor (1972) obtained closed form solutions to linearized dynamic equations. An eddy viscosity representation for shear stress was used which had an algebraically specified depth dependence. Analytical solutions to the nonlinear problem were obtained by Johns (1970) and Ianneillo (1977). They retained nonlinearities in the advective acceleration term of the equation of motion and in the continuity equation. Their results show that these nonlinearities induce residual currents. Hamilton and Bowden (1975), Vager and Kagan (1969, 1971) and Johns (1970) have proposed more detailed models having nonlinearities in the eddy viscosity representation as well. The eddy viscosity was taken to be proportional to mean current or to turbulent kinetic energy. Due to increased complexity, numerical solution was required. Analytical models for the vertical distribution of current and stress have been incorporated into 3-dimensional numerical models by Nihoul (1977) and Nihoul et al (1978). The theoretical work done in connection with tidal flow has features in common with models of wave boundary layers, such as those desscribed by Kajiura (1964, 1966), Smith (1977) and Grant and Madsen (1979), since both problems deal with the dynamics of oscillating boundary layers.

Experimental verification of these models is difficult because of the different types of specialized measurements required. Though comprehensive studies have been made of turbulence and Reynolds shear stress, a field program for validating models of the vertical distribution of current and stress should also include measurements of current profiles and pressure gradient. In investigations carried out by Swift et al (1979) and Reichard (1979), estimates of these variables were made in addition to turbulent shear stress. The data was used to compare eddy viscosity representations having different depth dependences. Though the data contained some scatter due to flow intermittency, a quadratic depth dependence was found to give better results than either a constant or a linear depth dependence.

Experimental studies of bottom turbulence have provided information on the structure of turbulence. Gordon (1974, 1975), Heathershaw (1974) and Gordon and Whiting (1977) have found that the turbulence which contributed to Reynolds stress is intermittent. This suggests that the important exchange of momentum is done by a few relatively large eddies. Experiments described by Smith and McLean (1977) indicate that spatial variation in bottom topography can alter stress and current significantly from values observed under uniform conditions.

These observations can have implications on how Reynolds stress should be modeled. Since the initial decay rate of turbulent eddies is inversely proportional to their length scale, larger eddies will decay more slowly. If the time scale for decay is long enough, current advection will have an effect. Production and dissipation will not be local. In this case, non-uniformity of the bottom (in the sense of a spatially varying turbulence production capacity) can take on added importance. Advection of turbulence under non-uniform conditions could change the local momentum exchange process. In this situation, the history of the flow will influence the local relationship between tidal current and shear stress.

In this paper, the effects of turbulence advection on current and stress will be examined when the turbulence production varies spatially. The analysis will focus on the outer flow response, and no attempt will be made to describe the near bottom flow over individual topographic features. Bottom non-uniformity will be treated as a spatially varying bottom stress representative of changing roughness or turbulence production capacity. It will not be necessary to consider the direct interaction of flow and topography because the non-uniform turbulence advection process affects flow remote from the point of generation. This distinguishes the problem considered here from studies done by Smith and McLean (1977) and Taylor and Dyer (1977).

The problem will be approached by developing a simple analytical model. In the model the effects of turbulence on the mean flow dynamics will be incorporated by modifying the eddy viscosity expression for stress. The eddy viscosity will be taken as a function of turbulent kinetic energy. The turbulent kinetic energy will be calculated from the turbulence energy equation in which the production term is spatially varying. This stress representation is essentially an adaptation of the Prandtl-Kolmogorov one-equation model as described by Launder and Spaulding (1972). Though turbulent kinetic energy dependent eddy viscosities have been used by Vager and Kagan (1969, 1971) and Johns (1979), the effects of non-uniform turbulence production were not considered.

MATHEMATICAL MODEL

The dynamic equations are developed for straight, narrow, well-mixed tidal channels having constant depth and breadth. Under these conditions the equation of motion reduces to

$$\frac{\partial U}{\partial t} + U \frac{\partial U}{\partial x} = - g \frac{\partial \eta}{\partial x} + \frac{\partial}{\rho \partial z} (R_{xz}) \tag{1}$$

In Eq. (1) x = channel axis coordinate; z = vertical coordinate; U = axial component of tidal current; g = gravitational acceleration; η = surface elevation above mean sea level; and R_{xz} = vertical plane Reynolds shear stress. For small tidal amplitude ($\eta/H \ll 1$), the vertically averaged continuity equation is

$$\frac{\partial}{\partial x} U_H = \frac{-1}{H} \frac{\partial \eta}{\partial t} \tag{2}$$

in which H is the mean depth and subscript H indicates a vertical average. An eddy viscosity expression is used for Reynolds stress and is written as

$$R_{xz} = \rho \nu_e \frac{\partial U}{\partial z} \tag{3}$$

where ν_e is the eddy viscosity.

The eddy viscosity is dimensionally equivalent to the product of a turbulent velocity scale and a length scale characteristic of the turbulent mixing. Near the bottom, the length scale should increase linearly with distance from the bottom due to the inhibiting effect of the solid boundary. Similar reasoning indicates that it should also diminish as the surface is approached. A simple mathematical form having these properties is a quadratic depth dependence. The turbulent velocity scale is taken as the square root of the turbulent kinetic energy per unit mass. Mathematically, the eddy viscosity is given by

$$\nu_e = u_t \ell$$

in which

$$\ell = K(\ell_0 + z) - \varepsilon z^2 \text{ and } u_t = \sqrt{k} \tag{4}$$

In the above expressions u_t = turbulent velocity scale; ℓ = turbulent length scale; k = turbulent kinetic energy per unit mass; ℓ_0 is the bottom roughness length, and K and ε are constants.

The turbulent kinetic energy is assumed to be well-distributed vertically and will therefore not be a function of depth. It is calculated from the vertically averaged turbulence energy equation which is

$$\frac{\partial k}{\partial t} + U_H \frac{\partial k}{\partial x} = \text{(Production)}_H - \text{(Dissipation)}_H$$

$$= \overline{(\tau/\rho \frac{\partial U}{\partial z})}_{z=h_p} - \frac{C_d k^{3/2}}{\ell_d}. \tag{5}$$

In Eq. (5), C_d is a dissipation coefficient and ℓ_d is a length scale associated with the decay of the energy containing eddies. The vertically averaged production is assumed equal to a characteristic production at height h_p. Since production is largely confined to the constant stress layer just above the bottom, stress at height h_p is approximately the bottom stress. Production is then given by

$$\text{(Production)}_H = \frac{\overline{(\tau/\rho)^2}_{\text{bottom}}}{\nu_e(z=h_p)} \tag{6}$$

Height h_p will be chosen so that the production balances dissipation under uniform conditions. The production term is time averaged to simplify the analysis. This approximation is satisfactory when the equations are used to calculate the first harmonic response at the semi-diurnal tidal frequency and residual currents. Time averaging the production term eliminates some higher harmonics so in this paper solutions for higher harmonic constituents will not be developed.

The mathematical model presented here is valid in the region where the total stress is transmitted by the Reynolds shear stress and the influence of topography is seen as effective bottom roughness. At the bottom, however, frictional shear is comprised of a skin friction component and a form drag part. Flow over individual roughness features must be considered, and sediment movement becomes a complicating factor. Thus a restriction should be placed on the equations so that it is clear that their range of applicability is above the near-bottom layer. The near-bottom layer is assumed thin relative to total depth and the origin of the z-axis will be placed immediately above it. The depth coordinate will be restricted to $0 \leq z \leq H$. The thickness of the omitted near-bottom layer will be denoted by h_0. It is chosen so that $\ell_0 \ll h_0 \ll H$, and ℓ_0 will be replaced by h_0 in the eddy viscosity expression.

Boundary conditons are necessary to completely specify the mathematical problem. Since velocity will not in general be zero at $z = 0$, the bottom boundary condition must be a stress condition. The bottom condition is written as

$$R_{xz} (z=0) = \tau_b = \rho \, \nu_e \, (z=0) \left. \frac{\partial U}{\partial z} \right|_{z=0} \tag{7}$$

where τ_b will henceforth be referred to as the "bottom stress". Wind stress will not be considered, so at the surface (z=H) stress is zero. Mathematically the surface condition is

$$R_{xz} (z=H) = \rho \, \nu_e \, (z=H) \left. \frac{\partial U}{\partial z} \right|_{z=H} = 0,$$

or

$$\left. \frac{\partial U}{\partial z} \right|_{z=H} = 0 \tag{8}$$

The dynamic equations will be applied to a channel segment having a length L much shorter than the wavelength of the channel's shallow water wave (that is, $L \ll T\sqrt{gH}$ in which T = tidal period). The bottom stress τ_b will be set equal to a part τ_1 nearly constant in x and a part $\alpha \, \tau_2$ which varies with x. Thus

$$\tau_b = \tau_1 + \alpha \, \tau_2$$

in which

$$\alpha = \text{non-dimensional parameter} \ll 1. \tag{9}$$

The stress τ_1 is the stress at z=0 if the channel had a uniform bottom. The stress $\alpha \, \tau_2$ is the variation in bottom stress due to non-uniformity of bottom roughness. It will cause two types of direct effects. One is a change in the mean flow dynamics due to alteration of the bottom friction characteristics. This influence is accounted for mathematically in the bottom boundary condition for the equation of motion. The second consequence is variable turbulence production which results in non-uniform effective viscosity.

These effects will be evaluated by obtaining an approximate solution to the nonlinear equations using a perturbation approach. Variables are expanded in terms of the small non-dimensional parameter α as written below:

$$U = U_1 + \alpha\, U_2 + \ldots \qquad\qquad u_t = u_{t1} + \alpha\, u_{t2} + \ldots$$

$$\eta = \eta_1 + \alpha\, \eta_2 + \ldots \qquad\qquad = \sqrt{k_1} + \alpha\, \frac{k_2}{2\sqrt{k_1}} + \ldots$$

$$k = k_1 + \alpha\, k_2 + \ldots \qquad\qquad \nu_e = \nu_1 + \alpha\, \nu_2 + \ldots \qquad\qquad (10)$$

Terms higher than second order will be neglected. Substituting these expansions into the dynamic equations and equating like powers of α results in the first and second order equations.

The first order equations are:

$$\frac{\partial U_1}{\partial t} = -g\, \frac{\partial \eta_1}{\partial x} + \frac{\partial}{\partial z}\, \{u_{t1}[K(h_o + z) - \epsilon\, z^2]\, \frac{\partial U_1}{\partial z}\}\ ,$$

$$\frac{\partial (U_1)_H}{\partial x} = \frac{-1}{H}\, \frac{\partial \eta_1}{\partial t}\ ,$$

and

$$0 = \frac{\overline{\tau_1}^2}{\rho^2\sqrt{k_1}\, K\, h_p} - \frac{C_d\, k_1^{3/2}}{\ell_d}$$

with boundary conditions

$$\frac{\partial U_1}{\partial z}\Big|_{z=H} = 0 \quad \text{and} \quad \rho K\, u_{t1}\, h_0\ \frac{\partial U_1}{\partial z}\Big|_{z=0} = \tau_1 \qquad\qquad (11)$$

In the above set of equations, the advective acceleration term has been neglected from the equation of motion. This is a reasonable approximation for small tidal amplitudes ($\eta/H \ll 1$). The first order response is therefore equivalent to the linear solution. The pressure gradient forcing and the bottom stress are taken to be harmonic at the semi-diurnal tidal frequency ($=\sigma$) so that

$$g\, \frac{\partial \eta}{\partial x} = R_e\, \{S_1\, e^{i\sigma t}\},$$

$$\tau_1 = R_e\, \{T_1\, e^{i\sigma t}\}$$

and consequently

$$U = R_e \{\hat{U}_1 \, e^{i\sigma t}\} \tag{12}$$

in which S_1, T_1 and $\hat{U}_1$ are complex amplitudes. The solution for the first order current is:

$$U = R_e \{\hat{U}_1 \, e^{i\sigma t}\} = R_e \{[\frac{S_1 i}{\sigma} + V_1] \, e^{i\sigma t}\}$$

where V_1 is the viscous part of the solution and is given by

$$V_1 = T_1 \, [C_1 \, P_\nu(\zeta) + C_2 \, Q_\nu(\zeta)]$$

in which P_ν and Q_ν are Legendre functions and

$$\zeta = [\frac{Kh_0}{\varepsilon H^2} + (\frac{K}{2\varepsilon H})^2]^{-1/2} (z/H - \frac{K}{2\varepsilon H}), \quad \nu = -1/2 + (\frac{1}{4} - \frac{i\sigma}{u_{t1}\varepsilon})^{1/2},$$

$$C_1 = \frac{B}{BC - AD} \qquad , C_2 = \frac{A}{AD - BC} \qquad ,$$

$$A = \zeta_H \, P_\nu \, (\zeta_H) - P_{\nu-1} \, (\zeta_H),$$

$$B = \zeta_H \, Q_\nu \, (\zeta_H) - Q_{\nu-1} \, (\zeta_H),$$

$$C = \frac{\rho K \, u_{t1} \, \nu}{2 \, \zeta_0} \, [\zeta_0 \, P_\nu \, (\zeta_0) - P_{\nu-1} \, (\zeta_0)].$$

$$D = \frac{\rho K \, u_{t1} \, \nu}{2 \, \zeta_0} \, [\zeta_0 \, Q_\nu \, (\zeta_0) - Q_{\nu-1} \, (\zeta_0)],$$

$$\zeta_0 = \zeta(z=0) \text{ and } \zeta_H = \zeta(z=H). \tag{13}$$

The turbulent velocity scale is given by

$$u_{t1} = \sqrt{\overline{k_1}} = \{(\frac{|T_1|}{\rho})^2 \, [\frac{\ell_d}{2C_d \, K \, h_p}]\}^{1/4}. \tag{14}$$

The eddy viscosity in the linear, first order solution should be consistent with values arrived at by the mixing length approach. This requirement is satisfied when $K \, h_p = \ell_d$ and $K/(2C_d)^{1/4} =$ von Karman's constant.

The second order dynamic equations are:

$$\frac{\partial U_2}{\partial t} + U_1 \frac{\partial U_2}{\partial x} = -g \frac{\partial \eta_2}{\partial x} + \frac{\partial}{\partial z} \{u_{t1} \, [K \, (h_0 + z) - \varepsilon z^2] \frac{\partial U_2}{\partial z}\}$$

$$+ \frac{\partial}{\partial z} \{u_{t2} \, [K \, (h_0 + z) - \varepsilon z^2] \frac{\partial U_1}{\partial z}\} ,$$

$$\frac{\partial (U_2)_H}{\partial x} = \frac{-1}{H} \frac{\partial \eta_2}{\partial t},$$

and

$$\frac{\partial k_2}{\partial t} + (U_1)_H \frac{\partial k_2}{\partial x} = \frac{2 \overline{\tau_1 \tau_2}}{\rho^2 u_{tl} K h_p} - \frac{\overline{\tau_1^2} u_{t2}}{\rho^2 u_{tl}^2 K h_p} - \frac{C_d}{\ell_d} (3/2 \sqrt{k_1}) k_2$$

with boundary conditions

$$\frac{\partial U_2}{\partial z}\Big|_{z=H} = 0 \quad \text{and} \quad (\nu_2 \frac{\partial U_1}{\partial z})_{z=0} + (\nu_1 \frac{\partial U_2}{\partial z})_{z=0} = \tau_2/\rho \tag{15}$$

The general solution for each of the second order variables will contain a residual (steady) current component and a time dependent part so that

$$U_2 = \overline{U}_2 + U_2',$$

$$\eta_2 = \overline{\eta}_2 + \eta_2',$$

and $k_2 = \overline{k}_2 + k_2'$ (16)

Second order bottom stress τ_2 must therefore have a steady component $\overline{\tau}_2$ in addition to the specified time dependent part (now denoted by τ_2'). The time dependent variables (primed quantities) are principally first harmonic corrections to the linear solution to account for non-uniform bottom conditions. Higher harmonic time variations are small and will not be considered. The Eq. (16) expansions are then substituted into Eqs. (15). Equations for residual currents are found by time averaging.

Note that advection terms have not been dropped as in the first order equations. Since the purpose of this paper is to determine the effects of non-uniform turbulence advection, the length scale for the bottom stress spatial variation will be chosen to be shorter than the channel segment length ($L_r < L$). In this case, horizontal gradients of second order variables are not negligible and advection must be taken into account.

The method of successive approximations will be used to deal with the advective term in the turbulence energy equation. The results yield a steady component of kinetic energy,

$$\overline{k}_2 = \frac{\overline{\tau_1 \tau_2'}}{\overline{\tau_1^2}} k_0 \tag{17}$$

and a time varying part,

$$k_2' = R_e \{K_2 e^{i\sigma t}\}$$

in which

$$K_2 = \frac{-\hat{U}_H \left[\frac{\partial}{\partial x} \left(\frac{\overline{\tau_1 \tau_2'}}{\overline{\tau_1}^2} \right) \right] k_1}{i\sigma + \frac{3}{2} \frac{C_d \sqrt{k_1}}{\ell_d}} \cdot \tag{18}$$

The time dependent contribution accounts for advection of the non-homogeneous turbulence field. Eq. (18) indicates that the fractional change in turbulent kinetic energy due to advection of non-uniformly generated turbulence is of order

$$\frac{\alpha(k_2')}{k_1} \sim \alpha \frac{U \, T_d}{L_r} \tag{19}$$

In the above expression, $T_d = \dfrac{\ell_d}{C_d \sqrt{k_1}}$ is the time scale for initial turbulence decay (assumed much shorter than the tidal period), and L_r is a length scale for the spatial variation in stress (due to non-uniform roughness). For advection induced variations to be important, therefore, eddies must be transported without significant decay over distances commensurate with changes in turbulence generation.

Time variation in turbulent kinetic energy implies fluctuations in the turbulent velocity scale ($u_{t2}' = \dfrac{k_2'}{2\sqrt{k_1}}$) and consequently the eddy viscosity as well.

This induces changes in the relationship between current and stress. The cyclic variation in the relationship between the periodic variables current and stress is a form of hystereses. Assymmetry in the stress/current relation contributes to the formation of residual currents.

The equation for the steady component of current is

$$\overline{U_1 \frac{\partial U_2'}{\partial x}} = -g \frac{\partial \overline{n_2}}{\partial x} + \frac{\partial}{\partial z} \left(\nu_1 \frac{\partial \overline{U_2}}{\partial z} \right) + \frac{\partial}{\partial z} \left(\frac{\nu_1}{u_{t1}} \overline{u_{t2}' \frac{\partial U_1}{\partial z}} \right). \tag{20}$$

Boundary conditions are no stress at the surface,

$$\frac{\partial \overline{U_2}}{\partial z} \bigg|_{z=H} = 0 \tag{21}$$

and a stress condition at the bottom,

$$\rho K \, h_o \, u_{t2}' \overline{\frac{\partial U_1}{\partial z}} \bigg|_{z=0} + \rho K \, h_0 \, u_{t1} \frac{\partial \overline{U_2}}{\partial z} \bigg|_{z=0} = \overline{\tau_2}. \tag{22}$$

To complete the second order equations, the steady bottom stress $\overline{\tau_2}$, is expressed using a linear drag coefficient approach. The value of the coefficient is taken to be the same as for the first order solution at maximum current, thus

$$\overline{\tau_2} = \rho C_\ell \overline{U_2} \quad (z=100 \text{ cm})$$

in which

$$C_\ell = \frac{(\tau_1) \text{ maximum current}}{\rho[U_1 (z=100 \text{ cm})]_{\text{maximum current}}}. \tag{23}$$

Eq. (20) shows that three mechanisms are responsible for forcing a steady current. The first is due to rectification between the time-dependent turbulent velocity scale and the first order velocity gradient. Physically, the flow experiences a variation in effective viscosity over the tidal cycle which biases the flow in one direction. Another source is the steady component of advective acceleration. Lastly, there is a contribution from the time-averaged pressure gradient. The pressure gradient arises because of continuity considerations. The continuity equation requires that

$$\frac{\partial}{\partial x} (\overline{U_2})_H = 0$$

or

$$(\overline{U_2})_H = \text{constant}$$
$$= 0 \text{ if there is a wall at the inland end of the channel} \tag{24}$$

The steady pressure gradient provides a return flow such that the net axial mass transport is constant. (Note that here the continuity equation is linearized and residual currents due to non-uniform bottom conditions are being treated. Additional contributions are possible from nonlinearities in the exact form of continuity. This problem has been treated by Ianneillo (1977)).

SINUSOIDALLY VARYING BOTTOM NON-UNIFORMITY

As an example, consider the case when the non-uniformity in bottom roughness is such that the first harmonic of the stress fluctuation, $\alpha \tau_2'$, is in phase with the first order bottom stress, τ_1, but varies sinusoidally in space. The amplitude ratio of the second order stress to the first order stress is α, so that τ_2' is written as

$$\tau_2' = R_e \{T_1 e^{i\sigma t}\} \sin k_r x. \tag{25}$$

In the above expression, $k_r = \frac{2\pi}{L_r}$ is a wavenumber based upon the characteristic length L_r of the roughness variation.

Eqs. (17) and (18) are used to evaluate the variation in turbulence kinetic energy. In this case the time-averaged second order turbulence kinetic energy is

$$\overline{k_2} = k_1 \sin k_r x, \tag{26}$$

and the time dependent second order turbulent kinetic energy is

$$k_2' = R_e \{K_2 e^{i\sigma t}\}$$

in which

$$K_2 = \frac{\hat{U}_H k_1 k_r \cos k_r x}{i\sigma + \dfrac{3 C_d k_1^{1/2}}{2\ell_d}}. \tag{27}$$

Eqs. (26) and (27) enable the second order turbulent velocity scale, $u_{t2} = \dfrac{k_2}{2\sqrt{k_1}}$, to be determined for substitution into the equation of motion for calculating currents.

The solution for the second order time dependent current has the form

$$U_2' = R_e \{[\frac{S_2 i}{\sigma} + V_2] e^{i\sigma t}\} \sin k_r x \tag{28}$$

in which $S_2 \sin k_r x$ is the complex amplitude of the pressure forcing term $(= g \dfrac{\partial \eta_2'}{\partial x})$. The two bracketed terms in Eq. (28) represent, respectively, the pressure gradient forcing and a correction to the linearized first harmonic solution due to changes in viscosity and the bottom stress boundary condition. The Eq. (28) solution must satisfy continuity,

$$H \frac{\partial}{\partial x} (U_2')_H = - \frac{\partial}{\partial t} \eta'_2 \tag{29}$$

with the result that

$$S_2 = \frac{i\sigma (V_2)_H}{1 - \dfrac{\sigma^2}{gHk_r^2}}$$

which becomes

$$= i\sigma (V_2)_H \tag{30}$$

for the $L_r < L \ll \dfrac{2\pi\sqrt{gH}}{\sigma}$ limitation introduced previously. When eddy viscosity values are sufficiently large, the tidal frequency first harmonic current pro-

files will approach vertical uniformity above the near-bottom layer and U_2'
becomes small. In this situation the advective acceleration term in the residual current equation can be neglected. This case will be considered here in
order to focus on the currents generated by fluctuations in turbulence kinetic
energy.

The residual current is then given by

$$\bar{U}_2 = -[\frac{R_e\{K_2\}\ R_e\{V_1\}}{4\ k_1} + \frac{I_m\{K_2\}\ I_m\{V_1\}}{4\ k_1}]$$

$$-\frac{\overline{S_2}}{u_{t1}\ \varepsilon}[\frac{r_1}{r_1-r_2}\ \ln\ (\frac{r_1-z}{r_1}) + \frac{r_2}{r_2-r_1}\ \ln\ (\frac{r_2-z}{r_2})]$$

$$+\frac{\overline{C_1}}{u_{t1}\ \varepsilon}[\frac{1}{r_2-r_1}\ \ln\ (\frac{r_1-z}{r_1}) + \frac{1}{r_1-r_2}\ \ln\ (\frac{r_2-z}{r_2})] + \overline{C_2} \tag{31}$$

in which $r_{1,2} = \frac{K}{2\varepsilon}\ \overset{-}{+}\ \sqrt{\frac{K\ h_0}{\varepsilon} + (\frac{K}{2\varepsilon})^2}$, $\overline{S_2} = g\ \frac{\partial\overline{\eta_2}}{\partial x}$, and $\overline{C_1}$, $\overline{C_2}$ are coefficients
to be evaluated. The coefficients $\overline{C_1}$, $\overline{C_2}$ and $\overline{S_2}$ are determined from Eqs. (22) –
(24), the boundary and continuity conditions.

APPLICATION OF THEORETICAL RESULTS

To evaluate the residual current solution, parameters associated with the
first order current must be specified. Table 1 gives parameter values obtained
from measurements made in the Great Bay Estuary, N.H., U.S.A. and reported by
Swift et al (1979). The channel in which the data was collected is shallow
[H=10 m] and well-mixed. Reichard (1979) found that the important dynamic
balance in this system is between pressure gradient and bottom stress.

$$|\frac{\partial\eta_1}{\partial x}| = 3.9 \times 10^{-5} \qquad\qquad \frac{|\tau_1|}{\rho g H} = 3.2 \times 10^{-5}$$

$$\frac{(U_1)_H}{(|\tau_1|/\rho)^{1/2}} = 8.1 \qquad\qquad C_D = 0.057$$

Table 1 - First Order Current Parameters from
Great Bay Estuary, N.H., U.S.A.

The steady current solution given by Eq. (31) is evaluated for the case when
$\varepsilon H/K = 0.9$ and $h_0/H = 0.01$. The results for locations having maximum current

are shown in Fig. 1

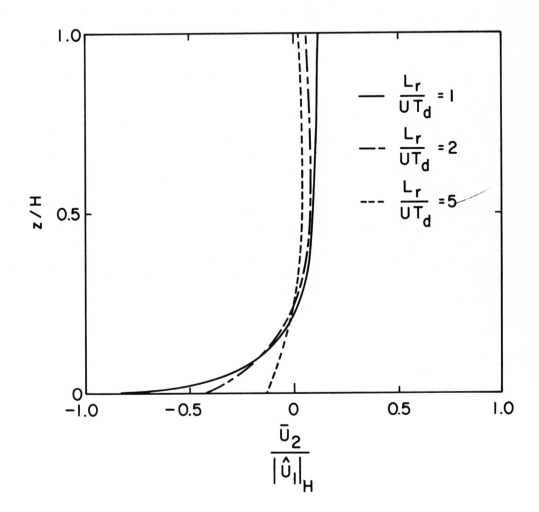

Fig. 1. Residual current profiles for $k_r x = 0, 2\pi, \ldots$

The profiles show that current magnitude increases as $L_r/U\,T_d$ approaches unity.
In all cases, however, current in the upper water column is relatively small.
The near bottom flow, on the other hand, is significant and may be a factor in
sediment transport processes.

300

CONCLUSION

The simple analytical model developed here shows that non-uniformity of bottom stress in tidal channels affects the outer flow region in several ways. Spatially changing turbulence production generates an inhomogeneous turbulent kinetic energy distribution. Advection of the turbulent field causes local fluctuations in the turbulent energy which depend on current direction as well as current magnitude. Since the effective viscosity of the flow is related to the turbulence, a cyclic time variation is introduced into the eddy viscosity. This results in hystereses of the stress/current relationship and a steady component to the tidal current.

Turbulence advection processes are important when the time scale of turbulence decay approximates the time necessary to advect eddies over a distance associated with the spatial variation in bottom conditions ($T_d = \ell_d/C_d\sqrt{k} \sim L_r/U$).

ACKNOWLEDGEMENT

The author appreciates the assistance given him by Dr. Ronnal Reichard of the University of Washington, U.S.A.

REFERENCES

Brown, W. and Trask, R., 1979. A study of estuarine bottom stress. (unpublished manuscript).

Gordon, C.M., 1974. Intermittent momentum transport in a geophysical boundary layer. Nature, 248:393-394.

Gordon, C.M., 1975. Period between bursts at high Reynolds number. Phys. Fluids, 18:141-143.

Gordon, C.M. and Witting, J., 1977. Turbulent structure in a benthic boundary layer. In: J.C.J. Nihoul (Editor), Bottom Turbulence. Elsevier, Amsterdam.

Grant, W.D. and Madsen, O.S., 1979. Combined wave and current interaction with a rough bottom. J. Geophys. Res., 84:1797-1808.

Hamilton, P. and Bowden, K.F., 1975. Some experiments with a numerical model of circulation and mixing in a tidal estuary. Est. and Coast. Mar. Sci., 3:281-301.

Heathershaw, A.D., 1974. Bursting phenomean in the sea. Nature, 248:394-395.

Ianniello, J.P., 1977. Tidally induced residual currents in estuaries of constant breadth and depth. J. Mar. Res., 35:755-786.

Johns, B., 1970. On the determination of the tidal structure and residual current system in a narrow channel. Geophys. J. Roy. Astro. Soc., 20:159-175.

Johns, B., 1978. The modeling of tidal flow in a channel using a turbulence energy closure scheme. J. Phys. Ocean. 8:1042-1049.

Johns, B. and Odd, N., 1966. On the vertical structure of tidal flow in river estuaries. Geophys. J. Roy. Astro. Sco., 12: 103-110.

Kajiura, K., 1964. On the bottom friction in an oscillating current. Bull. Earthquake Res. Inst. 42: 147-174.

Kajiura, K., 1966. A model of the bottom boundary layer in water waves. Bull. Earthquake Res. Inst. 46:75-123.

Launder, B.E. and Spalding, 1972. Mathematical Models of Turbulence, Academic Press, London and New York, 169 pp.

LeBlond, P.H., 1978. On tidal propagation in shallow rivers. J. Geophys. Res., 83:4717-4721.

McGregor, R.C., 1972. The influence of eddy viscosity on the vertical distribution of velocity in the tidal estuary. Geophys. J. Roy. Astro. Soc., 29:103-108.

Nihoul, J.C.J., 1977. Three-dimensional model of tides and storm surges in a shallow well-mixed continental sea. Dyn. Atmos. and Oceans, 2:29-47.

Nihoul, J.C.J., Runfola, Y. and Roisin, B., 1978. Non-linear three-dimensional modelling of mesoscale circulation in seas and lakes. In: J.C.J. Nihoul (Editor), Marine Forecasting. Elsevier, Amsterdam, pp. 235-259.

Reichard, R., 1979. The vertical structure of tidal flow in estuarine channels. Ph.D. dissertation, Univ. of New Hampshire, U.S.A.

Reichard, R. and Celikkol, B., 1978. Application of a finite element hydrodynamic model to the Great Bay estuarine system, New Hampshire, U.S.A. In: J.C.J. Nihoul (editor), Hydrodynamics of Estuaries and Fjords, Elsevier, Amsterdam, pp. 349-372.

Smith, J.D., 1977. Modelling of sediment transportation continental shelves. In: E.D. Goldberg, I.N. McCave, J.J. O'Brien, J.H. Steele (Editors), The Sea, Vol. 6. Wiley, New York, pp. 539-578.

Smith, J.D. and McLean, S.R., 1977. Spatially averaged flow over a wavy surface. J. Geophys. Res., 82:1735-1746.

Swift, M.R., Reichard, R. and Celikkol, B., 1979. Stress and tidal current in a well-mixed estuary. ASCE J. Hydraul. Div., 105:785-799.

Taylor, P.A. and Dyer, K.R., 1977. Theoretical models of flow near the bed and their implication for sediment transport. In: E.D. Goldberg, I.N. McCane, J.J. O'Brien, J.H. Steele (Editors), The Sea, Vol. 6. Wiley, New York, pp. 579-601.

Vager, B.G. and Kagan, B.A., 1969. The dynamics of the turbulent boundary layer in a tidal current. Izv. Akad. Nauk SSS R, Fiz. Atmos. Okeana, 5: 88-93.

Vager, B.G. and Kagan, B.A., 1971. Vertical structure and turbulence in a stratified boundary layer of a tidal flow. Izv Akad. Nauk SSS R, Fiz. Atmos. Okeana, 7:766-777.

A LASER-DOPPLER VELOCIMETER FOR SMALL SCALE TURBULENCE STUDIES IN THE SEA

C. VETH

Netherlands Institute for Sea Research, N.I.O.Z., Texel (The Netherlands)

ABSTRACT

    An in situ laser-doppler velocimeter is described, based on the so-called referen-
ce-beam configuration. The velocimeter is mounted in a twin type hull which permits
the simultaneous measurement of two orthogonal velocity components. The very small
measurement volume and high frequency response make it possible to measure turbulence
properties far into the dissipation range. The special hull has been tested by model
experiments in a flume for its influence on the turbulence spectrum of the flow.
An estimate of the Kolmogorov constant has been made from the isotropic part of the
turbulence spectra of the flume.

INTRODUCTION

    Since Yeh and Cummins (1964) introduced the laser-doppler technique as a tool for

measuring fluid flow, the method has become a common aid for determining turbulence

properties in laboratory flows.

    Several useful properties are combined in this method, such as a linear response

to velocity components, a fast response to velocity changes, a very small measuring

volume and, in principle, no influence of the measurement on the flow. The calibra-

tion of the system is determined by geometrical optics and the wavelength of the las-

er light only, so the instrument may be used to calibrate other types of current

meters.

    Several authers proposed or actually used already laser-doppler methods in the sea

(Kullenberg, 1978; Lespinard, 1978; Fowlis et al, 1974; Stachnik et al, 1977). We too

want to exploit the properties of the laser-doppler velocity measurement principle

in the sea, so we constructed an instrument system capable of determining aspects

of oceanic turbulence. The turbulence properties we aim at are the power and dissip-

ation spectra, the rate of dissipation, the turbulent intensity and the transport of

momentum and heat in several regions in the sea. Projects in which the laser-doppler

system will take a part, are a study of the seasonal thermocline in the Central North

Sea and the frontal zone between stratified and unstratified areas in the North Sea

and an investigation in the tidal streams of the Dutch Wadden Sea where very large

turbulent dissipation of tidal energy occurs.

    For these projects we need an instrument which measures two velocity components

simultaneously with a fast response to velocity changes. It seems that a laser-doppler system meets these requirements.

Special attention has been paid to the shape of the hull in order to be able of working in several modes of operation.

PRINCIPLE OF OPERATION

Various laser-doppler techniques are practised in flow research (Durst et al,1976; Durrani et al, 1977). The basic optical configurations are classified by the way the optical detector is located in the system. This location is important, because it determines in which way velocity components can be measured.

The arrangements most often used are shown in Fig.1. Many variations of these basic configurations are applied in practice, depending on the aspects of the flow one wants to measure.

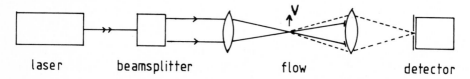

(a) forward_scatter fringe mode

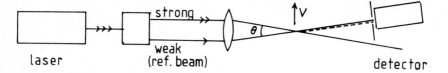

(b) forward_scatter reference beam mode

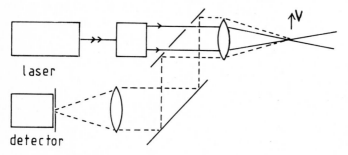

(c) backscatter fringe mode

Fig. 1. Three basic arrangements of laser-doppler velocimeters.
(- - - - - - = scattered light)

The arrangement we composed is derived from (b), the reference-beam mode. The reasons for this choise are:

- This mode of operation is not very sensitive to fluctuations in the number of scatterers in the water,
- One can use a simple photodiode as detector instead of the photomultiplier necessary in the other configurations,
- The system can be applied in daylight because the reference beam acts as a local oscillator in an optical heterodyne system. Such a system is only sensitive to light of frequencies near the frequency of the reference beam; daylight adds only a little to the noise,
- The alignment of this mode of operation is very simple in comparison with other modes.

The system we are using in our instrument has been designed at the Institute of Applied Physics (T.P.D.-T.N.O.), Technical University Delft, The Netherlands (Oldengarm et al, 1976).

The reference-beam mode of operation

A laser-doppler system senses velocity in the following way. The laser light (Fig. 1(b)) is split into a strong and a weak beam (reference-beam). The two beams intersect inside the flow at the point where we want to measure the velocity. The 'point' of intersection is in fact a finite volume due to the diffraction of the light of the beams. It is supposed that there are enough scatterers in the fluid inside this volume (the so-called measuring volume)(Stachnik et al, 1977). The photocell area receives two beams of light:

- direct light from the reference-beam and
- scattered light from the strong beam.

Only the scattered light from scatterers in the measuring volume takes a part in the following process, because other scattered light makes too great an angle with the reference beam to interfere noticeably.

The frequency of the scattered light is doppler-shifted when the scatterers are moving. The reference beam is not shifted in frequency. Because the photodiode responds to the square of the amplitude of the incident light field, the electronic signal at the output of the photocell contains the difference or beat frequency of the two light beams incident on the photocell area. This is shown, in a simplified way, as follows. The photocell area receives two parallel electromagnetic waves of frequency f_1 and f_2, where f_1 is the frequency of the reference beam and f_2 is the frequency of the scattered light. The total electric field E_t is given by

$$E_t = E_1 \cos 2\pi f_1 t + E_2 \cos 2\pi f_2 t \tag{1}$$

where E_1 and E_2 are the amplitudes of the individual incident waves. When the response of the photodetector is proportional to the intensity of the radiation, the output of the detector is proportional to the square of E_t:

$$r \propto E_t^2 = E_1^2 (\cos 2\pi f_1 t)^2 + E_2^2 (\cos 2\pi f_2 t) + E_1 E_2 \cos 2\pi (f_1 - f_2) t + E_1 E_2 \cos 2\pi (f_1 + f_2) t \quad (2)$$

Because the detector cannot follow the instantaneous intensity at light frequencies, it will respond to the average value of the first, second and fourth term in (2). The average values are $E_1^2/2$, $E_2^2/2$ and zero, respectively. However, it is assumed that the detector has a sufficiently high frequency response to follow the signal at the difference frequency $f_1 - f_2$. Thus the response of the detector to the two incident waves contains besides a d.c.-term the dopplershift:

$$f_d = f_1 - f_2 \quad (3)$$

The doppler-shift is proportional to the velocity component in the plane of the two original beams. The direction in this plane is indicated by the arrow in Fig. 1(b). Measured in this way, there is an ambiguity in the sign of the direction, but we return to this soon.

The relation between doppler-shift and velocity (of the component we measure) is easily found to be:

$$f_d = (2v \sin\tfrac{1}{2}\theta)/\lambda = k.v \quad , \quad (4)$$

where f_d is the doppler-shift, v the measured velocity, θ is the angle between the two original beams, λ is the wavelenght of the laser light and k is the optical transfer constant.

As one can see, k depends only on geometrical optics and the wavelength of the laser light, so the calibration is very easy.

To eliminate ambiguity in the sign of the determined velocity component v, a pre-shift is given to the reference beam. With respect to this preshift, negative frequencies are possible.

An additional advantage of this frequency shift is that low-frequency noise, which is abundant and has different sources, for example the discrete entrance of scatterers in the measuring volume, can be eliminated entirely by choosing a high enough preshift frequency and a high-pass electronic filter in the amplifier system of the photodetector to block the low-frequency noise.

As a preshift device a rotating radial grating is employed. This grating acts at the same time as a beam-splitter (Fig.2) (Durrani et al, 1977; Oldengarm et al, 1976).

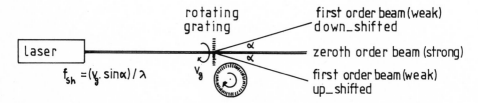

Fig. 2. Rotating radial grating used as preshift device and as beam-splitter.

Because two weak beams are generated with opposite shift with respect to the
strong beam, it is possible, by a displacement of the zeroth order beam, to construct
a two-component velocimeter system with orthogonal axes, as shown in Fig.3. This is
the system designed at T.P.D.-T.N.O. (Oldengarm et al, 1976).

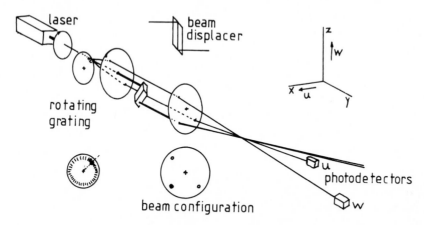

Fig.3. Two-component arrangement.

The detection of the doppler-shifted frequencies generated in the photodetectors
can be performed in different ways. In the case of a rather strong doppler signal,
as is the case when sea water is used as fluid, and when there are scatterers in
the measuring volume most of the time, a very good method is the application of
a frequency tracker which transforms the doppler-frequency into a voltage proportional to that frequency and hence proportional to the measured velocity component.
We use the frequency tracker built at the Institute of Applied Physics T.P.D.-T.N.O.
(type 1077). The electronic properties of the tracker determine the velocity range
and the frequency range of detectable velocity fluctuations (Table 1).

The dimensions of the measuring volume, determined by the diffraction of the
light beams, are given in Table 1. The velocity component that is determined this
way lies in the direction of the width d_x (Fig.4).

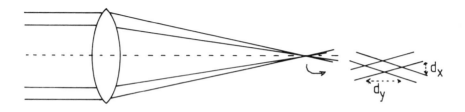

Fig.4. Measuring volume. Beam widths exaggerated.

308

TABLE 1

Technical details of the laser-doppler velocimeter system

Optical transfer constant	117	kHz/ms^{-1}
Tracker transfer constant	25	mV/kHz
Overall transfer constant	0.342	V/ms^{-1}
Preshift	760	$kHz \cong 6.5\ ms^{-1}$
Range tracker	-1.5 to $+1.5\ ms^{-1}$	
Lowest detectable velocity	$0.2\ 10^{-3}$	ms^{-1}
Tracking speed range (frequency)	0 – 2000	Hz
Tracking speed range (wavenumber) $v = 1\ ms^{-1}$	0 – 6000	m^{-1}
Width measuring volume	$0.2\ 10^{-3}$	m
length measuring volume	$5\ 10^{-3}$	m
Laser (He-Ne, Spectra Physics)	632.8 nm, 5 mW	

NOISE CONSIDERATIONS

We shall deal with a few specific aspects of noise concerning our laser-doppler system in this paper. Much more general information can be found in Durst et al (1976) and Buchhave et al (1976).

An important source of noise in our system is the rotating grating. Irregularities and slight eccentricity of the grating introduce a periodic noise signal in the pre-shift of the reference beams and hence a periodic noise signal in the velocity voltages. The frequency of this noise signal is in the frequency range expected for the turbulence spectra, so we have to eliminate it. With an extra photodetector the apparent velocity of an opal glass scatterer that is at rest is measured. This apparent velocity is subtracted from the fluid velocity components so, that the undisturbed velocity components remain. Fig.5 shows the optical configuration.

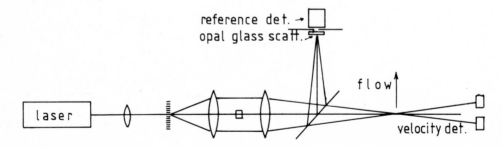

Fig.5. Reference detector for removal of periodic noise signal due to grating.

In this way, however, both the noise signal and the preshift are subtracted. To avoid this unwanted removal of the preshift, a very stable crystal oscillator has been applied to generate the wanted preshift. Fig. 6 shows how this is achieved.

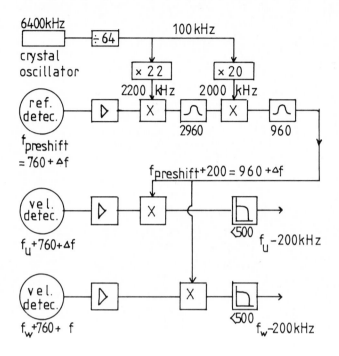

$\blacktriangleright$ amplifier

$\times$ mixer

$\sqcap$ low-pass filter

$\curvearrowright$ band filter

Fig. 6. Subtraction of periodic noise in the h.f. signal with preservation of a stable preshift.

The reference detector receives the preshift frequency generated by the rotating grating: $f_{preshift}$ = 760 kHz + Δf. Δf is the frequency of the periodic noise. The two velocity detectors receive the doppler-frequencies plus the preshift frequency: f_{signal} = $f_{u,w}$ + 760 kHz + Δf.

The crystal oscillator generates a stable 6400 kHz, which is transformed into 2200 kHz and 2000 kHz.

The output of the reference detector is after amplification, fed into a electronic mixer (X), together with the 2200 kHz signal. In the mixer the difference and sum frequencies are generated. The 2960 kHz bandfilter permits only the sum frequency 760 + Δf + 2200 kHz to pass. This signal is again mixed with 2000 kHz and the band-filter 960 kHz only transmits the difference frequency: 960 + Δf = $f_{preshift}$ + 200 kHz.

This resulting frequency is subtracted from the frequencies of the velocity detectors and the remaining frequencies are f_u - 200 kHz and f_w - 200 kHz, thus the doppler shifts of the two channels minus a stable 200 kHz preshift. The noise signal Δf is removed completely.

Ambiguity noise

Another well known source of noise is the fact that scatterers are in the measuring volume only for a limited time; new particles do not enter the volume in-phase with previous particles (George and Lumley, 1973). This results in ambiguity noise. In practice this noise influences the low intensity levels of turbulence spectra and thus the higher frequency ranges. If we have to measure in these contaminated frequency ranges, we will use a solution proposed by Van Maanen et al (1976).

In order to remove the major part of the ambiguity noise, we need both velocity channels to measure one velocity component. For that reason the beam displacer in Fig. 3 must be removed, and the optical system is rotated in the direction of the velocity component we want to measure. In this situation both channels are measuring in the same plane. Because we are working with very narrow angles between the light beams, the two channels measure almost the same velocity component. Also the measuring volume is almost the same, but the slight difference causes the fact that the velocity signals originate from different scatterers.

We obtain the turbulence power spectrum by a Fourier transform of the autocorrelation of the velocity signal under consideration. That spectrum contains the ambiguity noise. When we take the Fourier transform of the crosscorrelation of the almost identical velocity channels, the result is the same power spectrum but without a substantial part of the ambiguity noise. The scatterers in the two measuring volumes are not correlated.

Noise determinations we performed in a flume with the laser-doppler system in the autocorrelation (I) mode and the crosscorrelation (II) mode are shown in Fig.7. Hypothetical Kolmogorov spectra with different values of the dissipation rate are drawn in the same figure for comparison. The various dissipation rates, ε, cover almost the entire range one may expect in the sea:

$\varepsilon = 10^{-3}$ $m^2 s^{-3}$ in strong tidal currents and

$\varepsilon = 10^{-9}$ $m^2 s^{-3}$ in the ocean below the main thermocline

(Dillon et al, 1976; Grant et al, 1962; Gregg, 1977; Heathershaw, 1976; Williams et al, 1974; Osborn, 1974; Stewart et al, 1962; Webster, 1969)

These results show the possibility to use laser-doppler techniques for measurements in the dissipation range of turbulent spectra.

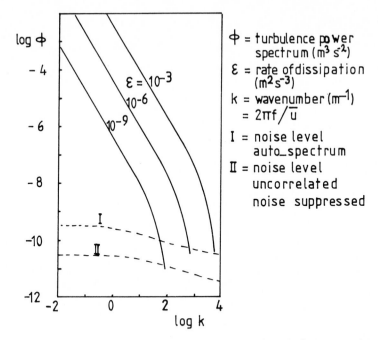

Fig. 7. Noise levels compared with hypothetical turbulence spectra.

UNDERWATER INSTRUMENT

A special hull has been designed in order to be able to use the reference beam mode. Fig. 8 shows the arrangement of the laser-doppler system schematically inside the hull. The velocity components which can be measured simultaneously are u (Fig.8) and w, perpendicular to the plane of the paper. The signals are treated for the periodic noise as described and transmitted to the ship by cable.

The noses of hull are ellipsoids of revolution. We expect a negligible influence of the shape of the instrument on the flow field at the measuring volume. This must be tested.

Just downstream the measuring volume a temperature sensor will be mounted for determinations of turbulent heat flux. The exact distance between the temperature sensor and the measuring volume has not jet been determined at this stage.

The underwater instrument is such that it can be used in a number of different ways. Towed from an anchored ship in a tidal stream or from a sailing ship, the system will be positioned with the noses forward. In this mode of operation the components u and w will be measured. A profiling mode is possible by hanging the system with the noses downward. To determine turbulence properties near the bottom, a tetrapod structure will be applied. In all modes of operation a cable must connect the system with the frequency tracker on board of a ship.

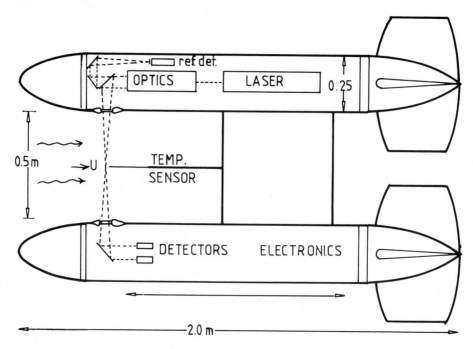

Fig.8. The underwater hull (top view).

To correct for rolling and pitching of the whole instrument, simple inclinometers are applied. An analog coordinate transformer translates the measured velocity components into velocity components parallel and normal to the horizontal plane.

Preliminary flume test of a model of the hull

Although one may expect that in a laser-doppler system the light beams will not affect the flow, the influence of the hull on the flow field is probably not negligible. To get a preliminary idea about such an influence, some simple tests have been performed with two models of the hull, a small one (scale 1:5) and a large one (scale 1:2). At the time these flume tests were done, the data processing was not at such a stage that two velocity components could be measured simultaneously, so the only thing which could be tested in the flume was the influence of the models of the hull on the longitudinal and transverse power spectrum.

For the test experiments we had to our disposal a flume of the Hydraulic Laboratory at Delft. The flume is 30 m long, 0.5 m wide and the water level is adjustable to a maximum hight of 0.65 m. The side walls of the flume are glass windows. The mean flow velocity is adjustable from 0 to 1 ms^{-1}.

To determine the influence of a model on the turbulence power spectra, we compare measurements with and without models in the flume. The experiment went as follows.

A laser-doppler velocimeter system was placed as shown in Fig.9. The laser beams go
through the glass windows of the flume. Over a distance of about three meters the
flow has been made very turbulent with blocks of concrete irregularly placed on the
bottom of the tank. About two meters downstream, where the turbulence was expected
to be almost isotropic, the lightbeams crossed at the centre of the flow.

The longitudinal and transverse power spectra of the flow were measured without
a model placed in the flow. These are the undisturbed spectra.

One of the models was placed in the flow in such a way that the light beams go
through holes in the model. Now the beams were in a configuration similar to that when
the laser-doppler system is placed inside the hull; compare Fig. 9 and Fig. 8.

The turbulence power spectra were measured again. The results were the disturbed
spectra.

The experiment has been repeated with the other model and different mean flow
velocities. The size of the large model was such that it was not feasible to place
it in the right way. Therefore it was put at right angles with the original arrangement (Fig.9).

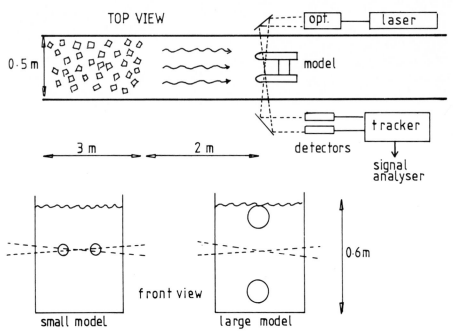

Fig. 9. Experimental set-up of the flume test of the underwater hull.

A comparison of the undisturbed and the disturbed spectra showed that the difference between these is negligible. Fig. 10 gives examples of the measured spectra.
The conclusion, based on the rather limited amount of data up till now, is that the
influence of the hull on the turbulent power spectra (longitudinal and transverse) is

negligible. In the near future, flume tests are planned with the prototype itself. The effect of the hull on other parameters will also be considered.

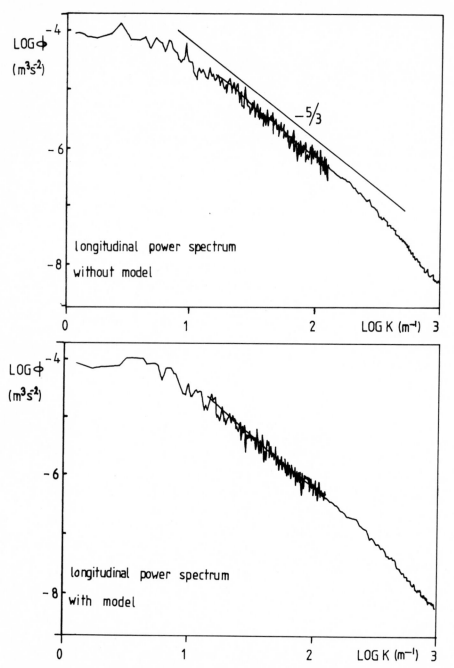

Fig. 10. Longitudinal turbulence spectra without and with the large model in the flume. The mean flow velocity was 0.32 ms^{-1}.

A DETERMINATION OF THE KOLMOGOROV CONSTANT FROM THE FLUME TEST DATA

Fig. 11 shows in one figure the longitudinal and the transverse power spectra of the flow in the flume. For about one and one-half decade the transverse spectrum Φ_{22} is 1.3 ± 0.2 times the longitudinal spectrum Φ_{11}. In the case of isotropic turbulence one would expect a factor 1.333 (Tennekes and Lumley, 1972). In the same wavenumber region the log.-spectra obey a $-5/3$ law. Hence the turbulence in the flume seems to be isotropic with an inertial subrange of about one and one-half decade.

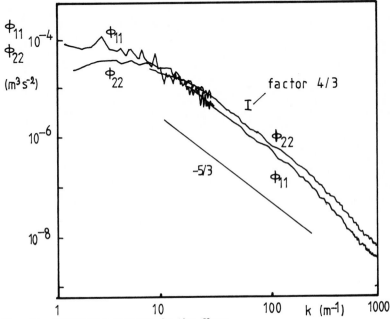

Fig. 11. Turbulence spectra in the flume.

With these spectra it is possible to make an estimate of the Kolmogorov constant α. We apply the method described by Grant et al (1962) using the dissipation spectrum.

In the inertial subrange, the longitudinal spectrum obeys (Tennekes and Lumley, 1972):

$$\Phi_{11} = (18/55)\alpha\varepsilon^{2/3} k^{-5/3} \tag{5}$$

ε can be found by determining the dissipation spectrum (Hinze, 1975):

$$\varepsilon = 15\nu \int_0^\infty k^2 \Phi_{11}(k)\,dk \tag{6}$$

where ν is the kinematic viscosity ($\nu = 1.1\ 10^{-6}\ m^2 s^{-1}$).

The Kolmogorov constant α is found as follows:

- Φ_{11} is determined by measurement,
- multiplication of Φ_{11} by k^2 yields the dissipation spectrum except for a constant

factor 15ν,

- integration gives ε,

- finally substituting ε into (5) at a representative k in the inertial subrange yields α.

The result of our measurements is:

$\varepsilon = 0.8 \ (\pm \ 0.1) \ 10^{-4} \ m^2 s^{-3}$

and thus:

$\alpha = 1.7 \pm 0.2$

The Reynolds number of the flow, based on the width of the flow and the mean velocity is $Re = 1.5 \ 10^5$.

The accuracy of the experiment is not very high because noise handling was not optimal during the flume tests and only a few measurements were available. The result is however in good agreement with a calculated value of Kraichnan (1966). His result is α = 1.77. A broad spectrum of values of α, determined in different experiments by several people, is given by Champagne (1978); our value seems not to disagree with these values (1.2 < α < 2.0).

CONCLUSIONS AND REMARKS

The laser-doppler velocimeter system which we built for measuring turbulence in the sea has properties which enable us to use it for spectral analysis of oceanic turbulence. The noise level is low enough to go far into the dissipation range and the fast response and small measuring volume allow this too. The two-component configuration enables us to determine Reynolds-stresses. Heat transport will be measureable when a fast temperature sensor is added to the system.

For data processing (correlations and spectra) a Hewlett-Packard 5420-A Signal Analyser is applied.

The measurements with the laser-doppler system will be supplemented with data from current meters, a thermo-chain, C.T.D., and other well-known hydrographic instruments.

The first measurements in the sea will take place in the winter season '79-'80.

ACKNOWLEDGEMENTS

I like to thank the people of the mechanical and electronic department at N.I.O.Z. for their indispensable work and the members of the Laser-Doppler Users Club for their contribution in the development and design of the instrument.

REFERENCES

Buchhave,P.(editor) ,1976. The Accuracy of Flow Measurements by Laser-Doppler Methods. *Proceedings of the LDA Symposium*, Copenhagen 1975 (Hemisphere Publ. Comp.).

Champagne, F.H.,1978. The fine-scale Structure of the Turbulent Velocity Field. *J.Fluid Mech.*,*86,1: 67-108.*

Dillon, T.M., Powell, T.M., 1976. Low Frequency Turbulence Spectra in the Mixed Layer of Lake Tahoe, California-Nevada. *J.Geoph.Res. 81,36:6421-6427.*

Durrani, T.S. and Greated, C.A., 1977. Laser Systems in Flow Measurements. *Plenum.*

Durst, F., Melling, A. and Whitelaw, J.H., 1976. Principles and Practice of Laser-Doppler Anemometry. *Academic Press.*

Fowlis, W.W., Thompson,J.D. and Terry, W.E., 1974. A Laser-Doppler Velocimeter with Ocean Applications. *J.Mar.Res. 32,1:93-102.*

George, W.K. and Lumley,J.L., 1973. The Laser-Doppler Velocimeter and its Application to the Measurement of Turbulence. *J.Fluid Mech., 60,2:321-362.*

Grant, H.L., Stewart, R.W. and Moillet, A., 1962. Turbulence Spectra from a Tidal Channel. *J.Fluid Mech. ,12:241-268.*

Gregg, M.C., 1977. Variations in the Intensity of Small-Scale Mixing in the Main Thermocline. *J.Phys.Oc., 7:436-454.*

Heathershaw, A.D., 1976. Measurements of Turbulence in the Irish Sea Bentic Boundary Layer. In: McCave, I.N. (editor), 1976. The Bentic Boundary Layer. *Plenum.*

Hinze, J.O., 1975. Turbulence 2nd ed. *McGraw-Hill.*

Kraichnan, R.H.,1966. Isotropic Turbulence and Inertial Subrange Structure. *Phys.Fluids, 9,9:1728-1752.*

Kullenberg, G., 1978. Preliminary Results of near Bottom Current Measurements in the Bothnian Sea. *Finn.Mar.Res., 244:42-51.*

Lespinard, G., 1978. Une Application importante de l'Anémométrie-Laser à l'Oceanographie: La Mesure des Courants. *Oceanologie, 9:73-94.*

Van Maanen et al, 1976. In Buchhave,1975.

Oldengarm, J. and Venkatesh, P., 1976. A simple Two-Component Laser-Doppler Anemometer using a Rotating Radial Grating. *J.Phys.E: Sc.Instr.,9:1009-1012.*

Osborn, T.R., 1974. Vertical Profiling of Velocity Microstructure. *J.Phys.Oc., 4: 109-115.*

Stachnik, W.J. and Mayo, W.T., 1977. Optical Velocimeters for use in Seawater. *Oceans '77. (MTS-IEEE).*

Stewart, R.W. and Grant, H.L., 1962. Determination of the rate of Dissipation of Turbulent Energy near the Sea Surface in the Presence of Waves. *J.Geoph.Res., 67,8: 3177-3180.*

Tennekes, H. and Lumley, J.L., 1972. A First Course in Turbulence. *The MIT Press.*

Webster, F., 1969. Turbulence Spectra in the Ocean. *Deep Sea Res. 16 suppl.:357-368.*

Williams, R.B. and Gibson, C.H., 1974. Direct measurements of Turbulence in the Pacific Equatorial Undercurrent. *J.Phys.Oc. 4: 104-108.*

Yeh, Y. and Cummins, H.Z., 1964. Localized Fluid Flow Measurements with a He-Ne Laser Spectrometer. *Appl.Phys.Lett., 4,10:176-178.*

ESTIMATION OF SHEARING STRESSES IN A TIDAL
CURRENT WITH APPLICATION TO THE IRISH SEA

J. WOLF

Institute of Oceanographic Sciences, Bidston Observatory,
Birkenhead, Merseyside, L43 7RA, England.

ABSTRACT

New estimates of the bed shear stress have been made from a set of current and
elevation data covering a period of a month at three stations in the Irish Sea.
Harmonic analysis of this data has made it possible to obtain various harmonic
components of the bottom frictional stress by solving the depth-integrated equation
of motion. Some unexpected results for the relation between friction stress and
bottom current have been obtained. There appears to be a phase difference between
stress and current at the bed, and the ratio of stress to the velocity squared law
appears to vary with frequency. In addition a 12-hour time series of vertical
profiles of current was available at one station. This has been used to calculate
time series of shearing stresses at each depth and obtain profiles of eddy viscosity. There is some evidence of a phase lag between the calculated stress at
any depth and the current shear at that depth.

1. INTRODUCTION

Three methods may be employed to make estimates of frictional parameters from

observation.

(a) Measure currents and surface elevations at a particular location and calcu-

late the size of the various terms in the equation of motion to deduce the friction

stress.

(b) Assume the classic logarithmic current profile in the bottom boundary layer

and fit measurements of current to this form, which gives estimates of the bottom

stress and a roughness length.

(c) Measure the Reynolds' stress directly by obtaining time series of the fluc-

tuating horizontal and vertical components of velocity and taking a time average

of their product.

Bowden's work has figured prominently in applying these methods e.g. Bowden and

Fairbairn (1952a); Bowden, Fairbairn and Hughes (1959); Bowden and Howe (1963).

However various attempts at applying method (a) have not been entirely successful,

mainly due to instrumental difficulties. Also, the values of the frictional parameters which have been calculated vary within quite wide limits. A review of the
work done on shear stress near the bed is included in Bowden (1978). Yet for
numerical modelling in particular, estimates of these parameters are required.
For vertically-integrated models an accurate parametrisation of bottom friction
is needed, and for 3-dimensional models some means of calculating stresses through
the vertical, such as a coefficient of eddy viscosity, must be used.

For the present work a month of elevation and current data was available from
3 stations in the northern Irish Sea. There were measurements of current at mid-
depth at the central station, and of elevation and bottom current at stations on
either side, aligned with the predominant tidal current. These were subjected to
a harmonic analysis so that the terms in the equation of motion could be examined
in detail for different tidal constituents. This had not been possible previously
with shorter data sets. Random errors in the measurements were reduced by this
means to insignificance. From this data, the harmonic constituents of bottom
stress were calculated and the relation between bottom stress and currents was
examined. The coefficient in the quadratic friction law was determined for each
constituent, and it appears to vary significantly with frequency.

The second part of the paper is concerned with an analysis of a time series of
current profiles through the vertical covering a 12-hour period. These were used
to examine the variation of shearing stress with depth and with time through a
tidal cycle. Then the shear stress was correlated with the vertical gradients of
velocity at each depth and also with their time derivatives to make estimates of
eddy viscosity.

2. EQUATIONS OF MOTION

The basic equations of motion for a given water particle were taken as

$$\frac{\delta u}{\delta t} + \frac{1}{2} \frac{\delta u^2}{\delta x} + v \frac{\delta u}{\delta y} - fv + g \frac{\delta \zeta}{\delta x} = -\frac{1}{\rho} \frac{\delta F_{zx}}{\delta z} \tag{1a}$$

and $$\frac{\delta v}{\delta t} + \frac{1}{2} \frac{\delta v^2}{\delta y} + u \frac{\delta v}{\delta x} + fu + g \frac{\delta \zeta}{\delta y} = -\frac{1}{\rho} \frac{\delta F_{zy}}{\delta z}, \tag{1b}$$

where x,y are the horizontal Cartesian coordinates aligned in east and north directions respectively,

  z   is the vertical coordinate, positive upwards, the origin being in the
      undisturbed water surface,

  t   is time,

  ζ is surface elevation above the undisturbed water surface,

  u,v are velocities in x,y directions respectively,

  ρ is the density of sea-water, assumed uniform and constant,

  g   is the acceleration due to gravity,

$f = 2\omega\sin\phi$ is the Coriolis frequency (where ω is the angular frequency of the Earth's rotation and ϕ is the latitude),

F_{zx} and F_{zy} are the components of shearing stress at depth z acting on the water column above that level, in the x and y directions respectively.

It has been assumed that the velocity and acceleration terms in the vertical direction are negligible, and that the direct tide-generating forces and horizontal shears are small. In general the convective terms involving horizontal gradients of u and v may be omitted as they tend to be much smaller than the other terms.

If equations (1) are integrated through depth, writing

$$\overline{u} = \frac{1}{h + \zeta}\int_{-h}^{\zeta} u \, dz, \quad \overline{v} = \frac{1}{h + \zeta}\int_{-h}^{\zeta} v \, dz,$$

where h is the mean water depth, we get

$$\frac{\delta\overline{u}}{\delta t} + \frac{1}{2}\frac{\delta\overline{u^2}}{\delta x} + \overline{v}\frac{\delta\overline{u}}{\delta y} - f\overline{v} + g\frac{\delta\zeta}{\delta x} = -\frac{1}{\rho(h + \zeta)}(F_{sx} - F_{bx}) \tag{2a}$$

$$\text{and } \frac{\delta\overline{v}}{\delta t} + \frac{1}{2}\frac{\delta\overline{v^2}}{\delta y} + \overline{u}\frac{\delta\overline{v}}{\delta x} + f\overline{u} + g\frac{\delta\zeta}{\delta y} = -\frac{1}{\rho(h + \zeta)}(F_{sy} - F_{by}), \tag{2b}$$

where $F_{zx} = F_{sx}$ and $F_{zy} = F_{sy}$ at $z = \zeta$,

and $F_{zx} = F_{bx}$ and $F_{zy} = F_{by}$ at $z = -h$.

It has been assumed that $\overline{u^2} \simeq \overline{u}^2$, $\overline{v^2} \simeq \overline{v}^2$ and $\overline{uv} \simeq \overline{u}\overline{v}$ i.e. the variation of velocity with depth is small. Except in very shallow water, $h \gg \zeta$ and $(h + \zeta) \simeq h$. Assuming this and ignoring the convective terms the linearised equations are

$$\frac{\delta u}{\delta t} - fv + g\frac{\delta\zeta}{\delta x} = -\frac{1}{\rho}\frac{\delta F_{zx}}{\delta z}, \tag{3a}$$

$$\frac{\delta v}{\delta t} + fu + g\frac{\delta\zeta}{\delta y} = -\frac{1}{\rho}\frac{\delta F_{zy}}{\delta z}, \tag{3b}$$

$$\frac{\delta\overline{u}}{\delta t} - f\overline{v} + g\frac{\delta\zeta}{\delta x} = \frac{F_{bx} - F_{sx}}{\rho h} \tag{4a}$$

$$\text{and } \frac{\delta\overline{v}}{\delta t} + f\overline{u} + g\frac{\delta\zeta}{\delta y} = \frac{F_{by} - F_{sy}}{\rho h}, \tag{4b}$$

which are equivalent to the equations used by Bowden, Fairbairn and Hughes (1959).

3. DATA

The observations were made during a joint I.O.S. Bidston/Liverpool University cruise in March and April 1977 in the northern Irish Sea. The stations of interest for the present application were:-

Station 10: $53^\circ46'N$, $03^\circ43'W$, h = 39.5 m.

Station 11: $53^\circ46'N$, $03^\circ55.5'W$, h = 41.1 m.

Station 12: $53^\circ46'N$, $04^\circ08'W$, h = 44.5 m.

as shown in Fig. 1.

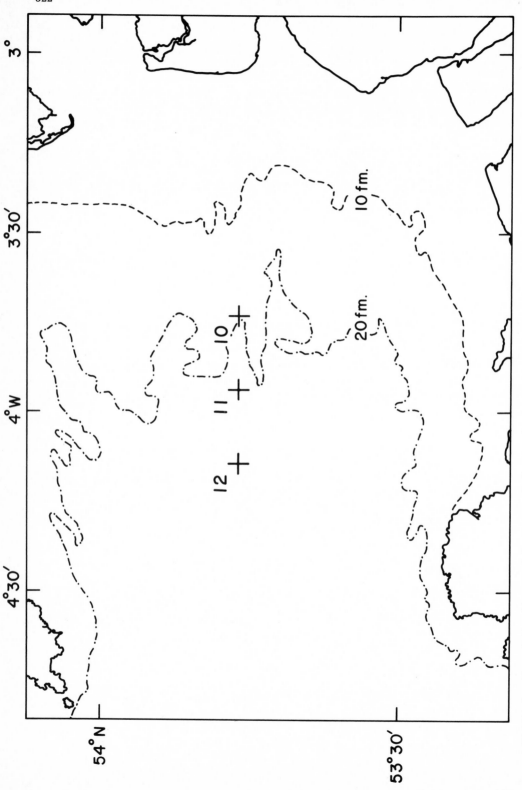

Figure 1. Map of the Liverpool Bay area of the northern Irish Sea, showing the locations of stations 10, 11 and 12.

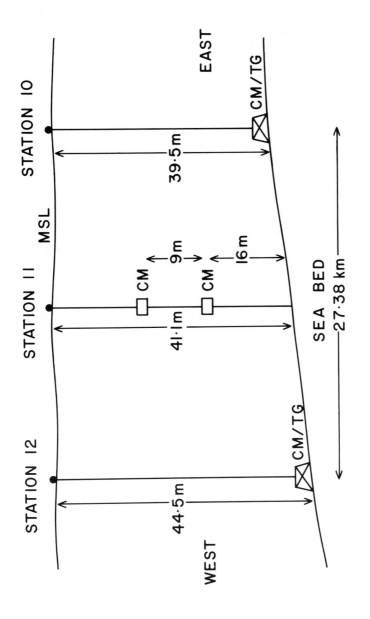

323

Figure 2. Vertical section through stations 10, 11 and 12, showing the layout of the instruments.

Fig. 2 shows the vertical layout of the instruments from which data was recovered in the section through stations 10, 11 and 12. At stations 10 and 12 bottom-mounted Aanderaa current meter/tide-gauges were deployed for one month from 16/3/77 to 17/4/77, a total of more than 29 days. Some data was lost from the beginning and end of the experiment at station 12, but the records were otherwise complete. An array of three current meters was moored at station 11 during the same period. No data was obtained from the meter nearest the sea-bed at station 11 and the middle current meter at 16 m above the bottom yielded only about 21 days of usable data, however the top current meter at 25 m above the bed gave a complete record. Hourly values and harmonic analyses of the pressure records from stations 10 and 12, and also of the north and east components of the current meter records, were available from Alcock and Howarth (1979).

During the same cruise, on 8-9/4/77, a series of velocity profiles through depth was made using a Braystoke direct-reading current meter at station 11. The profiles were taken at half-hourly intervals from 17.20 GMT on 8/4/77 to 07.59 GMT on 9/4/77 and from 20.02 GMT on 9/4/77 to 23.10 GMT on 9/4/77. Current velocity and direction were measured at six depths, nominally 3 m, 7 m, 13 m, 21 m, 29 m, and 37 m above the sea-bed, although the near-surface measurements were constrained to be about 5 m below the surface. Each profile took about 20 minutes to complete. A copy of this data was provided by Simon Ferguson of Liverpool University.

The following notation is used:

ζ_{10} and ζ_{12} are the surface elevations at stations 10 and 12 respectively,

$u_{10}^B, v_{10}^B, u_{12}^B$ and v_{12}^B are the u and v components of velocity near the sea-bed at stations 10 and 12 respectively,

$u_{11}^M, v_{11}^M, u_{11}^T$ and v_{11}^T are the u and v components of velocity at station 11, where the suffix M denotes the current meter at mid-depth and T that nearest the surface.

4. DETERMINATION OF BOTTOM STRESS FROM ONE-MONTH RECORDS

(a) Harmonic analysis and confidence limits

By means of a Fourier analysis, the components of velocity and stress, and the elevation, may be written as the sums of harmonic components.

$$\text{Let } \zeta = \sum_{i=0}^{\infty} R_i \cos(\sigma_i t - \delta_i),$$

$$u = \sum_{i=0}^{\infty} U_i \cos(\sigma_i t - \gamma_i),$$

$$v = \sum_{i=0}^{\infty} V_i \cos(\sigma_i t - \eta_i), \tag{5}$$

$$F_{zx} = \sum_{i=0}^{\infty} A_i \cos(\sigma_i t - \alpha_i)$$

$$\text{and } F_{zy} = \sum_{i=0}^{\infty} B_i \cos(\sigma_i t - \beta_i).$$

Then, inserting (5) in (3a), and equating terms in $\cos\sigma_i t$ and $\sin\sigma_i t$, gives

$$-\sigma_i U_i \cos\gamma_i - fV_i \sin\eta_i + g\frac{\delta R_i}{\delta x}\sin\delta_i + gR_i\frac{\delta\delta_i}{\delta x}\cos\delta_i = -\frac{1}{\rho}\frac{\delta}{\delta z}(A_i \sin\alpha_i),$$

$$\text{and } \sigma_i U_i \sin\gamma_i - fV_i \cos\eta_i + g\frac{\delta R_i}{\delta x}\cos\delta_i - gR_i\frac{\delta\delta_i}{\delta x}\sin\delta_i = -\frac{1}{\rho}\frac{\delta}{\delta z}(A_i \cos\alpha_i), \quad i=0,1,2\ldots$$

$$(6a)$$

or, equivalently,

$$\sigma_i U_i \sin(\gamma_i - \delta_i) - fV_i \cos(\eta_i - \delta_i) + g\frac{\delta R_i}{\delta x} = -\frac{1}{\rho}\frac{\delta}{\delta z}\{A_i \cos(\alpha_i - \delta_i)\},$$

$$\text{and } \sigma_i U_i \cos(\gamma_i - \delta_i) + fV_i \sin(\eta_i - \delta_i) - gR_i\frac{\delta\delta_i}{\delta x} = \frac{1}{\rho}\frac{\delta}{\delta z}\{A_i \sin(\alpha_i - \delta_i)\}, i=0,1,2\ldots$$

$$(6b)$$

Similarly, from (4a), assuming $F_{sx} = F_{sy} = 0$, i.e. the wind-stress is zero, we get

$$\sigma_i\overline{U_i}\sin(\gamma_i - \delta_i) - f\overline{V_i}\cos(\eta_i - \delta_i) + g\frac{\delta R_i}{\delta x} = \left(\frac{A_i\cos(\alpha_i - \delta_i)}{\rho h}\right)_B,$$

$$\text{and } \sigma_i\overline{U_i}\cos(\gamma_i - \delta_i) + f\overline{V_i}\sin(\eta_i - \delta_i) - gR_i\frac{\delta\delta_i}{\delta x} = -\left(\frac{A_i\sin(\alpha_i - \delta_i)}{\rho h}\right)_B, \quad i=0,1,2\ldots$$

$$(7)$$

Here $(\quad)_B$ indicates that the term inside the brackets is evaluated at $z = -h$. The overbar denotes the depth-averaged value of the term beneath it.

The current meter and pressure records from stations 10, 11, and 12, were subjected to harmonic analysis using a standard least squares analysis programme, TIRA, developed at I.O.S. Bidston, (Rossiter (1965)). A minimum of 21 days of hourly values was obtained from all the recording instruments, and in most cases 29 days were available, for which 35 constituents could be found. For the series which were much shorter than 29 days, 27 constituents were taken. The analysis required knowledge of the relation between the amplitudes and phases of some major constituents and those of other large constituents of nearly the same frequency which could not be resolved from 29 days of data. For this purpose, the known relations for tidal elevations from gauges at Hilbre, Heysham and Douglas were averaged and used as the best estimate. The largest amplitudes were found for the tidal constituents M_2, S_2, N_2, and M_4, which were used for the rest of this analysis.

After removing the coherent tidal part of the energy from each time series, a spectral analysis of the residuals was carried out for the purpose of estimating the standard errors on the harmonic components. It was assumed that, given a residual variance of S^2 in a frequency band of width dF cycles/unit time, then this was distributed equally among the fundamental frequencies which could be resolved within that band, i.e.

$$N^2 = S^2/T \, dF, \quad (8)$$

where N^2 is the residual noise at a particular frequency within the band and T is the length of the analysed record in time units. The sampling distributions of the amplitude and phase of the tidal component were taken to be approximately normal with standard errors $N/\sqrt{2}$ and $N/H\sqrt{2}$ respectively where H is the amplitude of the tidal constituent. Thus the 95% confidence limits on the harmonic analysis were given by ±2 standard errors. The background to this is set out in an appendix to Munk and Cartwright (1966).

TABLE 1

Amplitudes and phases of elevations and current components from a harmonic analysis, with standard errors

	M_2		S_2		N_2		M_4	
	H(cm)	$g(^{\circ})$	H(cm)	$g(^{\circ})$	H(cm)	$g(^{\circ})$	H(cm)	$g(^{\circ})$
ζ_{10}	263.64 ±0.62	317.8 ±0.1	84.27 ±0.42	359.6 ±0.3	50.66 ±0.62	293.7 ±0.7	11.35 ±0.25	203.6 ±1.3
ζ_{12}	237.70 ±0.33	316.6 ±0.1	75.91 ±0.27	357.3 ±0.2	46.48 ±0.33	291.3 ±0.4	7.12 ±0.11	203.9 ±0.9

	M_2		S_2		N_2		M_4	
	H(cm/s)	$g(^{\circ})$	H(cm/s)	$g(^{\circ})$	H(cm/s)	$g(^{\circ})$	H(cm/s)	$g(^{\circ})$
$^{B}u_{10}$	25.00 ±0.20	228.4 ±0.5	8.93 ±0.18	272.7 ±1.2	4.73 ±0.20	223.9 ±2.4	1.77 ±0.15	110.8 ±4.8
$^{B}u_{12}$	40.33 ±0.47	234.3 ±0.7	14.75 ±0.67	277.2 ±2.6	6.57 ±0.47	219.8 ±4.1	2.53 ±0.70	137.2 ±16.0
$^{M}u_{11}$	60.14 ±0.15	234.0 ±0.1	20.79 ±0.18	275.8 ±0.5	11.49 ±0.15	210.3 ±0.7	5.11 ±0.35	111.5 ±3.9
$^{T}u_{11}$	64.38 ±1.00	240.7 ±0.9	21.58 ±0.91	283.5 ±2.4	12.28 ±1.00	217.0 ±4.7	5.71 ±0.19	128.7 ±1.9
$\bar{u}_{11}$	58.42 ±1.02	237.3 ±0.9	19.92 ±1.10	279.8 ±1.6	11.06 ±1.01	214.6 ±5.0	4.94 ±0.76	121.8 ±7.9
$^{B}v_{10}$	10.71 ±0.13	315.1 ±0.7	2.97 ±0.25	355.6 ±4.8	1.75 ±0.13	279.5 ±4.4	0.93 ±0.11	178.3 ±7.0
$^{B}v_{12}$	4.44 ±0.07	248.5 ±0.9	1.45 ±0.08	307.6 ±3.3	1.12 ±0.07	247.4 ±3.6	0.46 ±0.07	159.3 ±8.0
$^{M}v_{11}$	4.27 ±0.12	347.9 ±1.6	0.75 ±0.07	1.4 ±5.7	0.82 ±0.12	324.2 ±8.6	1.20 ±0.13	136.5 ±6.2
$^{T}v_{11}$	2.85 ±0.15	311.1 ±3.1	1.52 ±0.11	14.2 ±4.2	0.52 ±0.15	348.1 ±17.2	0.53 ±0.12	181.1 ±13.8
$\bar{v}_{11}$	3.62 ±0.23	323.5 ±3.7	1.26 ±0.25	4.3 ±5.0	0.64 ±0.22	319.4 ±19.2	0.76 ±0.24	154.3 ±13.1

Table 1 lists the four main tidal constituents and their standard errors for all the current meter and pressure records. It also includes the calculated components of $\bar{u}_{11}$ and $\bar{v}_{11}$, the easterly and northerly components of the depth-mean current, using the method described in §4(b). In general the percentage errors in the currents are larger than the errors in the elevations. The northerly current components have much smaller amplitudes than the easterly components and the error percentages are higher. The smaller harmonic constituents have a higher percentage error than the larger ones.

(b) Calculation of harmonic components of bottom stress

In order to solve equations (7) with the available data it is necessary to assume that mean values of quantities across section 10-12 are representative of the values at station 11, and that gradients from 10 to 12 are uniform. Since the distance from station 10 to station 12 is much less than a tidal wavelength even for the highest frequencies considered, and the sea-bed appears to be fairly smoothly sloping, it is hoped that these requirements were fulfilled. Therefore the bottom velocities at station 11 were taken to be the mean of the bottom velocities at stations 10 and 12, i.e. $u_{11}^B = (u_{10}^B + u_{12}^B)/2$, $v_{11}^B = (v_{10}^B + v_{12}^B)/2$.

To find the harmonic components of the depth-mean velocity a parabolic velocity profile was fitted through the components of the observed velocity at various depths. Let the velocity take the form

$$u(z) = H(z) \cos(\sigma t - g(z)). \tag{9}$$

Then $\quad \bar{u} = \bar{X} \cos \sigma t + \bar{Y} \sin \sigma t, \tag{10}$

where $X = H \cos g$, $Y = H \sin g$

and $\quad \bar{u} = \dfrac{1}{h} \displaystyle\int_{-h}^{0} u \, dz$, etc.

Let H_j, g_j be the known values of H and g at $\xi = -z/h = \xi_j$,
so that we know $X = X_1$, $Y = Y_1$ at $\xi = \xi_1$,

$$X = X_2, \ Y = Y_2 \ \text{at} \ \xi = \xi_2,$$

$$\cdots\cdots\cdots\cdots\cdots\cdots \tag{11}$$

$$\text{and} \ X = X_n, \ Y = Y_n \ \text{at} \ \xi = \xi_n.$$

Given a general parabolic profile $X = A - B\xi^2$, the arbitrary constants A and B may be found by least squares regression, if H_j and g_j are known. Substituting the A and B thus determined in

$$\bar{X} = A - B/3$$

gives $\bar{X} = \dfrac{\displaystyle\sum_{j=1}^{n} \xi_j^2 \sum_{j=1}^{n} X_j \xi_j^2 - \sum_{j=1}^{n} X_j \sum_{j=1}^{n} \xi_j^4 - \sum_{j=1}^{n} X_j \xi_j^2 + \dfrac{1}{3} \sum_{j=1}^{n} X_j \sum_{j=1}^{n} \xi_j^2}{\left(\displaystyle\sum_{j=1}^{n} \xi_j^2\right)^2 - 3 \sum_{j=1}^{n} \xi_j^4}. \tag{12}$

$\bar{Y}$ may be calculated similarly. In this case the harmonically analysed velocities were known at three depths. At $\xi = 0.419$ we have u_{11}^T and v_{11}^T, at $\xi = 0.628$ we have u_{11}^M and v_{11}^M and at $\xi = 0.986$ we have u_{11}^B and v_{11}^B. From harmonic components of these were calculated the components of $\bar{u}_{11}$ and $\bar{v}_{11}$.

Each of the terms on the left-hand side of equations (7) were evaluated and $(A_i)_B$ and $(\alpha_i)_B$, the amplitude and phase of the easterly component of stress at the bed, F_{bx}, were then deduced. Including the estimates of errors in each harmonic component allowed the 95% confidence limits on the bed stress to be determined.

The quadratic friction law predicts that

$$F_{bx} = -k\rho u^B \sqrt{(u^B)^2 + (v^B)^2}, \text{ where k is some constant.} \tag{13}$$

Therefore the time series $u^B_{11} \sqrt{(u^B_{11})^2 + (v^B_{11})^2}$ was analysed harmonically and the confidence limits calculated so that a comparison could be made between the stress predicted by (13) and that calculated as above. The amplitude of the M_4 component of the calculated bottom stress appeared to be of similar magnitude to the M_2 component, whereas (13) predicted that it should be smaller, given a constant k. Since the M_4 component might have been affected to a greater extent by the omitting of the convective term, and the approximation of $(h + \zeta)$ by h, it was decided to include these effects and estimate their importance. For this purpose the time series of $(u^B_{10})^2$ and $(u^B_{12})^2$ were harmonically analysed. Numerical estimates gave

$$\overline{u} \simeq 2u^B \text{ so that } \frac{1}{2}\frac{\delta \overline{u^2}}{\delta x} \simeq \frac{2\left((u^B_{10})^2 - (u^B_{12})^2\right)}{\Delta x}$$ where Δx is the horizontal distance from

station 10 to station 12. No better estimate of the horizontal gradient of velocity was available unfortunately. The differences in friction stresses produced by including this term were approximately 30% in amplitude and phase in the M_4 constituent and on average about 10% for the semi-diurnal constituents. In order to estimate the effect of replacing $(h + \zeta)$ by h the time series of $hu^B_{11} \sqrt{(u^B_{11})^2 + (v^B_{11})^2}/(h + \zeta)$ was analysed. The differences caused by this change were quite small, reaching a maximum of about 5% in M_4 with only about 0.5% change in M_2. In analysing series such as $(u^B_{10})^2$ and $u^B_{11} \sqrt{(u^B_{11})^2 + (v^B_{11})^2}/(h + \zeta)$, the relations between constituents which could not be resolved in 29 days were found by generating year-long series of velocities and elevations from the previous analyses, and forming year-long time series for $(u^B_{10})^2$ and $u^B_{11} \sqrt{(u^B_{11})^2 + (v^B_{11})^2}/(h + \zeta)$ from these. Then each series was analysed for 103 constituents. The relations required were taken from this analysis. This method was found to reduce the magnitude of residuals, compared to using elevation relations.

Table 2 shows the final results for the amplitudes and phases of the four largest tidal constituents considered. The bottom friction stress calculated from the equation of motion, and two terms representing different parametrisations of quadratic friction are given. Bottom friction is frequently related to either $u^B \sqrt{(u^B)^2 + (v^B)^2}/(h + \zeta)$ using the near-bed velocity, usually at 1 m above the bottom, or $\overline{u}\sqrt{(\overline{u})^2 + (\overline{v})^2}/(h + \zeta)$ using the depth-averaged velocity. For each harmonic component the appropriate value of k to use in relating each of these terms to the stress is given by the ratio of the amplitude of the bottom stress component to the amplitude of the relevant term. These ratios and the phase differences between the stress and the velocity term, together with the confidence limits which may be applied to them, are shown in table 3.

It appears that, to 68% confidence at least, there are significant differences between the k which applies to M_2 and S_2, and that which applies to M_4, for both

TABLE 2. Harmonic components of bottom stress F_{bx} and quadratic friction terms $hu\overline{\sqrt{u^2 + v^2}}/(h + \zeta)$

Tidal constituent	East component of bottom stress, F_{bx}				$-\overline{u}^B\sqrt{(u^B)^2 + (v^B)^2}\,xh/(h+\zeta)$				$-\overline{u\sqrt{u^2 + v^2}}\,xh/(h+\zeta)$			
	Amplitude H_F (cm²/s²)	95% conf. limits	Phase g_F (°)	95% conf. limits	Amplitude H_{uB} (cm²/s²)	95% conf. limits	Phase g_{uB} (°)	95% conf. limits	Amplitude $H_{\overline{u}}$ (cm²/s²)	95% conf. limits	Phase $g_{\overline{u}}$ (°)	95% conf. limits
M_2	5.27	2.67 9.60	14	337 77	1080.6	1047.9 1113.3	52	50 55	3671.3	3596.7 3745.8	57	56 58
S_2	4.05	3.09 7.25	68	31 126	500.4	458.9 541.8	97	92 101	1574.3	1498.6 1653.0	101	100 101
N_2	1.39	0. 6.07	328	0 360	230.9	198.3 263.6	48	38 58	985.9	913.1 1058.7	29	25 33
M_4	4.15	0.23 7.94	228	198 284	153.1	125.1 181.0	297	287 308	384.6	306.8 462.3	290	283 298

TABLE 3. Bottom friction coefficient, k, and phase lags, for various tidal constituents

Tidal constituent	Amplitude ratio H_F/H_{uB}			Amplitude ratio $H_F/H_{\overline{u}}$			Phase lag $(g_{uB}-g_F)°$			Phase lag $(g_{\overline{u}}-g_F)°$		
	Mean value	68% conf. limits	95% conf. limits	Mean value	68% conf. limits	95% conf. limits	Mean value	68% conf. limits	95% conf. limits	Mean value	68% conf. limits	95% conf. limits
M_2	0.0049	0.0033 0.0069	0.0024 0.0092	0.0014	0.0010 0.0020	0.0007 0.0027	38	12 63	-27 78	43	17 66	-21 81
S_2	0.0081	0.0065 0.0115	0.0057 0.0158	0.0026	0.0021 0.0036	0.0019 0.0048	29	-1 54	-34 70	33	5 56	-26 70
N_2	0.0060	0. 0.0173	0. 0.0306	0.0014	0. 0.0039	0. 0.0066	88	-180 180	-180 180	61	-180 180	-180 180
M_4	0.0271	0.0131 0.0435	0.0013 0.0635	0.0108	0.0052 0.0175	0.0005 0.0259	69	36 93	3 110	62	30 84	-1 100

the depth-mean and the bottom current relation. The error bars on the N_2 component are much larger than on the other semi-diurnal components both because of its smaller size and because it falls in the same frequency band as M_2 in the spectral analysis. This means it is assigned the same magnitude of residual energy as M_2 although it might be expected to have less. Taken to 95% confidence the limits on k overlap, however there is an overall indication that k increases with frequency.

Phase differences between the stress and the quadratic velocity terms suggest that, at least to 68% confidence, the stress leads the velocity term. This phase difference remains positive to 95% confidence for M_4 but for the semi-diurnal constituents it then varies from positive to negative. The depth-mean velocity term lags behind the bottom velocity term for M_2 and S_2 by 4-5° which represents about 10 minutes at semi-diurnal frequency, but for M_4 the bottom velocity appears to lag behind the depth-mean term by about 7 minutes. The mean phase lead of the bottom stress with respect to the bottom velocity term is equivalent to about 1 hour. Part of this phase difference could perhaps be explained by the bottom velocity term, calculated at 60 cm above the sea-bed, lagging behind the velocity even nearer to the bed; but the phase difference seems rather large to be explained in this way.

From previous investigations a large scatter in the value of k has been found (see Bowden (1978)), the present results falling in the same range. The value of k is usually quoted for the relation between the bottom stress and the velocity at 1 m above the bottom. In this case the velocities were measured at 60 cm above the bottom. Assuming a logarithmic profile of velocity near the bed means that the values of k given here should be reduced by a factor of 0.87 to be equivalent to the values for 1 m, if the roughness length is of the order of 1 cm. This gives a mean value of k for M_2 of 0.0043. Bowden, Fairbairn and Hughes (1959) obtain a value of 0.0036 which is somewhat smaller. If the value of the friction coefficient is related in some way to frequency, as the present results to a certain extent suggest, this could explain the variability in previous measurements. A longer time series from which more components could be extracted with greater accuracy is required to confirm this.

The discrepancies between the calculated stress and the quadratic friction law are illustrated in Fig. 3. The thick line shows the time series from the sum of the M_2, S_2, N_2 and M_4 constituents of the calculated bottom stress plotted over a 15-day period. The thin line shows the quadratic friction law time series using the same four constituents with a value of k of 0.0075 in the bottom velocity squared term. This time series is in antiphase with the current, the stress opposing the current at each instant, so that it also represents the neap/spring current cycle. The large value of k was found to give the best overall agreement in magnitude. M_4 has a large amplitude in the stress calculated from observations which causes the strong asymmetry in the oscillations. It also causes the differences

331

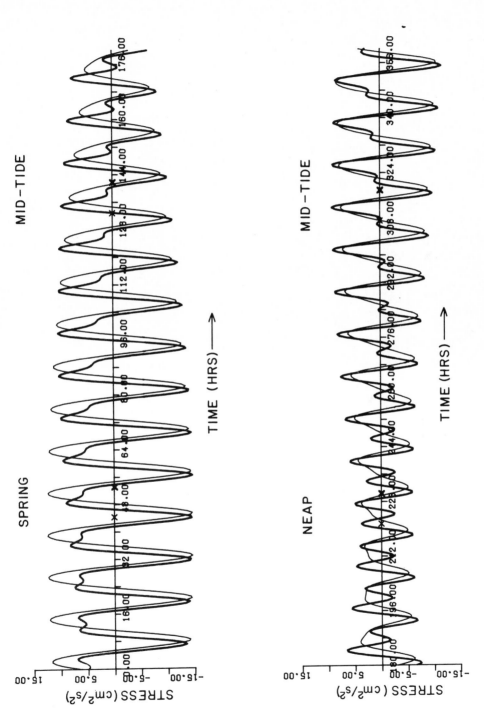

Figure 3. Time series of the bottom stress calculated from the equation of motion compared with the quadratic friction law, including components M_2, S_2, N_2 and M_4. —— Bottom stress from equation of motion. —— Quadratic friction stress = $-0.0075uB\sqrt{(uB)^2 + (vB)^2}$. ✗ ✗ Time of spring, neap and mid-tide for the respective stress series.

between the two series to be most exaggerated at neap tide. In general, the quadratic friction lags about an hour behind the observed bottom stress. This difference
in phase is minimised at about mid-way between neaps and springs, where best agreement is achieved. The suggestion is that for modelling neap or spring tides the
quadratic law would not perform as well as for an intermediate tide, or at least
the optimum k would vary with time.

5. ANALYSIS OF VERTICAL PROFILES OF CURRENT

(a) Calculation of shearing stresses through the water column

Equations (3a)-(4a) and (3b)-(4b) give

$$\frac{\delta(u - \bar{u})}{\delta t} - f(v - \bar{v}) = -\frac{1}{\rho}\frac{\delta F_{zx}}{\delta z} + \frac{F_{sx} - F_{bx}}{\rho h} \tag{14a}$$

and
$$\frac{\delta(v - \bar{v})}{\delta t} + f(u - \bar{u}) = -\frac{1}{\rho}\frac{\delta F_{zy}}{\delta z} + \frac{F_{sy} - F_{by}}{\rho h}, \tag{14b}$$

thus eliminating the surface elevation gradients. These equations can then be
solved for F_{zx} and F_{zy} using only velocity measurements, provided there are a
sufficient number through the vertical to give an estimate of the depth-mean velocity. Some assumption must also be made about the form of the stress at the sea-
surface and sea-bed. The stress at depth z is then given by:

$$F_{zx} = F_{sx}\left(1 + \frac{z}{h}\right) - F_{bx}\frac{z}{h} - \rho \int_{-h}^{z}\left(\frac{\delta(u - \bar{u})}{\delta t} - f(v - \bar{v})\right)dz \tag{15a}$$

and
$$F_{zy} = F_{sy}\left(1 + \frac{z}{h}\right) - F_{by}\frac{z}{h} - \rho \int_{-h}^{z}\left(\frac{\delta(v - \bar{v})}{\delta t} + f(u - \bar{u})\right)dz. \tag{15b}$$

Throughout the time when the current profiles were being recorded, the wind-speeds
were very low and the estimated surface stress was much less than 1 dyne/cm² in
general, possibly achieving a maximum of 1.5 dynes/cm². Therefore the surface
stress has been ignored.

Figs. 4 and 5 show the east and north components of velocity at each depth plotted
against time, for the majority of the current profiles. 7 more profiles were measured
after 20.00 on the 9th April 1977. These have been included in the calculations.
Some anomalous directions were recorded for the near-surface velocities where the
current meter was about 5 m below the surface. It seemed that there was an error
in direction of 180° mainly during the ebb. This was assumed to be caused by the
influence of the ship's magnetic field, and the directions of the near-surface
currents at these times were made equal to those of the level below.

In order to estimate the bottom stress it was first attempted to fit a logarithmic profile through the two current measurements nearest the bed, at any instant,
using the formula

$$U(z) = \frac{1}{\kappa}\left(\frac{F_b}{\rho}\right)^{\frac{1}{2}} \log_e \frac{z + h}{z_o} \tag{16}$$

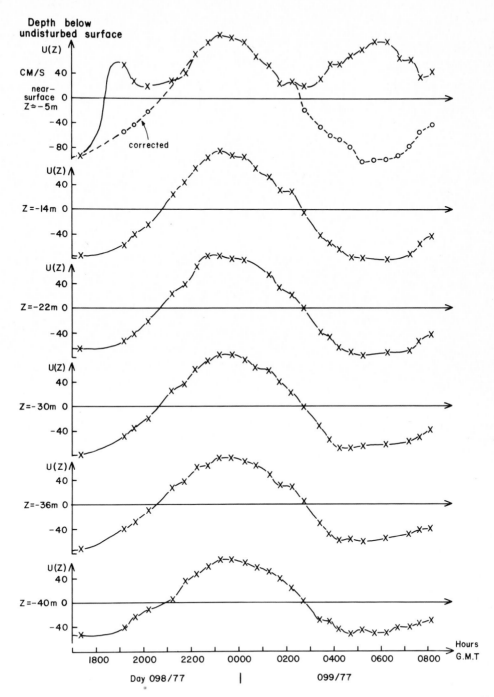

Figure 4. East component of current u(z) at depth z against time t, for station 11, days 098-099, 1977.

334

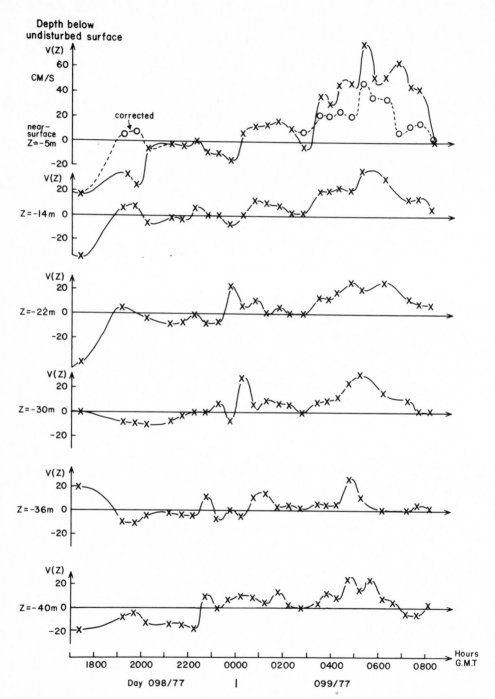

Figure 5. North component of current v(z) at depth z against time t, for station 11, days 098–099, 1977.

where U(z) is the magnitude of the velocity at height (z + h) above the bottom, F_b is the magnitude of the bed stress, κ is von Karman's constant = 0.4, and z_o is the roughness length. Using two simultaneous measurements of velocity at different heights, F_b and z_o may be determined, provided the logarithmic profile is applicable. In this case, measurements at 3 m and 7 m above the bottom were used, these being the only levels which might be considered to lie within the logarithmic layer. Several measurements within the bottom few metres should ideally have been used to check the shape of the profile and fit the best line through the points by regression. As it was, the results showed so much variability in F_b and z_o that it was decided to abandon the method in favour of the quadratic friction law. Comparing the estimates of bottom stress from the logarithmic profile method with the square of the velocity at 3 m above the bed gave values of the friction coefficient k ranging from 0.0003 to 0.043. In the absence of any better measure of k the bottom stress was taken to be

$$F_{bx} = -0.002\rho\bar{u}\sqrt{u^2 + v^2},$$ (17)
$$F_{by} = -0.002\rho\bar{v}\sqrt{u^2 + v^2}.$$

The depth-mean velocities $\bar{u}$ and $\bar{v}$ were calculated as

$$\bar{u} = \frac{1}{h} \sum_{z=-h}^{o} u(z)\Delta z$$ (18)

and $\bar{v} = \frac{1}{h} \sum_{z=-h}^{o} v(z)\Delta z.$

Then equations (15a) and (15b) were used to derive F_{zx} and F_{zy} at depths z = -10, -18, -26, -33, and -38 m, replacing the integral

$$\rho \int_{-h}^{z} \left(\frac{\delta(u - \bar{u})}{\delta t} - f(v - \bar{v})\right)dz \text{ by } \rho \sum_{z=-h}^{z} \left(\frac{\delta(u - \bar{u})}{\delta t} - f(v - \bar{v})\right)\Delta z$$ (19)

and $\rho \int_{-h}^{z} \left(\frac{\delta(v - \bar{v})}{\delta t} + f(u - \bar{u})\right)dz \text{ by } \rho \sum_{z=-h}^{z} \left(\frac{\delta(v - \bar{v})}{\delta t} + f(u - \bar{u})\right)\Delta z.$

The time derivatives were calculated using centered differences. Fig. 6 shows the east component of stress at each depth plotted against time. Figs. 7 and 8 show the east components of current and stress respectively, plotted against depth for selected times during the tidal cycle, starting at 20.45 GMT on 8th April and finishing at 07.38 GMT on 9th April. In general the stress decreases with increasing height above the bottom, except near slack water when the bottom stress tends to zero and the acceleration, or inertia, terms dominate in the upper part of the water column.

(b) Investigation of eddy viscosity

The shear stress F_{zx} is often related to the vertical gradient of current $\delta u/\delta z$ by means of an eddy viscosity N (which may be a function of x,y,z,t) such that

$$F_{zx} = -N\rho\frac{\delta u}{\delta z}$$ (20)

and, similarly, $F_{zy} = -N\rho\frac{\delta v}{\delta z}.$

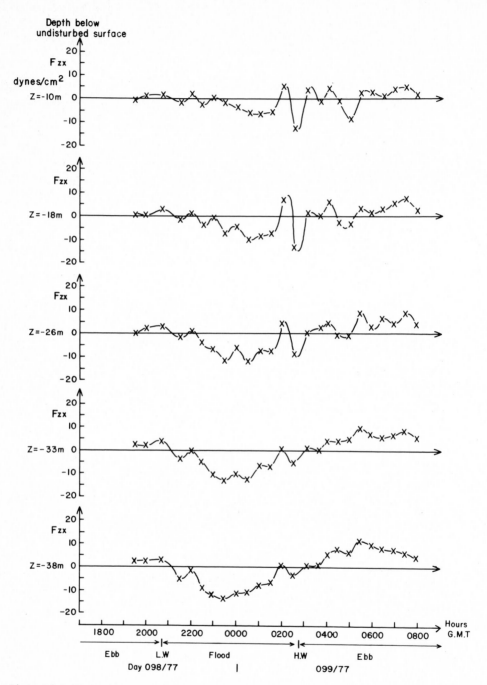

Figure 6.   East component of shearing stress, F_{zx}, at depth z against time t, at station 11, days 098-099, 1977.

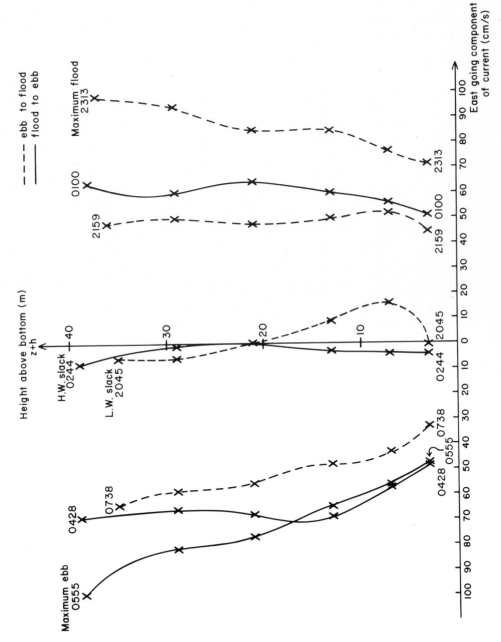

Figure 7. East component of velocity against height above sea-bed, at selected times during the tidal cycle.

338

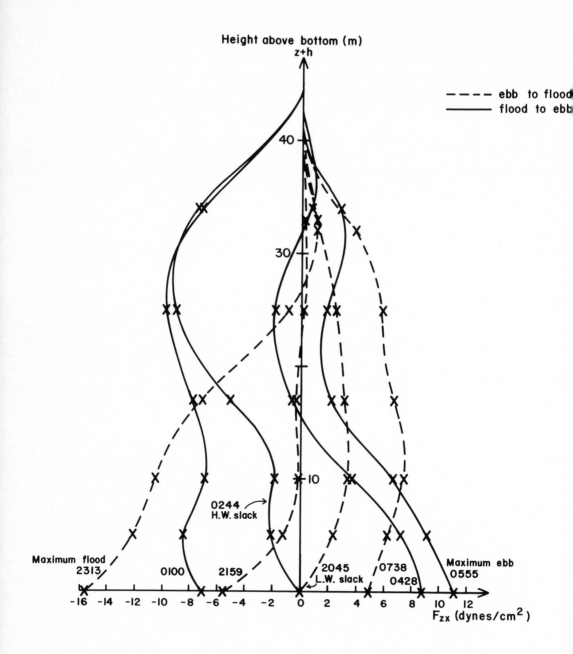

Figure 8. East component of shearing stress, F_{zx}, at depth z against height above sea-bed, for selected times during the tidal cycle.

TABLE 4

Correlation coefficients of F_{zx} with various terms involving velocity shear

Height above sea-bed (m)	Correlation coefficients of selected variables with F_{zx}						Multiple correlation coefficients	
	(1) $\dfrac{\delta u}{\delta z}$	(2) $\dfrac{\delta u}{\delta z}\sqrt{\left(\dfrac{\delta u}{\delta z}\right)^2+\left(\dfrac{\delta v}{\delta z}\right)^2}$	(3) $\dfrac{\delta u}{\delta z}\sqrt{u^2+v^2}$	(4) $(u^2+v^2)\dfrac{\delta u}{\delta z}$	(5) $\dfrac{\delta^2 u}{\delta z\delta t}$	(6) $\dfrac{\delta}{\delta t}\left[(u^2+v^2)\dfrac{\delta u}{\delta z}\right]$	(4) with (5)	(4) with (6)
5	-0.75	-0.68	-0.84	-0.88	0.22	0.20	0.90	0.91
10	-0.86	-0.86	-0.87	-0.86	0.11	0.23	0.86	0.89
17	-0.53	-0.44	-0.48	-0.46	0.36	0.53	0.61	0.71
25	-0.50	-0.41	-0.47	-0.44	0.51	0.47	0.69	0.58
33	-0.20	-0.19	-0.21	-0.20	0.75	0.71	0.78	0.72

To test this relation and find a value for N the series of F_{zx} and F_{zy} at each depth were correlated with the simultaneous values of $\delta u/\delta z$ and $\delta v/\delta z$ at the same depth, and a regression analysis carried out to find the optimum value of N, averaged over a tidal cycle. Other possible relations were also tested, for example, the mixing length hypothesis which suggests $F_{zx} \alpha - \frac{\delta u}{\delta z}\sqrt{\left(\frac{\delta u}{\delta z}\right)^2 + \left(\frac{\delta v}{\delta z}\right)^2}$. It has also been suggested that $N\alpha$ velocity $\times$ h (Bowden and Hamilton (1975)) so that $F_{zx} \alpha - \sqrt{u^2 + v^2}\,\delta u/\delta z$ was tested, and Davies (1979) has postulated $F_{zx} \alpha - (u^2 + v^2)\delta u/\delta z$. Since relation of the stress to the velocity shear has mainly been supported by evidence from steady flow, it was decided to include correlations with a term $\delta^2 u/\delta z\delta t$ to account for the effects of accelerations in tidal flow.

The correlations of F_{zx} with each of the above terms are tabulated in table 4, for each depth. Since the time derivative term was found to give significant correlation, various forms were tested, viz. $\sqrt{u^2 + v^2}\,\delta^2 u/\delta z\delta t$, $(u^2 + v^2)\delta^2 u/\delta z\delta t$, $\delta(\sqrt{u^2 + v^2}\,\delta u/\delta z)/\delta t$ and $\delta((u^2 + v^2)\delta u/\delta z)\delta t$. Only the last term gave an overall improvement in correlation compared with $\delta^2 u/\delta z\delta t$. The best overall correlation with a single variable was given by F_{zx} correlated with $\delta u/\delta z$, $\sqrt{u^2 + v^2}\,\delta u/\delta z$ or $(u^2 + v^2)\delta u/\delta z$. However, taking two terms in a multiple regression analysis the highest multiple correlation coefficients were obtained for combinations of $(u^2 + v^2)\delta u/\delta z$ with $\delta^2 u/\delta z\delta t$, and $\sqrt{u^2 + v^2}\,\delta u/\delta z$ with $\delta^2 u/\delta z\delta t$. Correlations of 0.7 to 0.9 were obtained except for the value at mid-depth. Terms in $\delta u/\delta z$ were all highly inter-correlated and therefore not included together but were almost independent of $\delta^2 u/\delta z\delta t$. The errors in F_{zy} were much larger than in F_{zx} due to the transverse velocities being much smaller than the velocities in the predominant direction of tidal flow and hence small errors in the measured direction of flow cause a much larger error in the v-component of velocity than in the u-component. For this reason correlations with F_{zy} are much less reliable and have been omitted.

The terms in $\delta u/\delta z$ tended to give a maximum correlation near the sea-bed, falling off almost to zero near the surface, while the correlations with $\delta^2 u/\delta z\delta t$ increased with height above the sea-bed, reaching a maximum near the surface.

A relation such as
$$F_{zx}(t) = -N_1\rho(u^2(t) + v^2(t))\frac{\delta u}{\delta z}(t) + N_2\rho\frac{\delta}{\delta t}\{(u^2(t) + v^2(t))\frac{\delta u}{\delta z}(t)\}, \tag{21}$$

may be looked upon as a truncated Taylor series so that it may be rewritten
$$F_{zx}(t) = -N_1\rho\{u^2(t - \Delta t) + v^2(t - \Delta t)\}\frac{\delta u}{\delta z}(t - \Delta t) = -N'\rho\frac{\delta u}{\delta z}(t - \Delta t), \tag{22}$$

where $\Delta t = N_2/N_1$ represents the time-lag between velocity-shear and the stress, and $N' = N_1(u^2(t - \Delta t) + v^2(t - \Delta t))$ is the effective eddy viscosity. Using terms $(u^2 + v^2)\delta u/\delta z$ and $\delta((u^2 + v^2)\delta u/\delta z)/\delta t$ in the multiple regression analysis with F_{zx} give a value for N_1, the coefficient relating the effective eddy viscosity to the magnitude of the current squared, and N_2/N_1, the time-lag, for each depth as shown in table 5. The eddy viscosity reaches a maximum between the bottom and mid-depth, while the time-lag increases with height above the bottom. The time-lag is of the order of 1 hour except near the surface. This anomalous value is probably due to the fact that the correlation of F_{zx} with $\delta u/\delta z$ is practically zero at this level, so that this estimate is not reliable.

TABLE 5

Effective eddy viscosity and time lags at different depths, including standard errors

Height above sea-bed (m)	Eddy viscosity/$(u^2 + v^2)$ (secs)	Time lag (hrs)	
5	0.114 ± 0.011	0.384	$\begin{cases} 0.222 \\ 0.583 \end{cases}$
10	0.119 ± 0.013	0.420	$\begin{cases} 0.231 \\ 0.655 \end{cases}$
17	0.102 ± 0.031	1.174	$\begin{cases} 0.664 \\ 2.128 \end{cases}$
25	0.043 ± 0.020	1.699	$\begin{cases} 0.661 \\ 4.541 \end{cases}$
33	0.005 ± 0.007	9.423	$\begin{cases} -\infty \\ +\infty \end{cases}$

Taking the terms which give the maximum multiple correlation, i.e. $(u^2 + v^2)\delta u/\delta z$ and $\delta^2 u/\delta z\delta t$, the values of the regression coefficient obtained for each, along with their 95% confidence limits, are plotted against depth in Fig. 9. There is also a small constant term to be added. This gives a relation

$$F_{zx} = A(u^2 + v^2)\frac{\delta u}{\delta z} + B\frac{\delta^2 u}{\delta z\delta t} + C, \tag{23}$$

where C is the constant term. A, B and C may be fitted empirically by functions of the depth, giving

$$F_{zx} = 1.204\frac{z^3}{h^3}\left(1 + \frac{z}{h}\right)(u^2 + v^2)\frac{\delta u}{\delta z} + 252.4 \times 10^4 \left(1 - \frac{z^2}{h^2}\right)\frac{\delta^2 u}{\delta z\delta t}$$

$$+ 1.106\left(\frac{z}{h} + .309\right)^2 - .505 \quad \text{(c.g.s. units).} \tag{24}$$

The predicted values of A and B from (24) are also plotted on Fig. 9 for comparison and fall well within the 95% confidence limits.

To sum up, the shearing stress at a depth z is correlated with $(u^2 + v^2)\delta u/\delta z$ and with $\delta^2 u/\delta z\delta t$. The first term gives maximum correlation near the sea-bed where the shear stress is most strongly determined by the bed shear stress. The contribution of the latter falls off with increasing height above the sea-bed. The

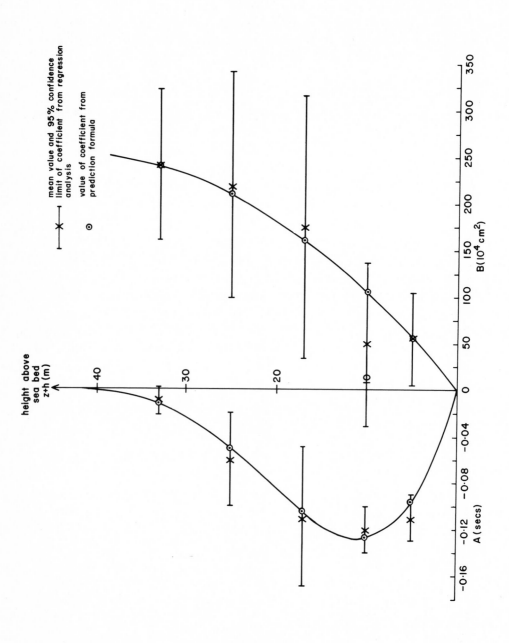

Figure 9. Regression coefficients A and B from the equation $F_{zx} = A(u^2 + v^2) \frac{\delta u}{\delta z} + B \frac{\delta^2 u}{\delta z \delta t}$, plotted against height above sea-bed.

correlation of the second term increases with height above the bed, reaching a maximum near the surface. This term represents inertial effects which are increasingly important further from the bottom. A limitation on the present work is the lack of an independent estimate of the bottom stress by either a logarithmic profile fitting or direct measurement of Reynolds' stress. Either of these would require a greater density of current measurements near the sea-bed. Current profiles through several tidal cycles would have also been a great advantage to check the consistency of the above results.

6. SUMMARY

(a) Harmonic analyses of a month's data seem to indicate that the coefficient of bottom friction in the quadratic friction law may be dependent on frequency. There appears to be a time-lag between the shear stress calculated from the equations of motion and that predicted by the quadratic friction law, the shear stress leading the velocity term. These effects should be investigated using even longer time series if possible.

(b) The shear stress through depth is correlated most strongly with a combination of the velocity squared times the velocity shear and a time derivative of the velocity shear. The latter introduces a time-lag or inertia effect increasing with height above the sea-bed. The effective eddy viscosity has a maximum just below mid-depth. Again, a longer time series, preferably covering several tidal cycles, would allow these effects to be clarified.

ACKNOWLEDGEMENTS

The author wishes to thank Mr M.J. Howarth and Mr G.A. Alcock of I.O.S. Bidston and Mr S. Ferguson of Liverpool University for providing the original data. Dr N.S. Heaps gave much advice and guidance during the work. Thanks are also due to Mr R.A. Smith for preparing the diagrams.

The work described in this paper was funded by a Consortium consisting of the Natural Environment Research Council, the Ministry of Agriculture, Fisheries and Food and the Departments of Energy and Industry.

REFERENCES

Alcock, G.A. and Howarth, M.J., 1979. I.O.S. Data Report No. 15.
Bowden, K.F., 1978. Geophysical Surveys, 3: 255-296.
Bowden, K.F. and Fairbairn, L.A., 1952a. Proc. R. Soc. A, 214: 371-392.
Bowden, K.F. and Fairbairn, L.A., 1956. Proc. R. Soc. A, 237: 422-438.
Bowden, K.F., Fairbairn, L.A. and Hughes, P., 1959. Geophys. J. R. astr. Soc., 2: 4, 288-305.
Bowden, K.F. and Hamilton, P., 1975. Estuarine & Coastal Mar. Sci., 3: 281-301.

344

Bowden, K.F. and Howe, M.R., 1963. J. Fluid Mech., 17, 2: 271-284.
Davies, A.M., 1979. Proceedings of 16th International Conference on Coastal Engineering, Hamburg, 1978.
Grace, S.F., 1936. Mon. Not. R. A. S. Geophys. Suppl., 3, 9: 388-395.
Grace, S.F., 1937. Mon. Not. R. A. S. Geophys. Suppl., 4, 2: 133-142.
Harvey, J.G. and Vincent, C.E., 1977. Estuarine & Coastal Mar. Sci., 5: 715-731.
Munk, W.H. and Cartwright, D.E., 1966. Phil. Trans. A, 259: 533-581.
Rossiter, J.R., 1965. Proceedings of the Symposium on Tidal Instrumentation and Predictions of Tides, Paris, 3-7 May 1965.

SHEAR EFFECT DISPERSION IN A SHALLOW TIDAL SEA

Jacques C.J. NIHOUL[1], Y. RUNFOLA and B. ROISIN[2]

Mécanique des Fluides Géophysiques, Université de Liège, Belgium.

INTRODUCTION

The hydrodynamics of shallow continental seas like the North Sea is dominated by long waves, tides and storm surges, with current velocities of the order of 1 m/s. The currents generate strong three-dimensional turbulence and vertical mixing, resulting, in general, in a fairly homogeneous distribution of temperature, salinity and concentrations of marine constituents over the water column.

Vertical gradients of concentrations may exist in localized areas where vertical mixing is partly (and temporarily) inhibited by stratification or during short periods of time - a few hours following an off-shore dumping, for instance - before vertical mixing is completed. However such cases are very limited in space and time and, in most problems, it is sufficient to study, in a first approach, the horizontal distribution of depth-averaged concentrations.

If c denotes the concentration of a given constituant, the three-dimensional "dispersion" equation, describing the evolution of c in space and time, can be written (e.g. Nihoul, 1975)

$$\frac{\partial c}{\partial t} + \underset{\sim}{\nabla}.(c\underset{\sim}{v}) = Q + I - \underset{\sim}{\nabla}.(\underset{\sim}{\sigma}c) + D \tag{1}$$

In eq. (1),

i) $\underset{\sim}{\nabla}.(c\underset{\sim}{v})$ represents advection and can be separated in two parts corresponding respectively to the horizontal transport $\underset{\sim}{\nabla}.(c\underset{\sim}{u})$ and to the vertical transport $\frac{\partial}{\partial x_3}(cv_3)$; $\underset{\sim}{u} = v_1 \underset{\sim}{e}_1 + v_2 \underset{\sim}{e}_2$ denoting the horizontal current velocity.

ii) Q represents the rate of production (or destruction) of the constituent by volume sources (or sinks).

---

1. Also at the Institut d'Astronomie et de Géophysique, Université de Louvain, Belgium.

2. Present address : Geophysical Fluid Dynamics Institute, Florida State University, Thallahassee, Florida, U.S.A.

346

(In most practical applications, inputs and outputs are located at the boundaries - in which case, they appear in the boundary conditions and not in Q -, or are localized quasi-instantaneous releases which may be conveniently taken into account in the initial conditions. In the following, one shall assume that this is the case and one shall set Q = 0).

iii) I represents the rate of production (or destruction) of the constituent by (chemical, ecological,...) interactions inside the marine system and I is, in general, a function of coupled variables c' , c",...

(A marine constituent is said to be passive when its evolution is not affected by such interactions. In the following, to simplify the formulation, one shall restrict attention to passive constituents and set I = 0 . The generalization of the theory to a system of interacting constituents presents no fundamental difficulty (Nihoul and Adam, 1977)).

iv) $- \nabla.(\sigma c)$ represents "migration". (sedimentation, horizontal migration of fish,..., e.g. Nihoul, 1975).

(Migration, at least in the most frequent case of sedimentation, can easily be taken into account (e.g. Nihoul and Adam, 1977). However, to avoid overloading the analysis, one shall assume, in the following, that the constituent is simply transported by the fluid and that the migration velocity σ is zero).

v) D represents turbulent diffusion and can be separated into a vertical turbulent diffusion and a horizontal turbulent diffusion.

(The horizontal turbulent diffusion is negligible as compared to the horizontal advection. The horizontal dispersion which is observed in the sea is mainly the result of the horizontal transport of the constituent by irregular and variable currents constituting a form of "pseudo horizontal turbulence" extending to much larger scales than the "proper" three-dimensional turbulence (e.g. Nihoul, 1975). In that case, D can be simply written

$$D = \frac{\partial}{\partial x_3} (\mu \frac{\partial c}{\partial x_3}) \tag{2}$$

where μ is the vertical turbulent diffusivity).

In the scope of the hypotheses made above, eq.(1) can be written, in the simpler form

$$\frac{\partial c}{\partial t} + \nabla.(cu) + \frac{\partial}{\partial x_3} (cv_3) = \frac{\partial}{\partial x_3} (\mu \frac{\partial c}{\partial x_3}) \tag{3}$$

The velocity field $\underset{\sim}{v} = \underset{\sim}{u} + v_3 \underset{\sim}{e}_3$ is given by the Boussinesq equations and in particular, one has

$$\underset{\sim}{\nabla} \cdot \underset{\sim}{u} + \frac{\partial v_3}{\partial x_3} = 0 \tag{4}$$

DEPTH-AVERAGED DISPERSION EQUATION

Let

$$\overline{c} = H^{-1} \int_{-h}^{\zeta} c \, dx_3 \quad ; \quad \hat{c} = c - \overline{c} \tag{5};(6)$$

$$\overline{\underset{\sim}{u}} = H^{-1} \int_{-h}^{\zeta} \underset{\sim}{u} \, dx_3 \quad ; \quad \hat{\underset{\sim}{u}} = \underset{\sim}{u} - \overline{\underset{\sim}{u}} \tag{7};(8)$$

with

$$\int_{-h}^{\zeta} \hat{c} \, dx_3 = 0 \quad ; \quad \int_{-h}^{\zeta} \hat{\underset{\sim}{u}} \, dx_3 = 0 \tag{9};(10)$$

and

$$H = h + \zeta \tag{11}$$

where h is the depth and ζ the surface elevation.

One has

$$\frac{\partial \zeta}{\partial t} + \underset{\sim}{u} \cdot \underset{\sim}{\nabla} \zeta = v_3 \quad \text{at} \quad x_3 = \zeta \tag{12}$$

$$\frac{\partial h}{\partial t} + \underset{\sim}{u} \cdot \underset{\sim}{\nabla} h = - v_3 \text{ at} \quad x_3 = - h \tag{13}$$

Integrating eqs.(1) and (4) over depth, inverting the order of integration with respect to x_3 and of derivation with respect to t, x_1 or x_2 and using eqs.(12) and (13) to eliminate the corrections due to the variable limits of integration, one obtains (e.g. Nihoul, 1975)

$$\frac{\partial}{\partial t} (H\overline{c}) + \underset{\sim}{\nabla} \cdot (H \overline{c} \, \overline{\underset{\sim}{u}}) + \underset{\sim}{\nabla} \cdot \int_{-h}^{\zeta} \hat{c} \, \hat{\underset{\sim}{u}} \, dx_3 = 0 \tag{14}$$

$$\frac{\partial H}{\partial t} + \underset{\sim}{\nabla} \cdot (H\overline{\underset{\sim}{u}}) = 0 \; . \tag{15}$$

In the right-hand side of eq.(14), one should have the difference between the fluxes of the constituent at the free surface and at the bottom. The hypothesis is made here that there is no exchange between the water column and the atmosphere and between the water column and the bottom sediments.

In this case, combining eqs.(14) and (15), one gets

$$\frac{\partial \bar{c}}{\partial t} + \bar{u}.\nabla \bar{c} = \Sigma \tag{16}$$

where

$$\Sigma = H^{-1} \nabla . \int_{-h}^{\zeta} (- \hat{c}\hat{u}) \ dx_3 \tag{17}$$

Σ contains the mean product of the deviations $\hat{c}$ and $\hat{u}$ around the mean values $\bar{c}$ and $\bar{u}$. The observations reveal that this term is responsible for a horizontal dispersion analogous to the turbulent dispersion but many times more efficient. This effect is called the "shear effect" because it is associated with the vertical gradient of the horizontal velocity u (e.g. Bowden, 1965 ; Nihoul, 1975).

PARAMETERIZATION OF THE SHEAR EFFECT

Substracting eq.(16) from eq.(1), one obtains

$$\frac{\partial \hat{c}}{\partial t} + \bar{u}.\nabla \hat{c} + \hat{u}.\nabla \hat{c} + \Sigma + v_3 \frac{\partial \hat{c}}{\partial x_3} + \hat{u}.\nabla \bar{c} = \frac{\partial}{\partial x_3} \left(\mu \frac{\partial \hat{c}}{\partial x_3} \right) \tag{18}$$

Because of the strong vertical mixing, one expects the deviation $\hat{c}$ to be much smaller than the mean value $\bar{c}$. This is not true for the velocity deviation $\hat{u}$ which may be comparable to $\bar{u}$; the velocity increasing from zero at the bottom to its maximum value at the surface. One may thus assume that the first four terms in the left-hand side of eq.(18) are negligible as compared to the sixth one $\hat{u}.\nabla \bar{c}$. The fifth term, representing vertical advection, is undoubtedly even smaller than the four neglected terms and eq.(18) reduces to

$$\hat{u}.\nabla \bar{c} = \frac{\partial}{\partial x_3} \left(\mu \frac{\partial \hat{c}}{\partial x_3} \right) \tag{19}$$

The physical meaning of this equation is clear : weak vertical inhomogeneities are constantly created by inhomogeneous convective transport and they adapt to that transport in such a way that the effects of advection and turbulent diffusion are in equilibrium for them.

Integrating eq.(19) with the condition that the flux is zero at the free surface, one obtains

$$H\hat{\underset{\sim}{r}} . \overline{\nabla c} = \mu \frac{\partial \hat{c}}{\partial x_3} \tag{20}$$

where

$$\hat{\underset{\sim}{r}} = H^{-1} \int_{\zeta}^{x_3} \hat{\underset{\sim}{u}} \, dx_3 \tag{21}$$

Integrating by parts and taking into account that $\hat{\underset{\sim}{r}} = 0$ at $x_3 = \zeta$ and $x_3 = -h$ (cfr eq.10), one gets

$$\Sigma = H^{-1} \underset{\sim}{\nabla} . (H \underset{\approx}{R} . \overline{\nabla c}) \tag{22}$$

where R is the shear effect diffusivity tensor, i.e. :

$$\underset{\approx}{R} = H \int_{-h}^{\zeta} \frac{\hat{\underset{\sim}{r}} \, \hat{\underset{\sim}{r}}}{\mu} \, dx_3 \tag{23}$$

To determine $\underset{\approx}{R}$, one must know the turbulent eddy diffusivity μ and the function $\hat{\underset{\sim}{r}}$, i.e. the velocity deviation $\hat{\underset{\sim}{u}}$.

VERTICAL PROFILE OF THE HORIZONTAL VELOCITY

The evolution equation for the horizontal velocity vector $\underset{\sim}{u}$ can be written, after eliminating the pressure (e.g. Nihoul, 1975)

$$\frac{\partial \underset{\sim}{u}}{\partial t} + \underset{\sim}{\nabla} . (\underset{\sim}{u} \, \underset{\sim}{u}) + f \underset{\sim}{e}_3 \wedge \underset{\sim}{u} + \frac{\partial}{\partial x_3} (v_3 \underset{\sim}{u}) = - \underset{\sim}{\nabla} \left(\frac{p_a}{\rho} + g\zeta \right) + \frac{\partial}{\partial x_3} \left(\nu \frac{\partial \underset{\sim}{u}}{\partial x_3} \right) \tag{24}$$

where f is equal to twice the vertical component of the earth's rotation vector, p_a is the atmospheric pressure, g the acceleration of gravity and ν the vertical turbulent viscosity.

In eq.(24), one has neglected the effect of the horizontal component of the earth's rotation vector (multiplied by $v_3 \ll u$) and the horizontal turbulent diffusion (because horizontal length scales are always much larger than the depth).

The observations indicate that, in shallow tidal seas, the turbulent viscosity ν can be written as the product of a function of t, x_1 and x_2 and a function of the reduced variable $\xi = H^{-1}(x_3 + h)$ (e.g. Bowden, 1965).

Let

$$\nu = H^2 \sigma(t, x_1, x_2) \lambda(\xi) \tag{25}$$

where σ and λ are appropriate functions.

The asymptotic form of ν for small values of ξ is well-known from boundary layer theory :

$$\nu = k \; u_*(x_3 + h) = k \; u_* H \xi \tag{26}$$

where k is the Von Karman constant and u_* the friction velocity given by

$$u_*^2 = \| \underset{\sim}{\tau}_b \| \qquad ; \qquad \underset{\sim}{\tau}_b = \left(\nu \; \frac{\partial \underset{\sim}{u}}{\partial x_3} \right)_{x_3 = -h} \tag{27};(28)$$

Hence

$$\sigma H = k u_* \tag{29}$$

and

$$\lambda(\xi) \sim \xi \qquad \text{for} \qquad \xi << 1 \; . \tag{30}$$

In a well-mixed shallow sea, where the Richardson number is small and the turbulence fully developed, it is reasonable (e.g. Nihoul, 1975) to take

$$\mu \sim \nu \; . \tag{31}$$

This hypothesis will be reexamined later.

It is convenient to change variables to (t, x_1, x_2, ξ) in eq.(24) In the final result (Nihoul, 1977), the non-linear terms combine with additional contributions from the time derivative to give three terms, related respectively to the gradients of velocity, depth and surface elevation. These terms are found negligible almost everywhere in the North Sea (Nihoul and Runfola, 1979). Thus although depth-integrated two-dimensional hydrodynamic models of the North Sea may not discard the non-linear terms[1], if one excludes localized singular regions like the vicinity of tidal emphydromic points, the "local" vertical distribution of velocity may be described, with a very good approximation, by a linear model.

---

1. It can be shown that these terms are essential in determining the residual circulation (Nihoul and Ronday, 1976b).

Then, the governing equation for the velocity deviation $\hat{u}$ can be written

$$\frac{\partial \underaccent{\tilde}{\hat{u}}}{\partial t} + f \underaccent{\tilde}{e}_3 \wedge \underaccent{\tilde}{\hat{u}} = \sigma \left\{ \frac{\partial}{\partial \xi} \left(\lambda \frac{\partial \underaccent{\tilde}{\hat{u}}}{\partial \xi} \right) - \frac{\underaccent{\tilde}{\tau}_s - \underaccent{\tilde}{\tau}_b}{\sigma H} \right\} \tag{32}$$

where

$$\underaccent{\tilde}{\tau}_s = \left(\nu \frac{\partial \underaccent{\tilde}{u}}{\partial x_3} \right)_{x_3 = \zeta} \tag{33}$$

is the wind stress (normalized with water density).

It is possible to find an analytical solution of eq.(32) giving $\hat{u}$ in terms of $\underaccent{\tilde}{\tau}_s$, $\underaccent{\tilde}{\tau}_b$ and their derivatives with respect to time ; the coefficients depending on the functions $s(\xi)$, $b(\xi)$ and $f_n(\xi)$ $(n=1,2,..)$ defined by (Nihoul, 1977)

$$s(\xi) = \int_0^\xi \frac{\eta}{\lambda(\eta)} \, d\eta \tag{34}$$

$$b(\xi) = \int_{\xi_0}^\xi \frac{1 - \eta}{\lambda(\eta)} \, d\eta \tag{35}$$

$$\frac{d}{d\xi} \left(\lambda \frac{df_n}{d\xi} \right) = - \alpha_n f_n \tag{36}$$

with

$$\int_0^1 f_n^2(\xi) \, d\xi = 1 \tag{37}$$

$$\lambda \frac{df_n}{d\xi} = 0 \quad \text{at} \quad \xi = 0 \quad \text{and} \quad \xi = 1 . \tag{38}$$

One should note here that, in the definition of $b(\xi)$, the lower limit of integration is not set equal to zero but to $\xi_0 = \frac{z_0}{H} \ll 1$ where z_0 is the "rugosity length". z_0 can be interpreted as the distance above the bottom where the velocity is conventionally set equal to zero, ignoring the intricated flow situation which occurs near the irregular sea floor and willing to parameterize its effect on the turbulent boundary layer as simply as possible. In the North Sea, the value of z_0, which varies according to the nature of the bottom, is of the order of 10^{-3}m ($\ln \xi_0 \sim - 10$) (e.g. Nihoul and Ronday, 1976a).

Although $\xi_o \ll 1$, it cannot be systematically put equal to zero because the linear variation of the vertical eddy viscosity near the bottom leads to a logarithmic velocity profile which is singular at $\xi = 0$. However, in the present description, the difficulty exists only for the function b and the eigenfunction $f(\xi)$ may be determined on the interval $(0, \xi)$.

In a shallow tidal sea like the North Sea, it is readily seen, comparing the orders of magnitude of the different terms, that one obtains a very good approximation with only the first two terms in the series expansion of $\hat{u}$, i.e. (Nihoul, 1977)

$$\hat{\underset{\sim}{u}} = \underset{\sim}{v}_s \big(s(\xi) - \bar{s}\big) + \underset{\sim}{v}_b \big(b(\xi) - \bar{b}\big)$$
$$- \Big(\frac{s_1}{\alpha_1 \sigma} \dot{\underset{\sim}{v}}_s + \frac{b_1}{\alpha_1 \sigma} \dot{\underset{\sim}{v}}_b\Big) f_1(\xi) \tag{39}$$

where $\bar{s}$ and $\bar{b}$ are the depth-averages of s and b, s_1 and b_1 two numerical coefficients ($s_1 = \int_o^1 s\, f_1\, d\xi$; $b_1 = \int_o^1 b\, f_1\, d\xi$) ; α_1 the eigenvalue corresponding to $f_1(\xi)$ and where

$$\underset{\sim}{v}_s = \frac{\underset{\sim}{\tau}_s}{\sigma H} \quad ; \quad \underset{\sim}{v}_b = \frac{\underset{\sim}{\tau}_b}{\sigma H} \tag{40};(41)$$

A dot denotes here a total derivative with respect to time

$$\Big(\dot{\underset{\sim}{v}}_s = \frac{d\underset{\sim}{v}_s}{dt} = \frac{\partial \underset{\sim}{v}_s}{\partial t} + f\, \underset{\sim}{e}_3 \wedge \underset{\sim}{v}_s \quad \text{and similarly for } \underset{\sim}{v}_b\Big) .$$

Knowing the function $\lambda(\xi)$, one can determine $\hat{\underset{\sim}{u}}$ by eq.(39), $\underset{\sim}{f}$ by eq.(21) and $\underset{\approx}{R}$ by eq.(23).

PARAMETERIZATION OF THE BOTTOM STRESS

The functions $\hat{\underset{\sim}{u}}$, $\underset{\sim}{f}$ and $\underset{\approx}{R}$ depend on the vectors $\underset{\sim}{v}_s$, $\underset{\sim}{v}_b$ and their derivatives. These can be determined by eqs.(27), (29), (40) and (41) from $\underset{\sim}{\tau}_s$ and $\underset{\sim}{\tau}_b$.

The surface stress $\underset{\sim}{\tau}_s$ can be calculated from atmospheric data, the bottom stress $\underset{\sim}{\tau}_b$ is not given and must be determined by the no-slip condition at the bottom, i.e.

$$\hat{\underset{\sim}{u}} = - \bar{\underset{\sim}{u}} \quad \text{at} \quad \xi = \xi_o \tag{42}$$

Eq.(42) provides a differential equation for $\tau_{\sim b}$ in terms of $\tau_{\sim s}$ and $\bar{u}_{\sim}$.

In shallow tidal seas like the North Sea, the terms including $\dot{v}_{\sim s}$ and $\dot{v}_{\sim b}$ are generally negligible and can only play a part during a relatively short time, at tide reversal (Nihoul and Runfola, 1979). The dominant term is, in fact, the term containing the bottom stress. The effect of the wind stress appears as a first order correction and the "memory" effect involving the derivatives $\dot{v}_{\sim s}$ and $\dot{v}_{\sim b}$ as a second order correction. One thus has

$$\bar{u}_{\sim} \sim \bar{b}\, v_{\sim b} \qquad\qquad \text{(zeroth order)} \qquad\qquad (43)$$

$$\bar{u}_{\sim} \sim \bar{b}\, v_{\sim b} + \bar{s}\, v_{\sim s} \qquad\qquad \text{(first order)} \qquad\qquad (44)$$

$$\bar{u}_{\sim} \sim \bar{b}\, v_{\sim b} + \bar{s}\, v_{\sim s} + (s_1 \dot{v}_{\sim s} + b_1 \dot{v}_{\sim b})\, \frac{f_{1,0}}{\alpha_1 \sigma} \quad \text{(second order)} \qquad (45)$$

Eq.(43) yields the well-known semi-empirical quadratic bottom friction law. Indeed, combining eqs.(27), (29) and (43), one finds

$$\| \tau_{\sim b} \| \sim \frac{\sigma H}{\bar{b}} \| \bar{u}_{\sim} \| \sim (\frac{\sigma H}{k})^2 \Rightarrow \sigma H \sim \frac{k^2}{\bar{b}} \| \bar{u}_{\sim} \| \qquad (46)$$

i.e.

$$\tau_{\sim b} = \frac{k^2}{\bar{b}^2} \| \bar{u}_{\sim} \| \bar{u}_{\sim} \qquad\qquad\qquad (47)$$

$\frac{k^2}{\bar{b}^2}$ is the so-called "drag coefficient".

At the first order, one gets another classical formula (e.g. Groen and Groves, 1966 ; Nihoul, 1975) :

$$\tau_{\sim b} = \frac{k^2}{\bar{b}^2} \| \bar{u}_{\sim} \| \bar{u}_{\sim} - \frac{\bar{s}}{\bar{b}} \tau_{\sim s} \qquad\qquad (48)$$

The second order parameterization is better understood if the last term in the right-hand side of eq.(45) is eliminated using eqs.(32), (34), (35), (36) and (39). One has, indeed

$$\frac{\partial \bar{u}_{\sim}}{\partial t} + f\, e_{\sim 3} \wedge \bar{u}_{\sim} = - \sigma \left(\frac{\partial}{\partial \xi} (\lambda \frac{\partial \hat{u}_{\sim}}{\partial \xi}) \right)_{x_3 = -h} + \frac{\tau_{\sim s} - \tau_{\sim b}}{H}$$

$$\sim - (s_1 \dot{v}_{\sim s} + b_1 \dot{v}_{\sim b})\, f_{1,0} \sim - \alpha_1 \sigma\, (\bar{u}_{\sim} - \bar{s}\, v_{\sim s} - \bar{b}\, v_{\sim b})$$

i.e., using (46) to estimate σH ,

$$\underset{\sim}{\tau}_b \sim \frac{k^2}{b^2} \| \overline{\underset{\sim}{u}} \| \overline{\underset{\sim}{u}} - \frac{\overline{s}}{b} \underset{\sim}{\tau}_s + \frac{H}{\alpha_1 \overline{b}} \left(\frac{\partial \overline{\underset{\sim}{u}}}{\partial t} + f \underset{\sim}{e}_3 \wedge \overline{\underset{\sim}{u}} \right)$$

With a typical drag coefficient of the order of $2 \ 10^{-3}$ (e.g. Nihoul and Ronday, 1976), one finds, using characteristic values for the North Sea,

$$\frac{k^2}{b^2} \| \overline{\underset{\sim}{u}} \| \overline{\underset{\sim}{u}} \sim 0 (2 \ 10^{-3} \overline{u}^2) \ ,$$

$$\frac{H}{\alpha_1 \overline{b}} \frac{\partial \overline{\underset{\sim}{u}}}{\partial t} \sim \frac{H}{\alpha_1 \overline{b}} f \underset{\sim}{e}_3 \wedge \overline{\underset{\sim}{u}} \sim 0 (10^{-4} \overline{u}) \ .$$

Thus, the "acceleration" terms containing $\frac{\partial \overline{\underset{\sim}{u}}}{\partial t}$ and $f \underset{\sim}{e}_3 \wedge \overline{\underset{\sim}{u}}$ (i.e. the terms arising from the time derivatives of $\underset{\sim}{v}_s$ and $\underset{\sim}{v}_b$) are not expected to play an important role except perhaps during relatively short periods of weak currents (when tides reverse, for instance). The total effect of the acceleration terms depends really on the local conditions. Obviously, (fig. 1) if the current velocity vector rotates clockwise during a tidal period, the two terms tend to oppose each other and their global contribution may remain always very small. On the other hand, if the current velocity vector rotates counter-clockwise (as it is often the case in the Southern Bight) the two terms reinforce each other and have a definite - although limited - effect on the velocity profile and the related relationship between the bottom stress and the mean velocity (fig. 2, fig. 3).

Most hydrodynamic models of shallow seas restrict attention to the determination of the depth-mean velocity $\overline{\underset{\sim}{u}}$. The two-dimentional time dependent evolution equation for $\overline{\underset{\sim}{u}}$ is obtained from eq.(24) by integration over depth. It includes the surface elevation ζ and the bottom stress $\underset{\sim}{\tau}_b$ and thus constitutes with eq.(15) and eq.(43) (or 44) a closed system for the depth-averaged circulation (e.g. Nihoul, 1975).

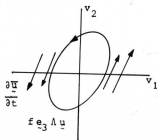
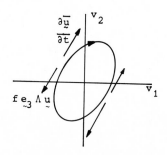

Fig. 1.

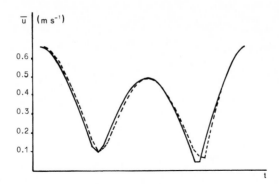

Fig. 2. Comparison between the magnitude of the mean velocity $\bar{u}$ computed at the test point 52°30'N, 3°50'E in the North Sea by a depth-averaged two-dimensional model using an algebraic parameterization of $\mathfrak{I}_b$ (full line) and by the three-dimensional model subject to the condition of zero velocity at the bottom (dash line) (Nihoul and Runfola, 1979).

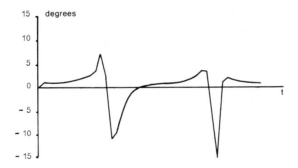

Fig. 3. Difference between the directions of the mean velocity vectors computed as in Fig. 2. (Nihoul and Runfola, 1979).

The determination of the vertical velocity profile can be carried simultaneously using eq.(39).

One can go a step further and devise a three-dimensional model based on the depth-averaged equation, the local depth-dependent equation for the velocity deviation $\hat{u}$ and the refined parameterization of the bottom stress given by eq.(49) ; the acceleration corrections being taken into account when required in the numerical calculation (Nihoul and Runfola, 1979).

The model gives H, $\bar{\underset{\sim}{u}}$, $\mathfrak{I}_b$, $\hat{\underset{\sim}{u}}$, $\underset{\sim}{f}$ and $\underset{\approx}{R}$.

Substituting in eq.(16) one obtains explicitly the dispersion equation for the mean concentration $\bar{c}$.

COEFFICIENTS OF THE SHEAR EFFECT DISPERSION

Shear effect dispersion is described by eq.(16) where Σ is given, in terms of the functions $\underset{\sim}{f}$ and μ, by eqs.(22) and (23). The functions $\underset{\sim}{f}$ and μ can be determined by eqs.(21), (25), (31) and (39) provided the function λ is known. Thus the parameterization of the shear effect - as well as the determination of the vertical profile of velocity - reduces to the choice of a single scalar function.
The function

$$\lambda = \xi(1 - 0.5\ \xi) \tag{50}$$

appears to cover a wide range of situations in the North Sea and other shallow tidal seas (e.g. Bowden, 1965 ; Nihoul, 1977).
In this case, the eigenfunctions and the eigenvalues of eqs.(36), (37) and (38) are given by

$$f_n = (4n + 1)^{1/2}\ P_{2n}(\xi - 1) \tag{51}$$

$$\alpha_n = n(2n + 1) \tag{52}$$

where P_{2n} is the Legendre polynomial of even order 2n.
Integrating eq.(39), one obtains, in this case,

$$\underset{\sim}{f} = \underset{\sim s}{v}\ S(\xi) + \underset{\sim b}{v}\ B(\xi) + \frac{\dot{\underset{\sim s}{v}} + 2\dot{\underset{\sim b}{v}}}{\sigma}\ F(\xi) \tag{53}$$

with

$$S(\xi) = 4\ \ln\ 2(\xi - 1) + 2(2 - \xi)\ln(2 - \xi) \tag{54}$$

$$B(\xi) = -2\ \ln\ 2(\xi - 1) + \xi\ \ln\ \xi - (2 - \xi)\ln(2 - \xi) \tag{55}$$

$$F(\xi) = \frac{5}{36}\ (\xi^3 - 3\ \xi^2 + 2\ \xi) \tag{56}$$

The shear effect diffusivity tensor can then be written, using eqs.(25) and (31)

$$R = \int_{\xi_O}^{1} \frac{\hat{\underset{\sim}{r}} \, \hat{\underset{\sim}{r}}}{\sigma \lambda} \, d\xi$$

$$= \frac{\gamma_{ss}}{\sigma} \underset{\sim}{v}_s \underset{\sim}{v}_s + \frac{\gamma_{sb}}{\sigma} (\underset{\sim}{v}_s \underset{\sim}{v}_b + \underset{\sim}{v}_b \underset{\sim}{v}_s) + \frac{\gamma_{bb}}{\sigma} \underset{\sim}{v}_b \underset{\sim}{v}_b$$

$$+ \frac{\gamma_{sf}}{\sigma^2} (\underset{\sim}{v}_s \dot{\underset{\sim}{v}}_s + 2 \underset{\sim}{v}_s \dot{\underset{\sim}{v}}_b + \dot{\underset{\sim}{v}}_s \underset{\sim}{v}_s + 2 \dot{\underset{\sim}{v}}_b \underset{\sim}{v}_s)$$

$$+ \frac{\gamma_{bf}}{\sigma^2} (\underset{\sim}{v}_b \dot{\underset{\sim}{v}}_s + 2 \underset{\sim}{v}_b \dot{\underset{\sim}{v}}_b + \dot{\underset{\sim}{v}}_s \underset{\sim}{v}_b + 2 \dot{\underset{\sim}{v}}_b \underset{\sim}{v}_b)$$

$$+ \frac{\gamma_{ff}}{\sigma^3} (\dot{\underset{\sim}{v}}_s + 2 \dot{\underset{\sim}{v}}_b)(\dot{\underset{\sim}{v}}_s + 2\dot{\underset{\sim}{v}}_b) \tag{57}$$

with

$$\gamma_{ss} = \int_{\xi_O}^{1} \frac{S^2}{\lambda} \, d\xi \sim 0.048 \qquad \gamma_{sf} = \int_{\xi_O}^{1} \frac{SF}{\lambda} \, d\xi \sim - 0.015 \tag{58},(59)$$

$$\gamma_{sb} = \int_{\xi_O}^{1} \frac{SB}{\lambda} \, d\xi \sim 0.090 \qquad \gamma_{bf} = \int_{\xi_O}^{1} \frac{BF}{\lambda} \, d\xi \sim - 0.031 \tag{60},(61)$$

$$\gamma_{bb} = \int_{\xi_O}^{1} \frac{B^2}{\lambda} \, d\xi \sim 0.196 \qquad \gamma_{ff} = \int_{\xi_O}^{1} \frac{F^2}{\lambda} \, d\xi \sim 0.005 \tag{62},(63)$$

APPLICATION TO THE SOUTHERN BIGHT OF THE NORTH SEA

In the Southern Bight of the North Sea, the depth is small and the bottom stress $\underset{\sim}{\tau}_b$, maintained by bottom friction of tidal currents, wind induced currents and residual currents is always fairly important. One can estimate that, in general, the characteristic time σ^{-1} is one order of magnitude larger than the characteristic time of variation of $\underset{\sim}{v}_s$ and $\underset{\sim}{v}_b$ (Nihoul and Ronday, 1976 ; Nihoul, 1977).

The terms of eq.(57) which contain the derivatives $\dot{\underset{\sim}{v}}_s$ and $\dot{\underset{\sim}{v}}_b$ - the coefficients of which are already smaller than the others - may then be neglected.

The shear effect diffusivity tensor reduces then to

$$\underset{\sim}{R} = \frac{H}{\|\underset{\sim}{v}_b\|} \left(\beta_1 \underset{\sim}{v}_b \underset{\sim}{v}_b + \beta_2 (\underset{\sim}{v}_s \underset{\sim}{v}_b + \underset{\sim}{v}_b \underset{\sim}{v}_s) + \beta_3 \underset{\sim}{v}_s \underset{\sim}{v}_s \right) \tag{64}$$

with

$$\beta_1 \sim 1.2 \qquad ; \qquad \beta_2 \sim 0.6 \qquad ; \qquad \beta_3 \sim 0.3 \tag{65)(66)(67}$$

In weak wind conditions ($v_b \leq 10^{-2} \bar{u}$), the first term in the bracket is largely dominant and, using eq.(43), one obtains, with a good approximation

$$\underset{\sim}{R} = \alpha \ \frac{H}{\bar{u}} \ \bar{\underset{\sim}{u}} \ \bar{\underset{\sim}{u}} \tag{68}$$

$$\Sigma = H^{-1} \underset{\sim}{\nabla} \cdot \left(\alpha \ \frac{H^2}{\bar{u}} \ \bar{\underset{\sim}{u}} (\bar{\underset{\sim}{u}} \cdot \underset{\sim}{\nabla} \bar{c}) \right) \tag{69}$$

with

$$\alpha \sim 0.14 \tag{70}$$

However, in weak wind conditions, the approximation which consists in neglecting the derivatives $\dot{v}_b$ and $\dot{v}_s$ is less justified and, furthermore, one may question the validity of eq.(50). If the wind is too weak to maintain turbulence in the sub-surface layer, one may expect, in some cases, a turbulent diffusivity which, instead of growing continuously from the bottom to the surface, instead passes through a maximum at some intermediate depth to decrease afterwards to a smaller surface value. This type of behaviour is described by the family of curves

$$\lambda = \xi(1 - \delta\xi) \tag{71}$$

Eq.(50) corresponds to the case $\delta = 0.5$. Values of δ from 0.5 to 1 correspond to lower intensity turbulence in the surface layer and the limiting value $\delta = 1$ would correspond to the case of an ice cover and the existence, below the surface, of a logarithmic boundary layer analogous to the bottom boundary layer.

In the Southern Bight of the North Sea, it is reasonable to assume that δ does not differ significantly from 0.5 and, in any case, never reaches extreme values close to 1. Nevertheless, to estimate the maximum error one can make on α , it is interesting to compute the coefficients β_1, β_2 and β_3 for some very different values of δ.
One finds

δ	0.5	0.7	0.9
β_1	1.2	1.5	2
β_2	0.6	0.8	1.3
β_3	0.3	0.5	1
α	0.14	0.17	0.23

The increase of the coefficients β_1, β_2, β_3 and α with δ is obviously associated with more important variations of u over depth, i.e. with larger values of $\hat{u}$.

One should note also that the existence of a vertical stratification, even a weak one, reduces the turbulent diffusivity ($\mu = \eta\nu$ with $\eta < 1$) and contributes similarly to increase the value of α (e.g. Bowden, 1965).

In the Southern Bight of the North Sea, eventual modifications of the magnitude ($\eta < 1$) or of the form ($\delta > 0.5$) of the turbulent diffusivity are not likely to be very important and eq.(68) can presumably be used with $\alpha = 0.14$ or some slightly higher value obtained by calibration of the model with the observations.

VERTICAL CONCENTRATION PROFILE

When $\bar{c}$ has been calculated, it is possible to compute the deviation $\hat{c}$ by eq.(20). Changing variable to ξ and using eqs.(25) and (31), one gets

$$\frac{\partial \hat{c}}{\partial \xi} = \frac{\hat{\underset{\sim}{r}}}{\lambda} \cdot \frac{\nabla \bar{c}}{\sigma} \tag{72}$$

with, from eq.(9),

$$\int_0^1 \hat{c}\ d\xi = 0 \ . \tag{73}$$

Restricting attention to the dominant terms, one finds

$$\hat{c}(\xi) = \left(H(\xi)\ \underset{\sim}{v}_s + G(\xi)\ \underset{\sim}{v}_b \right) \cdot \frac{\nabla \bar{c}}{\sigma}$$

where

$$H(\xi) = 4\ \{P(\xi) + \ln (2 - \xi)\ \ln \frac{\xi}{4} + \ln 2\ (4\ \ln 2 - 1 - \ln \xi) - 2\} \tag{75}$$

$$G(\xi) = - 2\ \{2\ L_2(\tfrac{1}{2}) + \ln \frac{2-\xi}{2}\ \ln \frac{\xi}{2} + \ln (2)\ \ln (2) + 2\} \tag{76}$$

and

$$P(\xi) = L_2(\tfrac{\xi}{2}) + 2\ L_2(\tfrac{1}{2}) - \bar{L}_2 \tag{77}$$

$$L_2(x) = \text{Dilo} (1-x) = \sum_{\nu=1}^{\infty} (-1)^\nu\ \frac{(x-1)^\nu}{\nu^2} \tag{78}$$

The functions H and G are shown in fig. 4. They are both negative near the surface and positive near the bottom. This is what one should expect from a physical point of view. Higher velocities near the surface carry water masses farther. If this transport is directed towards increasing mean concentrations the corresponding inflow of lower concentration fluid decreases the local concentration below the mean value $\bar{c}$. If the transport in the upper layer is directed towards decreasing mean concentrations, the corresponding inflow of high concentration fluid increases the local concentration above the mean value $\bar{c}$. The opposite situation occurs near the bottom. This is illustrated in fig. 5 showing the concentration profiles at two points situated downstream and downwind and respectively upstream and upwind on the same isoconcentration curve following a dumping.

The combination of eqs.(16) and (42) with a three-dimensional hydrodynamic model provides a three-dimensional dispersion model for the calculation of the evolution with time and the spatial - horizontal and vertical - distribution of any passive buoyant marine constituant.

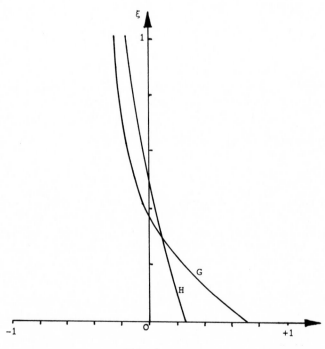

Fig. 4.

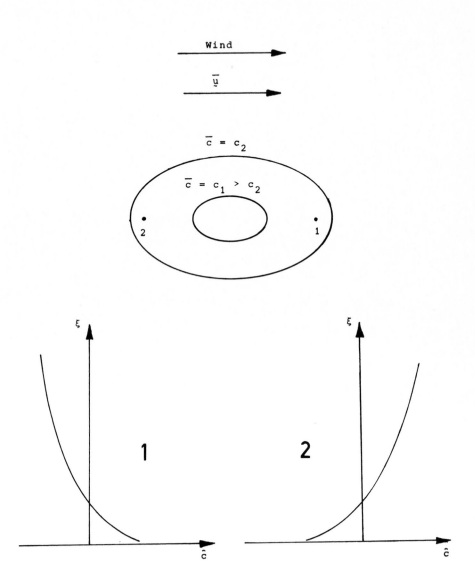

Fig. 5.

DISPERSION IN A TIDAL SEA

A. WARLUZEL and J.P BENQUE

Engineers at "Electricité de France" (French National Electricity Utility)
National Hydraulics Laboratory.

ABSTRACT

The theory which is presented in this paper definies the dispersion in tidal
flow by a tensor (2 x 2). Each component of this dispersion tensor is expressed
in terms of turbulent diffusion and variation of the horizontal velocity field
around its mean value along the depth. A mathematical model gives the minimum
values for each component of the dispersion tensor when the bottom is flat, the
sea infinite, without wind and when the turbulent diffusion is known. On the
other hand components of the dispersion tensor have been calculated from field
measurements.

INTRODUCTION

In many pollution problems, we are primarily interested in the average concentration field of the effluent in space. And so, in a linear model, our unknown

is the average concentration in a section. Daubert (1), adopting Taylor's ideas (2),

has shown that this concentration verifies a transport - diffusion equation where

the diffusion is due to heterogeneity in the velocity and concentration around the

mean (average) value in the section. This mechanism which is a much higher spacial

scale than the spatial scale for eddy diffusion, is called dispersion. The purpose

of this paper is to generalize these ideas and to provide figures for this dispersion

when the average flow is two-dimensional. This is the case of discharges into a

tidal sea where the eddy effect is assumed to be strong enough to make the concen-

tration almost homogeneous over a verticle.

EQUATION FOR THE MEAN CONCENTRATION

If we designate the pollutant concentration field at any instant $c(x,y,z,t)$, the equation for the evolution of this field is as follows (not taking into account the transport by horizontal diffusion) :

$$\frac{\partial c}{\partial t} + u \frac{\partial c}{\partial x} + v \frac{\partial c}{\partial y} + w \frac{\partial c}{\partial z} = \frac{\partial}{\partial z} \, d_t \, \frac{\partial c}{\partial z} \tag{1}$$

By separating out the fluctuating values from the mean values ($c = \bar{c} + c'$, $u = \bar{u} + u'$, $v = \bar{v} + v'$), by integrating this equation on the vertical, and by taking into account the impermeability of the free surface and the bottom, we obtain :

$$\frac{\partial h\bar{c}}{\partial t} + \frac{\partial h\bar{u}\bar{c}}{\partial x} + \frac{\partial h\bar{v}\bar{c}}{\partial y} + \frac{\partial h \,\overline{u'c'}}{\partial x} + \frac{\partial h \,\overline{v'c'}}{\partial y} = \phi_s + \phi_f \tag{2}$$

with :

$$\overline{u'c'} = \frac{1}{h} \int_{z_f}^{z} u'c'dz, \quad \overline{v'c'} = \frac{1}{h} \int_{z_f}^{z} v'c'dz$$

and ϕ_s and ϕ_f respectively designating the flow of the pollutant on the surface and at the bottom :

$$\phi_s + \phi_f = \left[d_t \, \frac{\partial c}{\partial z} \right]_{z_f}^{z}$$

Dispersion tensor

These equations show the divergence of an unknown vector $\vec{\psi}$ from components $\overline{u'c'}$, $\overline{v'c'}$. By analogy with what is done to solve certain eddy flow problems, we are tempted to write :

$$\vec{\psi} = \overline{\overline{K}} \; \overrightarrow{\text{grad}} \, \bar{c}$$

where $\overline{\overline{K}}$ is a tensor (2.2) called the dispersion tensor.

The purpose of this paper is to establish this result, show when it is valid, and estimate the tensor.

HORIZONTAL TRANSPORT BY VERTICAL DIFFUSION AND CONVECTION

The result which we seek depends essentially upon the comparison between two time scales : the one linked to horizontal convection and the one linked to vertical diffusion.

Time scales

If we call L a length which corresponds to the horizontal scale and H a length corresponding to the vertical scale, the time corresponding to the horizontal convection is expressed as follows :

$$T_{cv} = L/U_o$$

where U_o represents a speed which corresponds to the horizontal convection. In the same way, we define a time corresponding to the vertical diffusion as follows :

$$T_{diff} = H^2/D$$

where D represents a diffusion coefficient which corresponds to the vertical eddy effect. When the diffusion has the time to act on the convective exchanges, we have :

$$T_{diff}/T_{cv} = \nu \ll 1$$

When the effluent has exchanges with the atmosphere or the bottom, we define an exchange time scale. Assuming the bottom to be adiabatic, we have :

$$T_{ech} = HC_o/\phi_{so}$$

with ϕ_{so} being the value corresponding to the exchange on the surface, and C_o corresponding to the mean concentration variation on the vertical. Experience shows that in the usual situations (temperature), we have :

$$T_{cv}/T_{ech} = \nu' \ll 1$$

These hypotheses lead to simplifying the setting up of the later equations in this paper.

Placing in adimensional form

We intend to calculate the unknown terms in the equation (2) ($\overline{u'c'}$ and $\overline{v'c'}$) as a function of the mean values, the velocity fluctuations and the eddy diffusion.

With the following adimensional variables :

$\bar{c} = c_0 \, \bar{c}_+$ $\quad$ c_0 $\quad$ concentration of the reference introduced earlier

$\bar{u} = u_0 \, \bar{u}_+$ $\quad$ U_0 $\quad$ velocity of the reference introduced earlier

$u' = u_0 v'_+$
$v' = u_0 v'_+$ $\qquad \left\{ \begin{array}{l} \text{the velocity fluctuations around the vertical mean of the same} \\ \text{general magnitude as the mean magnitude} \end{array} \right.$
$w' = \nu u_0 \, Hw_+/L$ *

$c' = \nu c_0 \, c'_+$ $\qquad \left\{ \begin{array}{l} \text{the concentration fluctuations around the mean value are} \\ \text{considered small in relation to } c \end{array} \right.$

$\qquad\qquad\qquad$ D $\quad$ reference eddy diffusion coefficient introduced earlier. We

$d_t = D \, d_{t+}$ $\qquad$ often set $D = U_0 H$ so as to express that the eddy effect is

$\qquad\qquad\qquad\qquad$ strongly influenced by the horizontal convection and by the

$\qquad\qquad\qquad\qquad$ vertical scale.

$x = Lx_+$
$\qquad\qquad$ L $\quad$ horizontal reference length introduced earlier
$y = Ly_+$

$z = Hz_+$ $\qquad$ H $\quad$ vertical reference length

$t = T_{cv} \, t'_+$ $\qquad \left\{ \begin{array}{l} \text{the time corresponding to the temporal variations of } \bar{c} \text{ is linked} \\ \text{to the convection while the time corresponding to the } c' \text{ variations} \\ \text{is linked to the eddy diffusion} \end{array} \right.$
ou
$t = T_{diff} \, t_+$

Using the fact that

$\nu \sim \nu' \ll 1$

---

*this expression comes from the expression of the vertical velocity at the free surface.

and integrating the vertical, taking into account the conditions at the limits
and taking into account that the integrals of c'_+, u'_+, v'_+ are zero, the equation
(1) becomes :

$$\frac{\partial}{\partial t} \int_{Z_{f+}}^{Z_+} c'_+ \, dz \quad = \quad (Z_+ - Z_{f+}) \quad A \quad (x_+, \ y_+, \ t_+) \tag{3}$$

Term A, independent of Z in the second expression of (3) only contributes to the
mean of c'_+ and not to the transversal distribution along Z_+. The equation which
verifies C'_+ obliged to remain at zero mean, is therefore :

$$\frac{\partial c'_+}{\partial t_+} - \frac{\partial}{\partial z_+} \, d_{t+} \, \frac{\partial c'_+}{\partial z_+} = - \, u'_+ \, \frac{\partial \bar{c}_+}{\partial x_+} - v'_+ \, \frac{\partial \bar{c}_+}{\partial y_+} \quad , \quad d_{t+} \, \frac{\partial c'}{\partial z_+} = 0 \text{ en} \left. \begin{matrix} z_+ = Z_+ \\ z_+ = Z_{F+} \end{matrix} \right| \tag{4}$$

The search for the solution

Equation (4) is linear; the solution can be expressed based upon the Green
function of the operator at the partial derivatives. By omitting to write in the
symbol + for adminsionalizing and by choosing $c'(x, y, z, o) = o$, the general
solution to the equation (4) is expressed as follows :

$$c'(x, y, z, t) = \; - \int_o^t dt' \int_o^1 G \, (z, \, t/z't') \, u'(x,y,z;t') \, \partial \bar{c}/\partial x \, (x, \ y, \ t') \; dz'$$

$$- \int_o^t dt' \int_o^1 G \, (z, \, t/z't') \, v'(x,y,z;t') \, \partial \bar{c}/\partial y \, (x, \ y, \ t') \; dz'$$

By arrying this expression into the unknown flux terms and by considering the
problems as stationary,∗ it appears that flux $\vec{\psi}$ may be linked to the mean values
in the following way :

$$\vec{\psi} = - \, \bar{\bar{K}} \, \overrightarrow{\text{grad} \, \bar{c}}$$

$\bar{\bar{K}}$ is a tensor 2.2. This tensor is called the dispersion tensor.

∗this means choosing a time scale in (4) T_{cv} instead of Tdiff for the temporal
evolutions of C'.

By using the analytic expression of the Green function, we obtain the following for the components of the dispersion tensor :

$$Kij = - \int_o^1 u_i (z) \int_o^z dz'/d_t (z') \int_o^{z'} uj (u) \, du \, dz \qquad (5)$$

This expression is analogous to the one suppled by Elder, but generalized to the two-dimensional case; it can also be expressed as follows :

$$kij = - \int_o^1 dz'/d_t (z') \int_o^{z'} u_i (u) \, du \int_o^{z'} uj (v) \, dv \qquad (6)$$

It would seem that the latter expression is the most manageable when the mean flow is two-dimensional; it is actually easy to show, based on this expression, that tensor $\bar{\bar{K}}$ is associated with a definite and positive symmetrical quadratic form.

ESTIMATING THE DISPERSION TENSOR BASED UPON A SIMPLE MODEL

Mathematical formulation of the model

In this chapter, we give a pessimistic estimate of the dispersion based upon the expression given in the preceeding chapter, within the context of the following hypotheses : flat bottom, zero wind and infinite sea. Under these conditions, we can consider that the heterogeneity of the velocities on the vertical is less.

The forces affecting the flow are those due to eddy viscosity, to the pressure gradient, to temporal inertia and to the Coriolis force which is the only cause of a deviation in the velocity field in the direction perpendicular to the mean flow.

By considering the pressure to be hydrostatic, a mean sinusoidal current with amplitude U_o, by setting up :

$$u = u_o \, u_+ \quad t = h/u_o \, t_+ \quad f = f_+ \, u_o/h \quad \tau x/\rho = u_o^2 \, \tau_{x+}$$

$$z = hz_+ \quad \varepsilon = \varepsilon_+ \, u_o h \quad \omega = \omega_+ \, u_o/H \quad \tau y/\rho = u_o^2 \, \tau y_+$$

the water movement equations placed into adimensional form are expressed as follows :

$$\partial u_+/\partial t_+ = \partial/\partial z_+ \, (\varepsilon_+ \, \partial u_+/\partial z_+) + f_+ \, v_+ + \tau x_+ + \omega_+ \, \cos \omega_+ t_+$$

$$\partial v_+/\partial t_+ = \partial/\partial z_+ \, (\varepsilon_+ \, \partial v_+/\partial z_+) - f_+ \, u_+ + \tau y_+ + f_+ \, \sin \omega_+ t_+ \qquad (7)$$

With the conditions at the following limits :
$u_+ = v_+ = o$ at the bottom, $\partial u_+/\partial z = \partial v_+/\partial z_+ = o$ at the surface,

the determination of the velocity speed is obtained by solving the equations (7) after having specified the value of ε. We have adopted a vertical distribution of the eddy viscosity given by Bowden (3) and expressed as follows :

$$\varepsilon = \frac{k \ U \ m \ h\zeta^{1-\beta}(1 - \zeta)}{\beta(\beta + 1)}$$

It was obtained by beginning with a distribution of the main velocity in form $u = U \ \zeta^{\beta}$. U , this being the surface velocity, and β a coefficient which gives the profile shape - et was also successively designated at 0.2, 0.4, 0.6 - Figure 1 gives the pattern of the viscosity profiles obtained. It was evident that this expression included the values given by other authors.

Results

The calculation of the vertical distribution of the two velocity components beginning with the solution to system (7), then the calculation of the Kij coefficients of the dispersion tensor (formula 5 or 6) - considering d_t as equivalent to ε - was done on computer.

a) Steady flow :

Figures 2 to 7 give the adimensional form of the velocity distribution on the vertical and the dispersion tensor coefficients as a function of the single parameter upon which the steady flow phenomenon depends fhc^2/gu_m and for values of β of 0.2, 0.4 and 0.6. The analysis of these results shows that the transversal velocities as well as the dispersion increases with the draft or when the eddy viscosity diminishes, with a limit when ε is tending towards zero; the equations (7) then show that the flow becomes geostrophic.

The following table gives some dimensional values of the dispersion tensor coefficients. We see the size of the eddy viscosity corresponding to β and k (or C). We note that of the three coefficients, K_{22} is the largest : it corresponds to the dispersion in the direction across the flow.

	β	K	C	K_{11} m2/s	K_{12} m2/s	K_{22} m2/s
	0,2	2.10^{-3}	70	40	- 2,4	0,15
h = 20 m	0,2	4.10^{-3}	50	20	- 0,48	0,0128
	0,4	2.10^{-3}	70	160	-18	2,2
u_m = 1m/s	0,4	4.10^{-3}	50	84	- 4,2	0,24
	0,6	2.10^{-3}	70	420	-64	10
	0,6	4.10	50	220	-15,6	1,2

b) Tide flow :

The dispersion corresponds to the inertial effect and to the fact that the eddy viscosity becomes very weak at the reversals which makes K_{11}, K_{12}, K_{22} very large as we can see on figure 8 (U_o = 1 m/s, h = 20 m, C = 70, β = 0.2). Mean coefficients have been defined by eliminating the values of the coefficients obtained for longitudinal mean current velocities less than $U_o/3$. In the special case on figure 8, we have K_{11} = 27 m^2/s, K_{12} = 2.5 m^2/s, K_{22} = 0.5 m^2/s. The graphs on figures 9 and 10 give the values for the mean coefficients as a function of fh/U_o, C and β.

We come to the same conclusions as on steady flow : the dispersion increases with the draft and when the eddy viscosity diminishes. The following table gives some dimensional values for these coefficients; for the sake of comparison, it also gives the values of these coefficients in the case of steady flow ($U_m = U_o$).

We see that the tide flow is favorable to the cross dispersion corresponding to K_{22}.

Depth	β	C	Steady Flow			Tide Flow		
			$K_{11} m^2/s$	$K_{12} m^2/s$	$K_{22} m^2/s$	$K_{11} m^2/s$	$K_{12} m^2/s$	$K_{22} m^2/s$
20 m	0,2	70	40	2,4	0,15	27	2,5	0,5
	0,6	70	440	65	14	282	64	32
40 m	0,2	70	76	9,6	1,3	51	10	4
	0,6	70	760	214	68	511	190	152
100 m	0,2	70	160	50	15	111	44	37
	0,6	70	950	550	390	970	450	470

ESTIMATING THE DISPERSION TENSOR BASED UPON FIELD MEASUREMENTS

Based upon measurements of the vertical velocity distribution performed by E.D.F. (The French National Electricity Utility), just off Flamanville (Manche) with depths of 10 to 15 m, we have found (with β = 0.4 and C = 50) the following mean coefficients K_{11} = 12 m2/s, K_{12} = 5 m2/s, K_{22} = 3 m2/s; while those that we derived from the graph on figure 10 are K_{11} = 15 m2/s, K_{12} = 2.7 m2/s, K_{22} = 0.9 m2/s, which tends to prove that the model corresponds to a conservative evaluation of the dispersion.

CONCLUSION

Assuming that the vertical diffusion has time to act in relation to the convective exchanges, and that the time scale of the exchanges with the atmosphere is large in relation to the time scale for the convection, the theory which is described in this paper makes the diffusion appear in the form of a tensor (2 x 2) whose terms are a function of the vertical distribution of the velocities and of the eddy viscosity. A practical calculation of these dispersion coefficients was carried out using a mathematical model which assumes an infinite sea, a flat bottom and a zero wind. Since the sea flows are also disturbed by the form of the coastline, the relief of the bottom and the wind effects, it is evident that the results of this calculation, given in the form of graphs, are a conservative evaluation of the dispersion. This is confirmed by the results of the on-site measurements.

Notations :

u, v, w	velocity components in an orthonormal reference ox, oy, oz (oz ascendant vertical)
u', v', w'	velocity fluctuation components
$\bar{u}, \bar{v}, \bar{w}$	mean velocity components
c, c', $\bar{c}$	concentration, concentration fluctuation, mean concentration
h	depth $z = z/h$ Z datum level of the free surface Z_F datum level of the bottom
u_m	mean main current U_o amplitude of tide current
d_t	coefficient of vertical eddy diffusion
ε	coefficient of eddy viscosity ($d_t = \varepsilon$)
C	Chezy coefficient k friction coefficient (= $g/c2$)
Ω	instantaneous earth rotations w tide pulsation
φ	latitude $\tau_x = \varepsilon \dfrac{\partial u}{\partial z}\Bigg|$ $f = 2 \Omega \sin \varphi$
$\tau x \ \tau y$	bottom stresses $\tau y = \varepsilon \dfrac{\partial v}{\partial z}\Bigg|$ bottom ρ specific weight of the water

REFERENCES

Daubert, A., Mécanique des Fluides Appliqués - Chapitre 7, publié sous la direction de M. Michel HUG - Edition Eyrolles
Taylor, 1953 - pro. Soc. A. 219, 186
Bowden, K.F., Horizontal mixing in the sea due to a schearing current J. of Fluid Mechanic - 1965 - Vol. 21

FIGURES

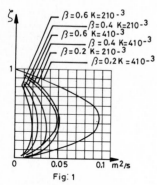

Fig. 1

Profil de viscosité turbulente
Vertical distribution of the
turbulent diffusion

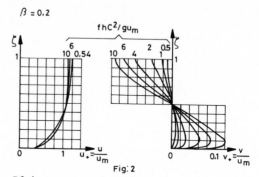

Fig. 2

Régime permanent : répartition verticale
des vitesses
Steady flow : vertical current distribution

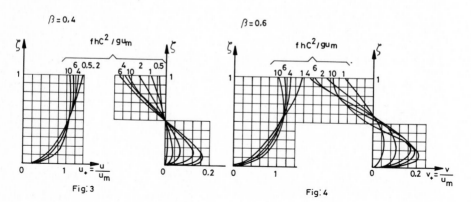

Fig. 3

Fig. 4

Régime permanent : répartition verticale des vitesses
Steady flow : vertical current distribution

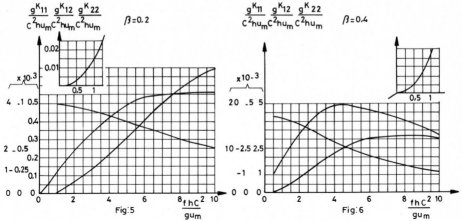

Régime permanent : coefficients du tenseur de dispersion
Steady flow : components of the dispersion tensor

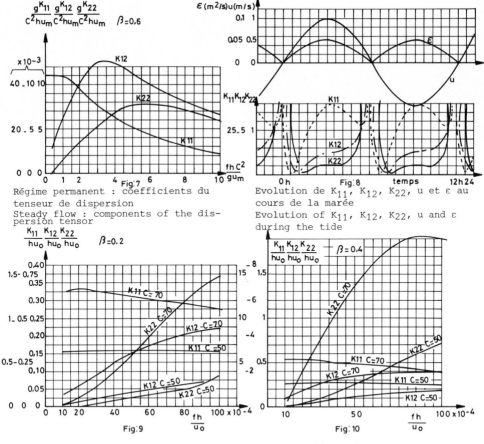

Régime permanent : coefficients du tenseur de dispersion
Steady flow : components of the dispersion tensor

Evolution de K_{11}, K_{12}, K_{22}, u et ε au cours de la marée
Evolution of K_{11}, K_{12}, K_{22}, u and ε during the tide

Régime de marée : coefficients moyens du tenseur de dispersion
Tidal flow : mean components of the dispersion tensor

SUBJECT INDEX